Beyond the Textbook: Designing Safe and Autonomous Control Systems

Furguson

TABLE OF CONTENTS

II Learning to Control Unknown Systems   136

Chapter 1

INTRODUCTION

Over the past few decades, the integration of our physical and digital worlds has deepened, with advances in machine learning, online optimization, and control theory playing pivotal roles. This progress, coupled with widespread access to computational power, is propelling us towards a future where interconnected smart systems, with humans in the loop, play an integral part in our lives. Understanding this complex closed-loop system and ensuring its reliability and safety, especially in safety-critical settings involving physical systems, is a paramount contemporary challenge. From a system engineering standpoint, designing control systems capable of operating in complex and unpredictable dynamic environments is an especially difficult problem. This naturally leads to a pertinent question regarding robust system design:

How do we design control systems for real world systems in one-shot, which can handle large model uncertainty online, and with worst-case guarantees?

Indeed, the central focus of this book seeks to address this pressing question, and contributes new approaches to the design of control systems that are capable of handling the inherent uncertainties that come with real-world dynamical systems. Part 1 of this book focuses on outlining a novel method of designing closed-loop systems that are robust to dynamic uncertainty, especially in settings involving nonlinear dynamics and complex control constraints. Part 2 introduces a general framework for learning-to-control algorithms that provide worst-case guarantees, even in scenarios with arbitrarily large dynamic uncertainty. By addressing these dually-important aspects, this work aims to advance our understanding and capabilities in designing control systems that can autonomously and effectively deal with uncertainty.

If you would be a real seeker after truth, it is necessary that at least once in your life you doubt, as far as possible, all things.

1.1 Definitions and Principles of Control System Design

In this section, we lay the groundwork by outlining fundamental concepts and terminologies, and offer a review of current control design m ethods. We differentiate these methods into two categories, namely "episodic" and "one-shot" strategies.

Every real-world control design problem begins merely with an unknown dynamical system and the aspiration to control it. As we venture into this, an almost philosophical question arises:

"How do we rationalize our certainty and uncertainty about the unknown?"

On the one hand, if we believe that the nature of the unknown system and the environment is predominantly random, then it would make sense to proceed with a probabilistic problem setup, i.e., viewing system and environment behavior as the realizations of random processes. That being said, it is oftentimes warranted to be more skeptical and prefer a more falsifiable, deterministic problem setup. In this book, we advocate for a deterministic view. Such perspective characterizes a control problem using three fundamental definitions: 1. Closed-Loop Complexity: The level of detail and class of models with which we want to describe the closed-loop system. 2. Model Hypothesis: Formulating a set of possible models. 3. Control Objective: Paraphrasing our objectives and requirements in control-theoretic terms. We next discuss these terms in detail.

Deterministic Formulation of a Control Problem

A deterministic setup consists of constructing a set of candidate models and phrasing our objectives as worst-case guarantees. Step one involves defining observation and action spaces, then choosing a representation to describe the closed-loop system. For example, one needs to choose whether to model the system as a linear or nonlinear dynamical system, how best to represent interactions with the environment, and how to impose constraints on the structure of the closed-loop system (i.e., constraints that may arise due to technological limitations: restricted and delayed communication, limited and distributed computational resources, limited sensing and actuation capabilities, etc.).

This first step in our decision-making process determines the level of *model complexity*. Essentially, we're choosing how detailed we want our representation to be of the system's behavior, its interaction with the controller, and the environment. This choice will set the complexity level for the entire closed-loop system dynamics.

Concurrently, or subsequently, we postulate our *model hypothesis*. This hypothesis comprises a collection of models, within which we expect at least one model to accurately depicting the dynamics of the unknown system. The size of this set mirrors the level of uncertainty (or confidence level) concerning our model hypothesis.

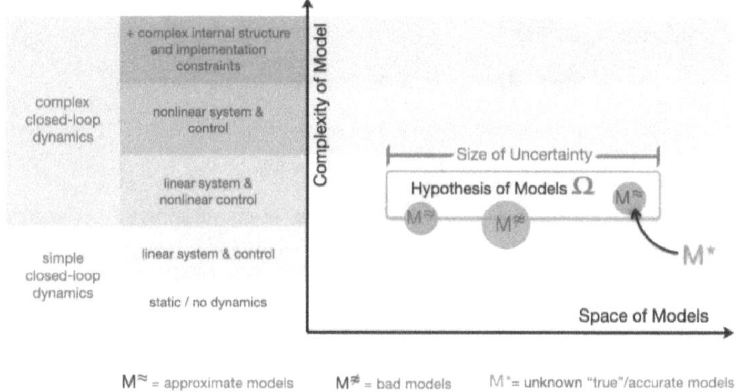

Figure 1.1: Deterministic problem formulation: Specifying a model class and stating a hypothesis set of possible models.

These two steps, choosing the model complexity and making a hypothesis, are illustrated in Figure 1.1. The box Ω represents a set of models forming our model hypothesis. The vertical axis indicates the level of complexity intrinsic to each of the models in the set, whereas the width of the box depicts the size of our set Ω, i.e., the breadth of uncertainty of our model hypothesis.

Note that determining whether or not our uncertainty is large or small can only be achieved once our control objectives are fully articulated. Flying a plane stably through turbulence, operating a manufacturing plant at peak efficiency, managing power distribution in a smart grid; many real world objectives and requirements can be phrased in terms of closed-loop stability conditions which should remain true even in worst-case scenarios. This type of control objective is often referred to as *worst-case closed-loop guarantees*. Some of the most common examples, which we will focus on in this work, are listed below:

- *ℓ_p-(internal)-stability*: A closed-loop system is ℓ_p-internally stable [49, 77, 139, 142] if its trajectories change continuously[1] under ℓ_p-bounded perturbations; that is, by adding small ℓ_p-bounded noise to every observed, controlled or computed signal.

- *Worst-Case Safety-Guarantees*: Most practical safety requirements are stated as verifiable and measurable conditions that should either always or never hold. Very often, this can be formally translated into some ℓ_∞-stability related condition [8, 45, 53], which can be expressed qualitatively as the following informal inequality

$$\max_t \{\text{Risk Score } \mathcal{R}_t \text{ at time } t\} \leqslant \text{Safety-Threshold},$$

  describing a worst-case ℓ_∞-bound on some scalar sequence $\{\mathcal{R}_t\}$ quantifying the risk or "unsafety" of the system at each time.

- *Worst-Case Cost-Performance Guarantees*: Many meaningful performance metrics can be phrased in terms of some cumulative sum of cost functions over time. Qualitatively, performance guarantees can be represented as an inequality of the type:

$$Total\ Cost = \sum_t \{\text{Cost } \mathcal{C}_t \text{ incurred at time } t\} \leqslant \text{Worst-Case Bound}.$$

  Guarantees of the above form can often be related to notions of ℓ_p-stability [49, 68, 85, 142][2] by expressing performance guarantees as worst-case ℓ_p-norm bounds on some suitable scalar cost sequence $\{\mathcal{C}_t\}$.

In summary, our ground zero for control system design is a deterministic problem formulation, which entails choosing a model representation of our closed-loop, stating our model hypothesis Ω, and translating our control objectives into suitable desired worst-case guarantees. Next, we review existing schools of thought and methods from literature. As illustrated in Figure 1.2, our overview will categorize literature by whether an episodic, "learning-then-control"-type, problem setting is assumed, or a one-shot problem setting is allowed.

---

[1] The stronger Lipshitz-continuity property corresponds with the definition of ℓ_p-finite-gain stability, and the Lipshitz constant is called *ℓ_p-gain*.

[2] Particularly common are parallels to ℓ_1- and ℓ_2-stability.

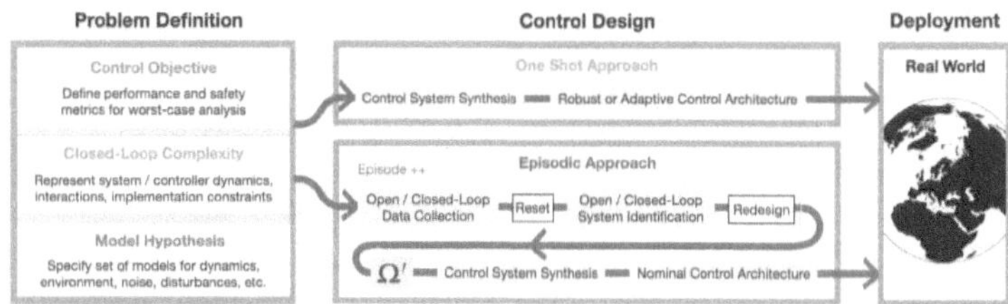

Figure 1.2: Overview of problem definition and different control design approaches.

One-Shot Approach to Control System Design

In the one-shot approach, we use the model hypothesis Ω and specified control objectives as our basis, and, without further input, task ourselves with designing a control algorithm that can be directly deployed on the unknown system; in other words, our goal is to design an algorithm capable of handling model uncertainty and feedback control decisions autonomously. In the classical control literature, this problem setting is the core subject of the fields commonly referred to as robust control theory (with the caveat that model uncertainty must be relatively small) [49, 141, 142] and adaptive control theory [15].

If the level of model uncertainty is *small* relative to our control objectives, we can apply methods of robust control theory literature. In robust control, synthesis processes primarily use the same internal structure or architecture of the feedback controller as in nominal control design. However, the feedback control law is developed to meet our control objective concurrently on all potential models of our assumption. Of course, these robust control design methods are only viable when the level of uncertainty is sufficiently small.

After a certain size of uncertainty, i.e, a certain set size of our model hypothesis, we have to drastically change the internal structure and architecture of our control algorithm, for it has to allow for incorporating online-learning in the closed loop. This is the canonical problem setting of the domain known as adaptive control [14, 69, 112].

Episodic Approach to Control System Design

If we have access to a controlled environment in which we can perform experiments and measurements of the unknown system (such as for the purpose of recording system trajectories, etc.), we can resort to an alternative approach to control design that leverages a two step process rather than just "one shot": a learning phase, and a controller synthesis phase. In the learning phase, we follow a process commonly known as system identification [16, 88], in which we perform repeated experiments, collect data, and apply learning algorithms to reduce the uncertainty of our initial model hypothesis Ω. Sometimes experiments are performed in closed-loop with a stabilizing feedback controller (assumed given as part of the problem formulation); this is referred to as closed-loop system identification [89] and is in general more difficult than open-loop system identification due to the distributional shift caused by the coupling of state and input signal. Particularly in the case of closed-loop system identification, it can also make sense to repeat the data collection and estimation process several times, wherein each round (also called *episode*), the experiment and stabilizing controller are redesigned based on the most recent data. The learning phase concludes once the system identification has made sufficient progress such that a new model hypothesis $\Omega' \subset \Omega$ can be formulated (refer to Figure 1.2) with minimal uncertainty. At this stage, we can synthesize the desired controller by choosing an arbitrary model within Ω' and designing a nominal controller specifically for it.

The episodic control design approach also commonly arises in model-based reinforcement learning [24, 57, 108]. Many modern learning and control algorithms also follow similar learning-then-control design approaches [3, 4, 62, 115].

Fundamental Prerequisites and Limitations to Episodic Design Optimally, the episodic control design is favored or paired with a single-shot method. However, the conditions for system identification may be too time-consuming or unattainable in reality. This includes the prerequisite of a strictly governed lab setting, essential for conducting safe experiments and affording the flexibility to pause or terminate the system when required. Particularly in real world applications, characterized by complex dynamical systems and uncertain environments, the former renders an episodic design approach impractical. Two main factors contribute to the impracticality of an episodic design approach in such settings. Firstly, system identification becomes quickly intractable for complex systems, particularly ones with nonlinear dynamics. Additionally, the potential for large model uncertainty in real-time operations necessitates the flexibility to adapt, further complicating the application of an episodic design approach.

An additional challenge is a fundamental theoretical obstacle inherent to the deterministic framework. Within episodic control design, worst-case error bounds on the complete system identification process are mandatory to ensure worst-case guarantees for the final control algorithm. However, achieving this is an intractable issue even for simpler linear systems [41]. This difficulty is evident in existing literature: probabilistic approaches offering closed-loop guarantees (stated with high probability) are plentiful [1, 28, 43, 50, 51]. Yet, deterministic approaches providing worst-case guarantees are scarce [35].

1.2 Major Challenges and Mission Statement

The main mission motivating this work is studying control design that can reliably and autonomously handle the complexity and large uncertainty inherent to real world dynamical systems. Tackling such a problem naturally calls for a one-shot design approach, and specifically, finding answers to the following central question:

How do we design control systems in one-shot, which provide worst-case guarantees in presence of large model uncertainty and complex closed-loop dynamics?

At its core there are three intertwined aspects to the problem: Complex Constraints and Dynamics, Worst-Case Guarantees, and Large Model Uncertainty. These issues present two critical challenges existing literature has yet to adequately address.

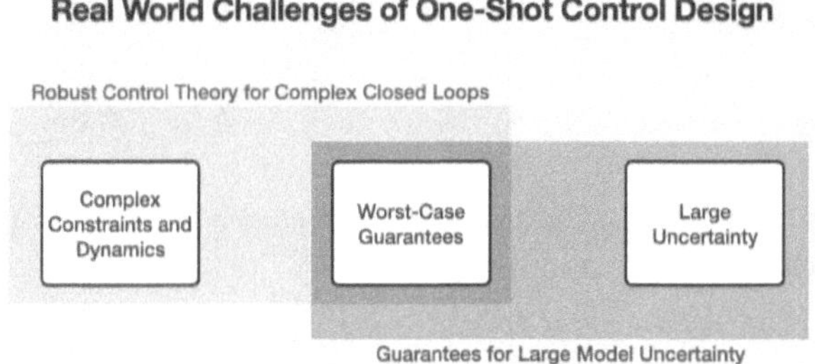

Figure 1.3: Overview of major challenges.

Challenge 1: Robust Control Theory for Complex Closed-Loop Dynamics
As alluded to previously, stability analysis and control design for nonlinear and complex systems is a core problem to consider.

Nonlinear dynamics in the system or controller result in overall nonlinear closed loops and require analysis and synthesis from the nonlinear control theory literature. Nonlinear stability analysis is based on some fundamental results in the classical literature [77, 101, 111, 117, 124], such as Lyapunov functions, passivity, the small-gain theorem, barrier functions [6], Poincaré maps, contraction analysis [90], and describing functions. Among these, the Lyapunov methods are by far the most widely used technique.

Methods for nonlinear control design can be separated into two groups: analytical and computational. Lyapunov stability analysis and the optimal control formalism [77, 85] have inspired most classical analytical methods such as feedback linearization, backstepping, sliding-mode control, and gain scheduling [77, 111, 117], among many others. With the technological breakthrough of computers in the 90s and the rapid advancement in computational capabilities and numerical optimization since then, computation-based control design approaches have also gained dominance. Prominent examples of such methods are Model-Predictive Control [30, 95] and reachability/viability-based approaches [20]. Another important line of work started with the sum-of-squares method (SOS) developed in [99, 103], which made it possible to compute Lyapunov functions through convex optimization and inspired new synthesis methods such as [104, 105, 133].

However, a common limitation in almost all approaches is that they do not scale well with system and controller complexity. We consider the following three aspects to be particularly important:

1. High-Dimensional Systems: Most optimal control or SOS-based synthesis procedures [19, 30, 67, 84, 85, 99, 103, 104] require solving (sometimes repeatedly [30]) optimization problems that are only tractable for small system dimensions.

2. Complexity of Controller Constraints: Many of the above synthesis methods cannot incorporate additional constraints on the controller. As mentioned above, technical limitations in practice can require us to impose structural constraints on the implementation of the realizing controller. These constraints are particularly common in large-scale systems, and some important ones include communication constraints between components used for sensing or actuation, limited actuation caused by saturation, and available computational resources that are possibly distributed.

3. Complexity of Controller Dynamics: Both stability analysis and control synthesis in a nonlinear setting primarily assume that the chosen controller is static or has a simple internal structure, such as an added state observer. Aside from a few exceptions (often analytical methods such as backstepping), many of the mentioned synthesis procedures, particularly computation-based ones, do not extend past static controllers. This is partially due to the fact that stability analysis of nonlinear closed-loop dynamics and controllers with rich internal structures is not well-studied. However, analysis and synthesis methods that are compatible with more complex and nonlinear control structures are increasingly needed in many modern control applications, especially in the rapidly growing area of learning and control [3, 5, 9, 37, 42, 59, 115]. In such problem settings, it is not unusual for controllers to have high-dimensional and complex internal dynamics. A common cause for this is that controller implementations require continuous and iterative solving of multiple (possibly interconnected and layered) tasks in closed loop, which potentially generate high-dimensional internal states with dynamics; such tasks include planning, machine learning, and online optimization.

Until recently [126], even for linear time-invariant systems, control design was challenging for the above problem settings. However, the system-level approach, as introduced in [126], enabled new efficient controller synthesis methods [66, 127] that allowed for localized, distributed, and scalable control design in large-scale linear systems. This was achieved by transforming constrained optimal linear control problems into convex optimization problems over achievable closed-loop maps that can be solved efficiently. A key component of the system-level synthesis (SLS) procedure is that once we have solved for the desired closed-loop map, there is a simple way to construct a controller that stably realizes this on the system.

We endorse the central idea of the system-level approach and use it in Part 1 of this book as our starting point, then go further to introduce a new framework for nonlinear control design and system analysis dedicated to addressing the aforementioned additional challenges.

Challenge 2: One-Shot Worst-Case Guarantees vs. Large Model Uncertainty

One-shot control design, in the case of arbitrarily large model uncertainty, naturally has to embed some kind of online learning process into the closed-loop system, i.e., data collection and inference is performed in closed loop, at the same time as the system is being controlled. This design approach is commonly taken in modern "learning and control" literature [1, 3, 4, 37, 43, 58, 62, 115] and in the more classical "adaptive control" literature [14, 15, 52, 69, 71, 72, 83, 112, 121, 132]. Exactly how these two fields integrate online interference into the closed-loop, however, remains quite different.

One-shot control design approaches in modern learning and control literature focus exclusively on problem settings wherein the system is either *perfectly* known [3, 4, 37, 58, 62, 115], or the model hypothesis enjoys small uncertainty. Moreover, said literature primarily focuses on the task of regret-optimal (with respect to some cost function) online control [1, 3, 4, 37, 42, 42, 43, 51, 62]. Furthermore, aside from a few exceptions ([28, 76] for example), the online learning for control literature mostly focuses on linear systems [35, 43, 62, 116] and convex cost functions. Linear-quadratic (LQ) costs have been of particular interest: [1, 2, 42, 43, 62] study algorithms with sublinear regret bounds, while [58, 115] present online control algorithms with competitiveness guarantees. However, compared to the adaptive control literature, even for linear systems, modern learning and control approaches that provide worst-case guarantees in the large model uncertainty setting are almost non-existent ([35][3]).

In contrast, adaptive control algorithms are primarily focused on providing stability guarantees in the presence of large model uncertainty, and less on cost performance guarantees. Despite a long history dating back to the 1960s [23], the field of adaptive control has not matured as much as other areas of control literature. From a conceptual point of view, the central problem of adaptive control, that is robustness to large model uncertainty, is clearly relevant and of interest in almost any control application, and one would naturally expect that methods and principles of adaptive control can be applied in a way that builds on top of existing control methods of other areas of control; however, this is not the case. Despite some efforts to address this [8–10], there is still no unifying framework that allows us to bridge this gap. As an example, theoretical tools, particularly concerning robustness and stability analysis, which are considered standard in other areas of control [6, 90, 139, 141], cannot be applied (or have no analog) in the adaptive control literature. In particular, theory and design methods which come with safety and cost performance under adversarial disturbances and noise, are almost non-existent ([68][4] is a rare exception).

---

[3]Proves sublinear regret and stability guarantees for a linear system with large model uncertainty.

[4]$\mathcal{L}_1$-adaptive control provides ℓ_∞-stability guarantees for linear systems and a small class of nonlinear systems

Reviewing the current literature, old and new, we are still very far from a unifying theory for designing control systems with worst-case guarantees in the presence of large model uncertainty. Part 2 of this book is dedicated to closing this gap (at least partially) by introducing new theory and frameworks for control design.

1.3   Main Contributions of the book

In this book, we aim to address the previous challenges by investigating the problem of one-shot control design from two complementary perspectives.

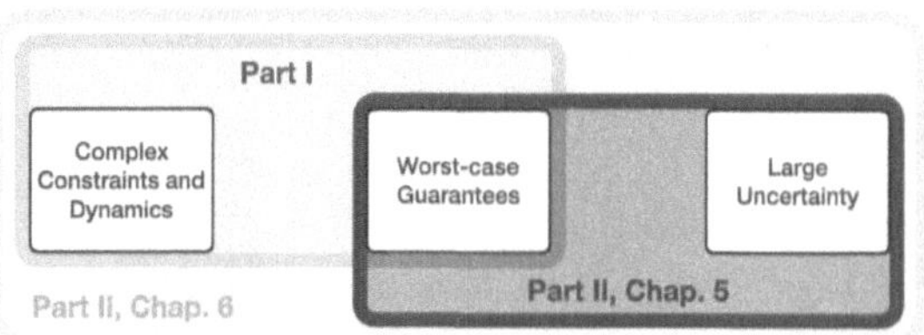

Figure 1.4: book overview.

In Part 1, we restrict ourselves to the small uncertainty setting and focus on developing new theory and methods for robust control design and system analysis of closed-loop systems with nonlinear dynamics, complex constraints and complex internal structure. In the Part 2, we develop new theory and algorithms for the large model uncertainty setting. In particular, we introduce a first general and modular one-shot control design framework ("PixSel") with worst-case cost- and safety performance guarantees in the presence of arbitrarily large uncertainty in our model hypothesis. The following sections will delve into these topics, shedding light on our novel contributions to this emerging interdisciplinary intersection.

Synopsis of Part I: Nonlinear Closed Loops and System Level Control

In Part 1 of the book we introduce a new framework for nonlinear control design and system analysis dedicated to addressing the first set of challenges of Section 1.2. In Chapter 2, we show that for a very general class of nonlinear systems, there is a universal connection between the closed-loop map, the external representation of a closed-loop as a map between input and output sequences (see Figure 1.5), and the operators corresponding to the feedback controller and system dynamics, which describe the behavior of the closed-loop from an internal point of view. We discuss important equivalence relations between different representations of the closed-loop system via a set of characterizing operator equations. These equations and their solution space provide useful insights for system analysis and control design, which form the foundation of the *nonlinear system level approach*.

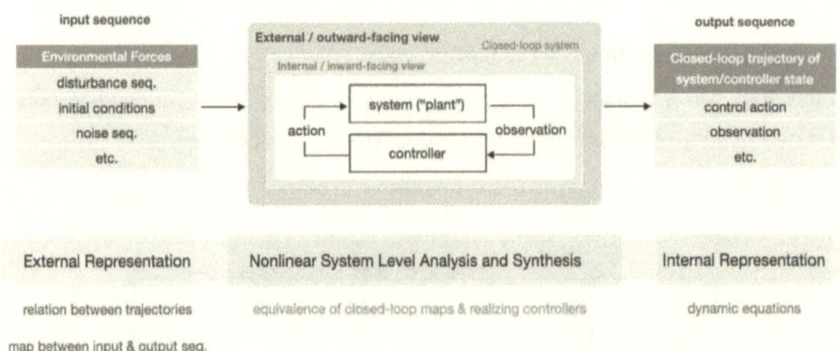

Figure 1.5: Part 1: Nonlinear system level approach

The solutions, called closed-loop maps (CLM) of the characterizing operator equations, lead to concrete implementations of the realizing controller. This enables us to reformulate control design problems equivalently in terms of CLM operators, and realize feedback control as a structured dynamic system parameterized by the desired CLM operator. We refer to this special control structure as the "system-level" implementation, which turns out to offer more benefits than its intended original purpose. In fact, parameterizing a system-level implementation with approximate solutions of the operator equation provides a stable closed-loop system, as long as the approximation error is small enough in an appropriate sense. This leads to a result for robust stability analysis of nonlinear closed-loop systems, which is both instructive and can be leveraged for robust controller synthesis.

We then proceed to investigate closed-loop maps in the special setting where we have a linear system and nonlinear dynamic control. We define a particular class of "blended" closed-loop maps, which are particularly suited for this problem setting, and offer new insight into the dynamics of such closed-loops. The corresponding system-level implementations of these blended closed loops lead to a promising synthesis approach, which is capable of "blending" multiple linear controllers into one nonlinear controller, thus allowing us to combine desirable properties of each individual component.

In Chapter 3, we explore some first implications of these results, such as robust discrete-time trajectory tracking controllers for continuous-time nonlinear systems. Second, we investigate some first application s cenarios, n amely, d istributed constrained LQR and distributed anti-windup control, where this technique naturally provides significant b enefits in th e la rge-scale sy stem se tting ov er ex isting me thods. We describe a synthesis procedure for blended SLS controllers that outperforms any optimal linear controller for the constrained LQR problem [36, 86, 91, 140]. We then discuss how the blended SL approach provides a natural remedy for controller windup in a way that is easily scalable for use in large-scale control systems. We discuss the efficacy of the methods with simulations and show that synthesis and implementation enjoy the same benefits as previous SLS synthesis methods: both are distributed, handle delays, sparse actuation, and allow for localized disturbance rejection.

In Chapter 4, we begin to shift our focus towards the problem of learning-to-control in closed-loop; the core topic of the second part of the book. We investigate a common feature found in many learning and control algorithms across various problem settings: At each time step, we estimate a model of the system dynamics and then switch to a control law designed for that model; i.e., we pretend to have found a model that will remain accurate going forward. This design approach is often referred to as "certainty equivalent" control.

We show that this general design principle imposes a certain structure on the closed-loop maps, which we can leverage for stability analysis and algorithm design. This leads to a sufficient condition for closed-loop stability, which admits a natural interpretation within our framework. Decomposing the dynamics of the lumped disturbance reveals that three factors are important for robust and stable adaptation in closed-loop:

1. Model Sensitivity of Nominal Control: Small changes in models should cause small changes in nominal controllers (the control law selected for each model).

2. Consistency of Models: Models should be consistent (or consistent up to some bounded error) with our observations.

3. Select Models Efficiently: The posited model should change as little as possible.

This observation is a manifestation of a deeper concept, which we explore in Part 2, Chapter 6 of the book. Learning-to-control can be decomposed into control via *Robust Oracle Design* (addressing factor 1) and learning via *Consistent Model Chasing* (addressing factors 2 and 3). This idea serves as the foundation for many of the results in Part 2, where we develop the theory and framework for learning to control largely uncertain dynamical systems with guarantees.

In the second part of the Chapter 4, we explore applications of this result for problem settings concerned with online learning of optimal controllers. To this end, we focus on the setting of linear time-invariant systems and linear-quadratic costs; a problem setting that has received immense recent attention in the learning and control literature.

Guided by the theoretical findings in Part 1, we follow the principle of certainty equivalence to design a learning-to-control scheme with LQ-optimal system-level controllers as our basis of nominal control laws. We perform perturbation analysis of the solutions for the LQ-optimal control problem and show that the solutions are Lipschitz-continuous with respect to changes in the system matrices of the linear system (over parameter sets of equal degree of controllability). This partial result in itself is new and characterizes the sensitivity of LQ-optimal closed-loop maps for linear time-invariant (LTI) systems in terms of system-theoretic properties such as controllability and observability. With this result, we analyze the closed-loop stability of the learning-to-control scheme and provide conditions for model selections that are sufficient for closed-loop st ability. We revisit these conditions again in Part 2 in the context of consistent model chasing.

In summary, Part 1 of this book discusses new fundamental connections between closed-loop maps, realizing controllers, and system-level implementations that open up new possibilities for the analysis and design of closed-loop control systems, allowing for a broad range of nonlinear dynamics in the system and controller. The strength of this approach, compared to existing ones, is that it is particularly well-suited for the handling of complex systems, such as large-scale systems, as well as complex control structures caused by high-dimensional internal dynamics (such as learning and optimization in closed-loop) and/or restrictive implementation constraints (spatially and temporally structured sensing, actuation, communication, and computation constraints). This has proven to be a catalyst for new control approaches, some of which we outline in the next section.

Impact, Current Work and Outlook. The results presented in Part 1 of the book were inspired by our earlier work [66], which for the first time, extended the

SLS-theory to the linear time-varying case and noticed its importance for online learning of feedback control: [66] introduced a framework for distributed adaptive control via system-level controllers, which allowed for incorporating online learning in large-scale closed-loop systems with delay and communication constraints. In later work, we leverage some of the ideas in [7] to develop sub-optimal adaptive system-level controllers that are robust to communication dropouts. With [1], we were the first to extend the system level analysis approach to general nonlinear systems, and then demonstrated its use-case for nonlinear control design in the large-scale system setting in [8]. [8] uses the CLM blending approach to formulate a nonlinear control synthesis method that outperforms linear controllers for the constrained LQR problem; notably, CLM blending is also shown to provide a solution for distributed and localized anti-windup. In [6] we introduced a new way for robust control design through a decomposition into a problem of CLM design and robust controller implementation. The theory developed in [1] and presented in Chapter 2 has served as a foundation for new results on nonlinear controller synthesis: [39] presents synthesis methods for nonlinear system-level control for dynamical systems with polynomial dynamics. [56] applies the theory to provide a characterization of the space of nonlinear stabilizing controllers and introduces a new framework, called neural system-level synthesis, to learn stabilizing nonlinear system-level controllers via neural networks.

Synopsis of Part II: Learning to Control Unknown Systems

Part 2 of the book focuses on addressing the second set of challenges, a conundrum we referred to as "One-Shot Worst-Case Guarantees vs. Large Uncertainty" earlier in Section 1.2.

In contrast to Part 1, which focuses on the "small uncertainty" setting, Part 2 focuses on the more general "large uncertainty" scenario, where the set of possible models is bounded (in an appropriate sense), but is allowed to be arbitrarily large. Thus, in a practical sense, we assume that the dynamics of the system are almost entirely "unknown".

Chapter 5 explores possible trade-offs and difficulties associated with our overall objective of handling completely unknown systems. It is clear that with larger model uncertainty comes degradation in robustness; hence, there has to be a limit at which model uncertainty renders possibly desired guarantees impossible to achieve. Surprisingly, there are no clear and general answers to this problem, even for simple linear systems. In Chapter 5, we focus on a simple class of linear systems and explore the following questions:

At what size of model uncertainty does "learning-to-stabilize" become an impossible task? Is there a minimal model assumption?

The results presented in Chapter 5 are based on the work published in [65] and develop new methods for stability analysis and control design without the need for a model. In particular, we demonstrate the first instance of an all-model-free formulation of controller, closed-loop dynamics, and robust stability analysis. We end up with a positive answer for the considered class of systems. We present a simple model-free control algorithm that is able to robustly learn and stabilize an unknown discrete-time linear system with full control and state feedback subject to arbitrary bounded disturbance and noise sequences. The controller does not require any prior knowledge of the system dynamics, disturbances, or noise, yet it can guarantee robust stability and provide asymptotic and worst-case bounds on the state and input trajectories. To the best of our knowledge, this is the first model-free algorithm that comes with such robust stability guarantees without the need to make any prior assumptions about the system. Moreover, the theory and results developed provide a first set of tools that allow for an entirely model-free formulation of controller, closed-loop dynamics, *and* robust stability analysis. The simulation results also show that despite the generality and simplicity, the controller demonstrates good closed-loop performance: fast convergence, small learning transients, and almost optimal asymptotic gain.

Investigating this problem led us to a new convex geometry-based approach towards robust stability analysis, which served as a key enabler in our results. A distinguishing feature of the approach is that stability conditions can be phrased entirely in terms of data and there is an intuitive geometric way of quantifying and analyzing model uncertainty, which can be phrased in terms of metric entropy, the absolute convex hull of the observed data, and the disturbance. This perspective allows us to conduct stability analysis independent of the system matrix and the size of the disturbance and noise. The key idea that led to these results was to describe the reduction of uncertainty obtained from new observations in a set-theoretic way and study the geometric relation between the data and the uncertainty. The insight gathered from this approach inspired us to rethink the general problem of learning to control as a whole from a new angle, which led us to a framework introduced in Chapter 6.

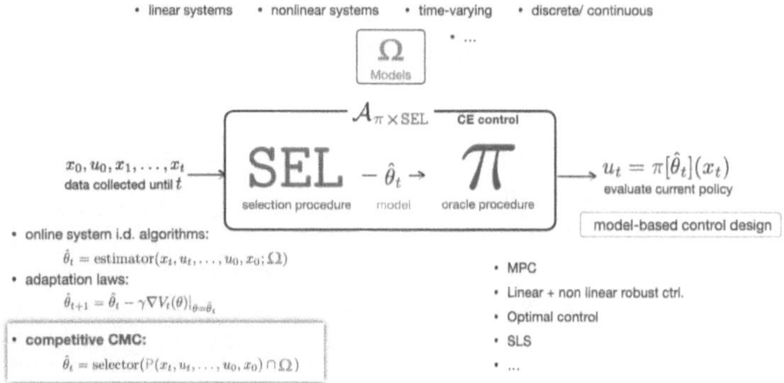

Figure 1.6: Chapter 6: PixSel Framework for adaptive systems analysis and control design

In Chapter 6, we introduce a new modular framework for one-shot control design that is particularly well suited for learning-to-control problem settings that require robust safety and cost performance guarantees in the presence of potentially large model uncertainty. Our approach is based on decomposing the problem into two sub-problems: online learning, named "consistent model chasing", and the underlying control problem in the absence of model uncertainty, called "oracle design". Each of these sub-problems can be addressed separately, and their solutions (a control oracle and a model chaser) are used to instantiate a certainty-equivalent learning-to-control scheme. This scheme, in a symbiotic way, inherits both control- and learning-theoretic guarantees, certifying the robustness of the closed-loop, even for large model uncertainty in the system dynamics. The range of closed-loop guarantees that we can obtain through this inheritance is fairly broad. In this chapter, which is based on the work in [5], we present worst-case performance guarantees in terms of $\{0, 1\}$ costs (commonly referred to as finite mistake or mistake bound guarantees in online learning) and represent worst-case safety guarantees as bounds on the ℓ_∞-norm of the trajectories of the closed-loop. As we show later, this way of phrasing closed-loop "safety" and "performance" is expressive enough to represent many other guarantees of stability, convergence, or even set invariance.

A defining feature of this new approach is the inheritance relation between the learning-to-control algorithm ("PixSel") and its instantiating sub-procedures, which builds a new bridge between the fields of online learning and robust control theory. This allows us to rigorously merge online learning with traditional control algorithms for the purpose of learning to control uncertain dynamical systems.Moreover, it opens up an important connection between two, so far mostly separated, problem settings in control theory: small model uncertainty and large model uncertainty. As discussed earlier, the classical (adaptive control) and modern (learning and control) literature do not provide a simple way to scale robust controllers designed for the setting of small uncertainty to the more general setting of large model uncertainty. Within the "PixSel" framework, we accomplish this by merging the controller with a consistent model chaser, thereby extending the robustness guarantees of the original controller to the large-model uncertainty setting. As far as we are aware, there are no other existing approaches that enable this in such a general and straightforward way.

As an instructive example and to demonstrate the practicality of our approach, we show how to instantiate this framework for general robotic systems for common tasks such as stabilization or trajectory tracking. In addition to providing theoretical guarantees, empirical results show that our framework is a promising approach to designing efficient algorithms for learning and control in practice. We apply our approach to the problem of swinging up a cartpole with large parametric uncertainty in a realistic and highly challenging setting and show that it consistently achieves good performance over 900 experiments with different parameter settings. Despite its popularity, the cart-pole swing-up problem presents many fundamental challenges (underactuated, non-minimum phase, nonlinear dynamics) for control design, and the majority of existing design approaches in the adaptive control literature [73, 83, 117] are not applicable. There does not seem to be any empirical evidence of other design methods capable of tackling this learning-to-control problem in a large-model uncertainty setting.

In conclusion, the main motivation behind the research presented in this part was to establish a theoretical foundation for a systematic approach to online learning of feedback control in the presence of uncertainty and constraints. The objective was to develop a theory for "Online Learning of Feedback Control with Robustness to Large Uncertainty." Chapter 5 introduced new theoretical tools for a model-free approach to the problem, formulating controller synthesis and analysis in a model-free and model-agnostic way.

Chapter 6 addressed the core problem of learning to control unknown systems from a model-based perspective, introducing the "PixSel" framework. This framework modularly combines online learning and control design methods to provide robust safety and cost performance guarantees in the presence of large model uncertainty.

The framework establishes an inheritance relation between online learning and control procedures, merging the fields of online learning and robust control theory. It also bridges the gap between small and large model uncertainty settings, providing a straightforward way to extend robustness guarantees to the more general setting.

Impact, Current Work and Outlook. The theory developed in Part 2 has served as a foundation for recent progress in this problem space. The work of [135] applies the PixSel framework to provide an efficient solution to the problems of robust voltage control under uncertain grid topology. In [9], we adapt the PixSel framework for the distributed case and combine it with the system-level approach to provide the first distributed, localized adaptive control approach, both in learning and control, that provides worst-case safety and stability guarantees for arbitrarily large bounded uncertainties. The algorithms which we presented in [9], form a symbiotic marriage of the theoretical results developed in Chapter 6 and Chapter 4 and provide a tremendous improvement over our early adaptive system-level control framework [66] in terms of scalability (more efficient distributed learning), computational cost ([66] relied on solving a robust optimization problem, which, compared to [9], scales very poorly with local model complexity), generality, and guarantees ([9] provides worst-case stability bounds for arbitrarily large uncertainty, while in [66] we can only provide guarantees for the small uncertainty setting).

More recently, the work presented in [138] applies the PixSel framework to the special case of LTV systems and develops an approach to online stability of unknown linear time-varying systems by reformulating the corresponding problem of consistent model chasing into a convex body chasing problem.

Part I

Nonlinear Closed-Loops and System Level Control

OVERVIEW

In this part of the book, we introduce a new operator-theoretic framework for the design and analysis of complex closed-loop systems. This framework allows for general nonlinear dynamics in the plant and controller, robust control synthesis and stability analysis, and complex constraints on the controller implementation. We refer to this framework as the *nonlinear system-level approach*. Chapter 2 introduces this theory, while the following two chapters demonstrate how this theory opens up new methods for control synthesis and system analysis.

Chapter 2 is based on the work presented in [64] and begins with an introduction to the required mathematical concepts from operator theory. It also reviews old and new lemmas for operator stability analysis. The first core result of this chapter characterizes the space of achievable closed-loop systems once feedback control is introduced to a given nonlinear system F. By representing closed-loop systems as nonlinear causal operators mapping between inputs and outputs, we identify the space as solutions to an operator equation, the *Closed-Loop-Map-Equation*, parameterized by F. Moreover, any operator Ψ that solves the CLM equation, called a closed-loop map (CLM) of the system F, can be realized by means of a particular controller structure, the system level controller $\mathrm{SL}(\Psi)$. The internal stability of this realization is the topic of the second main theorem, which discusses sufficient and necessary conditions for stability. The final part of this chapter investigates closed-loop systems consisting of nonlinear controllers and linear plant dynamics. We show that in this case, the set of nonlinear CLMs is closed under a special type of combination, which we call *CLM-blending*, and this allows us to construct more expressive CLMs from simpler ones in a hierarchical way.

The ramifications of this result are explored in Chapter 3, which is based on the work presented in [8, 64, 66]. As a first application, we use blending to develop a scalable method for robust control design of linear time-varying dynamic controllers for trajectory tracking in nonlinear continuous-time systems. We then discuss how CLM-blending allows for control design with improved safety and cost-performance trade-offs in control applications considering large-scale linear systems subjected to controller constraints and actuator saturation. We discuss this in the context of two control problems, constrained LQR and distributed anti-windup.

The first half of Chapter 4 applies the nonlinear system level framework to derive new results for stability analysis of closed-loop systems with online model estimation/selection and certainty-equivalent control in the loop. We introduce an operator-theoretic characterization of the space of closed-loop systems such as tuples $(\Omega, \hat{F}, \Psi, S)$ consisting of a compact parameter space Ω of possible system models ω with dynamics $\hat{F}[\omega]$, a parameterization of nominal CLMs $\Psi : \omega \mapsto \Psi[\omega]$ underlying the certainty-equivalent control policies, and an operator S representing an algorithm for online model estimation. Our main theorem states that if the parameterizations of dynamics $\hat{F} : \omega \mapsto \hat{F}[\omega]$ and desired closed-loop behavior $\Psi : \omega \mapsto \Psi[\omega]$ are continuous and the algorithm S guarantees convergent and consistent model selection, then the overall closed-loop dynamics are stable.

Guided by these theoretical findings, in the second half of the chapter, we follow the certainty equivalence principle to design a learning-to-control scheme with LQ-optimal system-level controllers as the basis of nominal control laws. Most of the results are based on work published in [9, 66]. We perform perturbation analysis of the solutions for the LQ-optimal control problem and show that the solutions are Lipschitz-continuous with respect to changes in the system matrices of the linear system (over parameter sets of equal degree of controllability). This partial result is novel and characterizes the sensitivity (i.e., analytic bounds on the Lipschitz constant) of LQ-optimal closed-loop maps for LTI systems in terms of system-theoretic properties such as controllability and observability. With this result, we analyze the closed-loop stability of the proposed learning-to-control scheme and provide conditions for model selection that are sufficient for closed-loop stability, which we revisit in Part 2 in the context of consistent model chasing.

Chapter 2

NONLINEAR CLOSED LOOP MAPS FROM THE INSIDE AND OUTSIDE

There is a universal connection between the achievable closed-loop dynamics and the corresponding feedback controller that produces it, which shows promise of leading to new methods for robust non-linear control in discrete time. In this chapter, we derive that given a causal nonlinear discrete-time system and controller, the resulting closed loop is a solution to a nonlinear operator equation. Conversely, any causal solution to the nonlinear operator equation is a closed loop that can be achieved by some causal controller. Moreover, solutions can be substituted into a simple dynamic controller structure, which we refer to as a *system level* controller, to obtain an implementation of the unique corresponding feedback controller. System-level controllers are a promising approach for robust nonlinear control, as we show that even when they are parameterized with approximate solutions to the operator equation, they can still produce robustly stable closed loops. We provide theoretical results that state how the degree of approximation and robust stability of the closed loop are related and show that this relationship can be leveraged for controller synthesis.

2.1 Introduction

Compared to linear control theory, there are fewer mathematical tools for tackling controller synthesis of general nonlinear systems. Nonlinear stability analysis is based on some fundamental results in the classical literature [77, 101, 111, 117, 124], such as Lyapunov functions, passivity, the small gain theorem, barrier functions [6], Poincare maps, contraction analysis [90] and describing functions. Among these, the Lyapunov methods are by far the most widely used technique.

Methods for non-linear control design can be separated into two groups: analytical and computational. Lyapunov stability analysis and the optimal control formalism [85], [77] have inspired most classical analytical methods such as feedback linearization, backstepping, sliding-mode control, gain scheduling [77, 111, 117], and many others. Nevertheless, with the recent explosion of available computational resources and progress in the optimization and control theory community, significant progress has been made toward achieving a more generalized, data-driven approach to nonlinear control design. With the sum of squares methods (SOS) [103], [99], it became possible to compute Lyapunov functions for stability analysis through convex optimization. SOS-based controller synthesis methods are presented in [104, 105], and [133] for the continuous-time (CT) and discrete-time (DT) settings, respectively. Examples of computational methods based on approximating solutions of the Hamilton-Jacobi-Bellman type of equations are found in [120], [67], [130]. Other more recent works [84] (CT), [61] (DT) provide alternative formulations of optimal controller synthesis through occupation measures.

Inspired by the recently developed system-level approach to linear control theory [126], we present a new insight into nonlinear discrete-time systems that enables new synthesis methods for nonlinear discrete-time systems. The system-level approach, as introduced in [126], enabled new efficient controller synthesis methods [66, 127] that allow for localized, distributed, and scalable control design in large-scale systems. This is achieved by transforming constrained optimal control problems as convex optimization problems into achievable closed-loop maps that can be solved efficiently. A key component of the system-level synthesis (SLS) procedure is that once we have solved for the desired closed-loop map, there is a simple way to construct a controller that stably realizes this on the system.

In this chapter, we show that this connection between closed-loop maps and their corresponding realizing controller is not merely a phenomenon of linear systems, but rather a surprisingly universal control principle that extends to general nonlinear discrete-time systems. We demonstrate that given any feasible nonlinear closed-loop map from disturbance to state and input, we can construct an internally stable dynamic controller that realizes it. Specifically, we characterize the space of all feasible closed-loop maps as solutions to a nonlinear operator equation and define a dynamic controller that realizes them. This controller structure, which we refer to as a system-level (SL) controller, is parameterized by the solutions of the operator equation. We further show that even approximate solutions of this equation, with a small enough error, can still yield stabilizing controllers when parameterized into the SL controller structure. We characterize the internal stability of the nonlinear closed loop and discuss a simple sufficient closed-loop stability condition based on the small-gain theorem.

The presented approach motivates new paths towards nonlinear control synthesis: 1) finding approximate solutions to the closed-loop operator equation and 2) obtaining a stabilizing controller by parameterizing an SL controller with the approximate solutions. In the latter part of the chapter, we also discuss a method for synthesizing nonlinear dynamic controllers based on *blending* of closed-loop maps. These ideas and their importance for control applications will be explored in Chapter 3. To formulate the framework rigorously, we first review some background on operator theory and stability and introduce necessary preliminaries such as notation and basic definitions.

Preliminary Definitions and Notation

Binary Relations. A subset $R \subset \mathcal{X} \times \mathcal{Y}$ of the Cartesian product of two sets $\mathcal{X}$ and $\mathcal{Y}$ is called a binary relation. $do(R) = \{x \in \mathcal{X} \mid \exists y \in \mathcal{Y} : (x, y) \in R\}$ is called the domain and $\mathcal{Y}$ is called the codomain, denoted $co(R)$ of the relation R. One says x relates to y to state $(x, y) \in R$. The image $R(\mathcal{X}')$ of a subset $\mathcal{X}' \subset do(R)$ refers to the set $\{y \in \mathcal{Y} \mid \exists x \in \mathcal{X}' : (x, y) \in R\}$ and similarly $R^{-1}(\mathcal{Y}') := \{x \in \mathcal{X} \mid \exists y \in \mathcal{Y}' : (x, y) \in R\}$ is the preimage of a subset $\mathcal{Y}' \subset \mathcal{Y}$. A composition $R_1 \circ R_2$ or $R_1 R_2$ of two relations $R_1 \subset \mathcal{X} \times \mathcal{Y}$ and $R_2 \subset \mathcal{Y} \times \mathcal{Z}$ refers to the relation $\{(x, z) \mid \exists y \in \mathcal{Y} : (x, y) \in R_1, (y, z) \in R_2\} \subset \mathcal{X} \times \mathcal{Z}$ and, similarly, $z = R_1 R_2(x)$ expresses that there exists a $y \in \mathcal{Y}$ such that $y = R_2(x)$ and $z = R_1(y)$.

Relations and Functions. A relation R is said to be functional if for any two pairs $(x, y) \in R, (x', y) \in R$ holds $y = y' \implies x = x'$. We write $f : \mathcal{X} \to \mathcal{Y}$ as a shorthand to say that f is a functional relation with domain $\mathcal{X}$ and codomain $\mathcal{Y}$ and write $y = f(x)$ or $f : x \mapsto y$ to express $(x, y) \in f$. A *restriction* of a function $f : \mathcal{X} \to \mathcal{Y}$ to a subset $\mathcal{X}' \subset$ is a function $g : \mathcal{X}' \to \mathcal{Y}$ such that $g(x) = f(x), \forall x \in \mathcal{X}'$ and we refer to the function g with the notation $f|_{\mathcal{X}'}$ or $f|\mathcal{X}'$. On the other hand, an *extension* of a function $f : \mathcal{X} \to \mathcal{Y}$ to $\mathcal{X}_e \supset \mathcal{X}$ refers to the function $g : \mathcal{X}_e \to \mathcal{Y}$ such that $g(x) = f(x), \forall x \in \mathcal{X}$; *co-restriction* and *co-extension* of $f : \mathcal{X} \to \mathcal{Y}$ refer to functions $g_1 : \mathcal{X} \to \mathcal{Y}'$ for $\mathcal{Y}' \supset f(\mathcal{X})$ and $g_2 : \mathcal{X} \to \mathcal{Y}_e$, $\mathcal{Y}_e \supset \mathcal{Y}$ such that $g_1(x) = g_2(x) = f(x), \forall x \in \mathcal{X}$.

Sequences. A sequence s of elements in $\mathcal{S}$ over the time horizon $\mathcal{T}$ is a function $s : \mathcal{T} \to \mathcal{S}$ where the domain $\mathcal{T} \subset \mathbb{N}$ is a subset of natural numbers $\mathbb{N} = \{0, 1, \dots\}$ and the codomain is $\mathcal{S}$; the space of all such sequences is denoted by $\mathcal{S}^\mathcal{T}$. If we say s is a sequence of $\mathcal{S}$ and do not specify $\mathcal{T}$ or just write $s \in \mathcal{S}^\mathbb{N}$, it is implied that $\mathcal{T} = \mathbb{N}$. We will often use s_t to refer to the value $s(t)$ of a sequence s in t. As defined for general functions above, we will write, for example, $s_{[t_1, t_2]}$ or $s_\mathcal{T}$ for some $\mathcal{T} \subset \mathbb{N}$, to refer to the *restrictions* of s to an interval $[t_1, t_2] \subset \mathbb{N}$ or subset $\mathcal{T}$; *extensions* of a sequence $s \in \mathcal{S}^\mathcal{T}$ to time horizon $\mathcal{T}_e \subset \mathbb{N}$ are defined analogously as in the context of general functions. A sequence $s \in \mathcal{S}^{\mathcal{T}}$ over some finite horizon $\mathcal{T} \subset \mathbb{N}$, i.e., $|\mathcal{T}| < \infty$, is called a *finite sequence*.

Remark. *It is important to note that a finite sequence $s_{[\tau,\bar{\tau}]}$ entails more information than merely the vector $[s_\tau, s_{\tau+1}, \ldots, s_{\bar{\tau}}]$, because in the latter all information about the domain is lost. To point out this difference formally: the space of all finite sequences $s : \mathcal{T} \to \mathcal{S}$ with domain of size n (that is: $|\mathcal{T}| = n$) is infinite dimensional and <u>not</u> simply an n-dimensional vector space of $\mathcal{S}$.*

Normed Vector Spaces and their Sequence Spaces   We will use caligraphic font variables - most commonly just $\mathcal{X}$ and $\mathcal{U}$ - to denote two fixed finite-dimensional vector spaces equipped with norms $|\cdot|_{\mathcal{X}}$ and $|\cdot|_{\mathcal{U}}$. For convenience, we assume that there is some norm $|\cdot| : \mathcal{X} \times \mathcal{U} \mapsto \mathbb{R}_0^+$ such that $|\cdot|_{\mathcal{X}} : x \mapsto |(x,0)|$ and $|\cdot|_{\mathcal{U}} : u \mapsto |(0,u)|$, and therefore we drop the dependence on the space and let $|\cdot|$ represent a default norm for any finite-dimensional vector spaces. If not otherwise specified, the reader can assume that $\mathcal{X} = \mathbb{R}^n, \mathcal{U} = \mathbb{R}^m$ for some n and $m < n$, and $|\cdot|$ represent the standard Euclidean norm. We will use $\ell^{\mathcal{X}}$, $\ell^{\mathcal{U}}$, and $\ell^{\mathcal{X} \times \mathcal{U}}$ to denote the vector space of sequences $\mathcal{X}^{\mathbb{N}}, \mathcal{U}^{\mathbb{N}}$, and $(\mathcal{X} \times \mathcal{U})^{\mathbb{N}}$. Similarly, we define the ℓ_p norms in these spaces as

$$\|\boldsymbol{x}\|_p := \left(\sum_{k=0}^{\infty} |x_k|^p \right)^{1/p} \qquad\qquad \|\boldsymbol{x}\|_\infty := \sup_{k \geq 0} |x_k|$$

where $|\cdot|$ denotes the corresponding norm in finite-dimensional vector spaces. Consequently, define the subspaces of ℓ_p bounded sequences in $\ell^{\mathcal{X}}$ as $\ell_p^{\mathcal{X}} \subset \ell^{\mathcal{X}}$, $\ell_p^{\mathcal{X}} := \{\boldsymbol{x} \in \ell^{\mathcal{X}} | \|\boldsymbol{x}\|_p < \infty\}$ and define $\ell_p^{\mathcal{U}}, \ell_p^{\mathcal{X} \times \mathcal{U}}$ analogously. Sequences that have finitely many nonzero elements span a linear subspace of $\ell^{\mathcal{X}}$ which we refer to as $\ell_0^{\mathcal{X}}$. Furthermore, the subset of sequences $\boldsymbol{x} \in \ell_0^{\mathcal{X}}$ for which $\boldsymbol{x}(k) = 0, \forall k \geq t$, i.e., the sequence takes on zero values after time t, form a $(t+1)\dim(\mathcal{X})$-dim. subspace of $\ell_0^{\mathcal{X}}$, which we refer to as $\ell_{[0,t]}^{\mathcal{X}}$.

Remark. *Unless it is crucial for the discussion, we will not distinguish between the two spaces $\ell_{[0,t]}^{\mathcal{X}}$ and $\mathcal{X}^{[0,t]}$. However, it should be noted that strictly speaking, $\mathcal{X}^{[0,t]}$ and $\ell_{[0,t]}^{\mathcal{X}}$ are only isomorphic to each other, since $\ell_{[0,t]}^{\mathcal{X}}$ is an embedding of $\mathcal{X}^{[0,t]}$ into space $\ell^{\mathcal{X}}$.*

A truncation $\boldsymbol{x}'$ of a sequence $\boldsymbol{x} \in \ell^{\mathcal{X}}$ is a sequence $\boldsymbol{x}' \in \ell_{[0,t]}^{\mathcal{X}}$ such that $x_k' := x_k$ for $k \in \{0, \ldots, t\}$ and some t, that is, a sequence formed by truncating all but the first of the elements t of $\boldsymbol{x}$.

Tuples and Concatenations of Sequences We will use small $\mathtt{tt}$-font styled variables $\{\mathtt{x}, \mathtt{u}, \mathtt{w}, \hat{\mathtt{w}}, \mathtt{y}, \mathtt{z}\}$ to form index sets for sequence tuples or partial operators. For example, if we write $\tau = \{\tau_i\}_{i \in \mathcal{I}} \in \times_{i \in \mathcal{I}}(\mathcal{V}_i)^{\mathbb{N}}$, $\mathcal{I} = \mathtt{x}, \mathtt{u}, \mathtt{w}$, then we mean that τ is a tuple $(\tau_{\mathtt{x}}, \tau_{\mathtt{u}}, \tau_{\mathtt{w}})$ of sequences $\tau_{\mathtt{x}} \in (\mathcal{V}_{\mathtt{x}})^{\mathbb{N}}$, $\tau_{\mathtt{u}} \in (\mathcal{V}_{\mathtt{w}})^{\mathbb{N}}$, and $\tau_{\mathtt{w}} \in (\mathcal{V}_{\mathtt{w}})^{\mathbb{N}}$; $\times_{j \in \mathcal{J}} \mathcal{S}_j$ generally denotes the Cartesian product of a family of sets $\{\mathcal{S}_j\}_{j \in \mathcal{J}}$. Occasionally, we match the variable names of the sequences with the labels and will write $\tau_{\mathtt{xuw}} = (\boldsymbol{x}, \boldsymbol{u}, \boldsymbol{w})$ to suggest assignment $(\tau_{\mathtt{x}}, \tau_{\mathtt{u}}, \tau_{\mathtt{w}}) = (\boldsymbol{x}, \boldsymbol{u}, \boldsymbol{w})$. The concatenation of a sequence $\boldsymbol{x} \in \ell^{\mathcal{X}}$ and $\boldsymbol{u} \in \ell^{\mathcal{U}}$ is a new sequence $\boldsymbol{z} \in \ell^{\mathcal{X} \times \mathcal{U}}$, such that $z_t = (x_t, u_t)$, $\forall t \in \mathbb{N}$; we refer to sequence $\boldsymbol{z}$ by writing $\left[\begin{smallmatrix} \boldsymbol{x} \\ \boldsymbol{u} \end{smallmatrix}\right]$. With slight abuse of notation, we use $\top$ to identify $(\boldsymbol{x}, \boldsymbol{u})^{\top}$ with $\left[\begin{smallmatrix} \boldsymbol{x} \\ \boldsymbol{u} \end{smallmatrix}\right]$ and $\left[\begin{smallmatrix} \boldsymbol{x} \\ \boldsymbol{u} \end{smallmatrix}\right]^{\top}$ with $(\boldsymbol{x}, \boldsymbol{u})$, i.e., $(\cdot)^{\top} : (\boldsymbol{x}, \boldsymbol{u}) \mapsto \left[\begin{smallmatrix} \boldsymbol{x} \\ \boldsymbol{u} \end{smallmatrix}\right]$ and $[\cdot]^{\top} : \left[\begin{smallmatrix} \boldsymbol{x} \\ \boldsymbol{u} \end{smallmatrix}\right] \mapsto (\boldsymbol{x}, \boldsymbol{u})$, and define this correspondence similarly for any tuple of sequences; hence, $(\cdot)^{\top}$ is the canonical isomorphism between $\times_{i \in \mathcal{I}}(\mathcal{V}_i)^{\mathbb{N}}$ and $(\times_{i \in \mathcal{I}} \mathcal{V}_i)^{\mathbb{N}}$.

2.2 Fundamentals of Operator Theory

Operators will be denoted in bold capital letters $\boldsymbol{A}$ and will represent maps between vector sequence spaces $\ell^{\mathcal{X}} \to \ell^{\mathcal{U}}$. An operator will be called causal if for any pair of input $\boldsymbol{x} \in \ell^{\mathcal{X}}$ and the corresponding output $\boldsymbol{y} = \boldsymbol{A}(\boldsymbol{x})$, the values of y_t do not depend on future input values x_{t+k}, $k \geqslant 1$. More precisely, we define $\boldsymbol{A} : \ell^{\mathcal{X}} \to \ell^{\mathcal{U}}$ to be a *causal* operator if there are functions, $A_t : \mathcal{X}^{[0,t]} \to \mathcal{U}$ that allow $\boldsymbol{A}$ to be equivalently represented as

$$\boldsymbol{A}(\boldsymbol{x}) = (A_0(x_0), A_1(x_1, x_0), \ldots, A_t(x_{t:0}), \ldots). \tag{2.1}$$

If in addition, the functions A_t satisfy $A_t(x_{t:0}) = A_t(0, x_{t-1:0})$, i.e., are constant in their first parameter, then $\boldsymbol{A}$ will be called *strictly causal*. The functions $\{A_t\}$ fully characterize a causal operator and will also be called *component functions* or just *components* of $\boldsymbol{A}$. Notice that every component function A_t has $t + 1$ arguments which are populated in reverse-chronological order in Definition (2.1). For notational convenience, for an interval $\mathcal{I} \subset \mathbb{N}$, $\mathcal{I} = [i, j]$, $i \leqslant j$ we will use $A_{\mathcal{I}} : \mathcal{X}^{[0:j]} \to \mathcal{U}^{\mathcal{I}}$ (or $A_{i:j}$) to refer to the mapping defined below: $x_{j:0} \ni \mathcal{X}^{[0,j]} \mapsto (A_i(x_{i:0}), A_{i+1}(x_{i+1:0}), \ldots, A_j(x_{j:0})) \in \mathcal{U}^{[i,j]}$ for $j \geqslant i$ as

$$A_{i:j}(x_{j:0}) := (A_i(x_{i:0}), A_{i+1}(x_{i+1:0}), \ldots, A_j(x_{j:0})).$$

An operator with the same domain and co-domain, for example: $\boldsymbol{A} : \ell^{\mathcal{X}} \to \ell^{\mathcal{X}}$, is called square. Define the space of all causal and strictly causal operators $\ell^{\mathcal{X}} \to \ell^{\mathcal{Y}}$ as $\mathcal{C}(\ell^{\mathcal{X}}, \ell^{\mathcal{Y}})$ and $\mathcal{C}_s(\ell^{\mathcal{X}}, \ell^{\mathcal{Y}})$, respectively. Similarly, define $\mathcal{LC}(\ell^{\mathcal{X}}, \ell^{\mathcal{Y}}) \subset \mathcal{C}(\ell^{\mathcal{X}}, \ell^{\mathcal{Y}})$

and $\mathcal{LC}_s(\ell^{\mathcal{X}}, \ell^{\mathcal{Y}}) \subset \mathcal{C}_s(\ell^{\mathcal{X}}, \ell^{\mathcal{Y}})$ the spaces of all linear causal and strictly causal operators. The corresponding spaces of square (linear)-(strictly)-causal operators on $\ell^{\mathcal{X}}$ will be denoted by $\mathcal{C}(\ell^{\mathcal{X}})$, $\mathcal{C}_s(\ell^{\mathcal{X}})$, $\mathcal{LC}(\ell^{\mathcal{X}})$, $\mathcal{LC}_s(\ell^{\mathcal{X}})$. The right-shift operation $\mathcal{S}^+ : \mathcal{C}(\ell^{\mathcal{X}}, \ell^{\mathcal{Y}}) \mapsto \mathcal{C}_s(\ell^{\mathcal{X}}, \ell^{\mathcal{Y}})$ is defined to map any causal operator $A \in \mathcal{C}(\ell^{\mathcal{X}}, \ell^{\mathcal{Y}})$, to a strictly causal one $\mathcal{S}^+[A] := A^+ \in \mathcal{C}_s(\ell^{\mathcal{X}}, \ell^{\mathcal{Y}})$ by shifting all component functions to the right, that is, $T_0^+(x_0) := 0, \quad T_t^+(x_{t:0}) := T_{t-1}(x_{t-1:0})$.

Addition and Multiplication of Operators

Sums and products of operators are defined as binary operations on the space of causal operators where

$$A + B : x \mapsto A(x) + B(x)$$
$$AB \text{ or } A \circ B : x \mapsto A(B(x)).$$

It is crucial to remember that for general operators, the above defined multiplication is not commutative and is **only** right-distributed over the summation but **not** left-distributed, i.e.,

$$(A + B)C = AC + BC \text{ \underline{but} } C(A + B) \neq CA + CB.$$

Moreover, for two operators $A \in \mathcal{C}(\ell^{\mathcal{X}})$, $B \in \mathcal{C}(\ell^{\mathcal{X}}, \ell^{\mathcal{U}})$ with matching domain, $\left[\begin{smallmatrix} A \\ B \end{smallmatrix}\right]$ refers to the operator $C \in \mathcal{C}(\ell^{\mathcal{X}}, \ell^{\mathcal{X} \times \mathcal{U}})$ with components C_t defined each t and sequence $x \in \ell^{\mathcal{X}}$ as

$$C_t(x_{t:0}) := (A_t(x_{t:0}), B_t(x_{t:0})).$$

Truncation Operator

Sequences that have finitely many nonzero elements span a linear subspace of $\ell^{\mathcal{X}}$ which we refer to as $\ell_0^{\mathcal{X}}$. Furthermore, for a fixed subset $\mathcal{I} \subset \mathbb{N}$, the subset of sequences $x \in \ell_0^{\mathcal{X}}$ such that $x_k = 0$ for all $k \notin \mathcal{I}$ forms a subspace of $\ell_0^{\mathcal{X}}$, which we refer to as $\ell_{\mathcal{I}}^{\mathcal{X}}$; for example, $\ell_{[0,t]}^{\mathcal{X}}$ represents the subspace of sequences with elements equal to 0 after the first $t + 1$ terms. A truncation x' of a sequence $x \in \ell^{\mathcal{X}}$ is a sequence $x' \in \ell_{[0,t]}^{\mathcal{X}}$ such that $x_k' := x_k$ for $k \in \{0, \ldots, t\}$ and some t, that is, a sequence formed by truncating all but the first t elements of x. For each $t \in \mathbb{N}$, the correspondence $x \mapsto x'$ between sequences $x \in \ell^{\mathcal{X}}$ and their truncations $x' \in \ell_{[0,t]}^{\mathcal{X}}$ defines a unique linear projection map $P_n^t : \ell^{\mathcal{X}} \to \ell^{\mathcal{X}}$ onto $\ell_{[0,t]}^{\mathcal{X}}$, commonly called a truncation operator. Below, we define the truncation operators $\{P_n^t \mid t \in \mathbb{N}\}$ in terms of the broader – however less standard – family of operators $\{P_n^{\mathcal{I}} \mid \mathcal{I} \subset \mathbb{N}\}$:

Definition 2.1. *[Truncation Operators] For a subset $\mathcal{I} \subset \mathbb{N}$ and its complement $\mathcal{I}^c := \mathbb{N} \setminus \mathcal{I}$, define the operator $P_n^{\mathcal{I}} : \ell^{\mathcal{X}} \to \ell^{\mathcal{X}}$ and $\overline{P_n^{\mathcal{I}}} : \ell^{\mathcal{X}} \to \ell^{\mathcal{X}}$ for each $x \in \ell^{\mathcal{X}}$ as:*

$$P_n^{\mathcal{I}}(x) := \begin{cases} x_k & \text{for } k \in \mathcal{I} \\ 0 & \text{else} \end{cases} \qquad \overline{P_n^{\mathcal{I}}}(x) := \begin{cases} x_k & \text{for } k \in \mathcal{I}^c \\ 0 & \text{else.} \end{cases}$$

Moreover, we use the following shorthand notation for $P_n^{\mathcal{I}}$ and $\overline{P_n^{\mathcal{I}}}$ if $\mathcal{I}$ is an interval or a singleton:

- *For $t \in \mathbb{N}$, $P_n^t := P_n^{\mathcal{I}}$ and $\overline{P_n^t} := \overline{P_n^{\mathcal{I}}}$ where $\mathcal{I} = [0, t] \subset \mathbb{N}$.*
- *For $t, t' \in \mathbb{N}: t < t'$, $P_n^{[t,t']} := P_n^{\mathcal{I}}$ and $\overline{P_n^{[t,t']}} := \overline{P_n^{\mathcal{I}}}$ where $\mathcal{I} = [t, t'] \subset \mathbb{N}$.*
- *For $t \in \mathbb{N}$, $P_n^{[t]} := P_n^{\mathcal{I}}$ and $\overline{P_n^{[t]}} := \overline{P_n^{\mathcal{I}}}$ where $\mathcal{I} = \{t\}$.*

The truncation operators are linear projection maps in sequence space and have several important properties listed in Cor. 1, all of which are trivial consequences of the previous definition:

Corollary 1. *For any $\mathcal{I} \subset \mathbb{N}$, the truncation operator $P^{\mathcal{I}}$ satisfies the following identities:*

(i) $P_n^{\mathcal{I}}(x) \in \ell_p^{\mathcal{X}}$ *for any x and finite set $\mathcal{I}$.*   (ii) $P_n^{\mathcal{I}}(x + y) = P_n^{\mathcal{I}}(x) + P_n^{\mathcal{I}}(y)$.

(iii) $P_n^{\mathcal{I}} P_n^{\mathcal{I}} = P_n^{\mathcal{I}}$ (iv) *If $x \notin \ell_p^{\mathcal{X}}, (p \in [1, \infty])$ then $P_n^{\mathcal{I}}(x) \notin \ell_p^{\mathcal{X}}$ or $\overline{P_n^{\mathcal{I}}}(x) \notin \ell_p^{\mathcal{X}}$.*

(v) $\|x\|_p^p = \|(I - P_n^{\mathcal{I}})(x)\|_p^p + \|P_n^{\mathcal{I}}(x)\|_p^p = \|\overline{P_n^{\mathcal{I}}}(x)\|_p^p + \|P_n^{\mathcal{I}}(x)\|_p^p$ *holds for all $p < \infty$ and $x \in \ell_p^{\mathcal{X}}$.*

(vi) $\|x\|_\infty = \|(I - P_n^{\mathcal{I}})(x)\|_\infty \vee \|P_n^{\mathcal{I}}(x)\|_\infty = \|\overline{P_n^{\mathcal{I}}}(x)\|_\infty \vee \|P_n^{\mathcal{I}}(x)\|_\infty$ *holds for all $x \in \ell_\infty^{\mathcal{X}}$. ($a \vee b := \max\{a, b\}$).*

In light of this more general definition, the standard truncations P_n^t we discussed in the beginning are identified with the intervals $\mathcal{I} = [0, t]$. Furthermore, the operation $P_n^{[t]}(x)$ truncates all terms of x_k, except $k = t$. To familiarize with the introduced notation, let $t, t' \in \mathbb{N}$, $t < t'$ and notice the following relationships:

$$P_n^t + \overline{P_n^t} = I_n \qquad\qquad P_n^{[t]} = P_n^t - P_n^{t-1}$$

$$P_n^t = \sum_{k=0}^{t} P_n^{[k]} \qquad\qquad P_n^{[t,t']} = P_n^{t'} - P_n^{t-1}.$$

Given that we live in the space $\mathcal{LC}(\ell^{\mathcal{X}}, \ell^{\mathcal{X}})$, as long as $\mathcal{I}$ is a finite subset of $\mathbb{N}$, it is clear that $P_n^{\mathcal{I}}$ is a finite-rank linear projection and $\operatorname{rank}(P_n^{\mathcal{I}}) = n|\mathcal{I}|$.

For finite sets $\mathcal{I}$, we can factor $\boldsymbol{P}_n^{\mathcal{I}}$ into the familiar form $\boldsymbol{P}_n^{\mathcal{I}} = \boldsymbol{U}^{\mathcal{I}}\boldsymbol{U}^{*\mathcal{I}}$, by letting the linear maps $\boldsymbol{U}^{\mathcal{I}} : \mathbb{R}^N \to \ell^{\mathcal{X}}$ and $\boldsymbol{U}^{*\mathcal{I}} : \ell^{\mathcal{X}} \to \mathbb{R}^N$ be extensions of the canonical isomorphisms between $\mathbb{R}^N$, where $N = n|\mathcal{I}|$, and the $\ell_0^{\mathcal{X}}$-subspace $\{\boldsymbol{x} \in \ell_0^{\mathcal{X}} \mid x_k = 0, \forall k \notin \mathcal{I}\}$. To clarify, consider $\mathcal{I} = [1]$ as an example. Then, for any $x \in \mathbb{R}^n$, $\boldsymbol{y} = \boldsymbol{U}^{[1]}(x)$ is the sequence $\boldsymbol{y} = (0, x, 0, \dots) \in \ell_{0|1}^{\mathcal{X}}$ and, vice versa, for any sequence $\boldsymbol{x} = (0, y, 0, \dots) \in \ell_{0|1}^{\mathcal{X}}$ with $y \in \mathbb{R}^n$, we have $\boldsymbol{U}^{*[1]}(\boldsymbol{x}) = y$. To match our previous notation, we also abbreviate $\boldsymbol{U}_n^{[0,t]}$ and $\boldsymbol{U}_n^{*[0,t]}$ as $\boldsymbol{U}_n^t$ and $\boldsymbol{U}_n^{*t}$, respectively, which allows us to factor the standard truncation operators as $\boldsymbol{P}_n^t = \boldsymbol{U}_n^t \boldsymbol{U}_n^{*t}$.

Causal Operators and Truncations

There are several important relationships between truncations and causal operators, one being that the causality of an operator $\boldsymbol{Q}$ can be equivalently defined in terms of its interaction with truncations.

Lemma 2. *For each $t \in \mathbb{N}$, let $\boldsymbol{P}_n^t$ and $\boldsymbol{P}_m^t$ denote the t-truncation on $\ell^{\mathcal{U}}$ and $\ell^{\mathcal{U}}$. For any operator $\boldsymbol{Q} : \ell^{\mathcal{X}} \mapsto \ell^{\mathcal{U}}$, the following hold:*

$$\boldsymbol{Q} \in \mathcal{C}(\ell^{\mathcal{X}}, \ell^{\mathcal{U}}) \quad \Leftrightarrow \quad \forall t \in \mathbb{N} : \ \boldsymbol{P}_m^t \boldsymbol{Q} = \boldsymbol{P}_m^t \boldsymbol{Q} \boldsymbol{P}_n^t$$
$$\boldsymbol{Q} \in \mathcal{C}_s(\ell^{\mathcal{X}}, \ell^{\mathcal{U}}) \quad \Leftrightarrow \quad \forall t \in \mathbb{N} : \ \boldsymbol{P}_m^t \boldsymbol{Q} = \boldsymbol{P}_m^t \boldsymbol{Q} \boldsymbol{P}_n^{t-1}.$$

Using the above definition and right-distributiveness, we can decompose a causal operator $\boldsymbol{Q} : \ell^{\mathcal{X}} \mapsto \ell^{\mathcal{U}}$ as follows:

$$\boldsymbol{Q} = (\boldsymbol{I}_m - \boldsymbol{P}_m^t)\boldsymbol{Q} + \boldsymbol{P}_m^t \boldsymbol{Q}$$
$$= \overline{\boldsymbol{P}_m^t}\boldsymbol{Q} + \boldsymbol{P}_m^t \boldsymbol{Q} \boldsymbol{P}_n^t.$$

As mentioned in Cor. 1, a truncation operator exhibits some properties that resemble orthogonal projections, but without requiring a specific inner product. In particular, in any normed space $\ell_p^{\mathcal{U}}/\ell_\infty^{\mathcal{U}}$, from the above decomposition and the properties mentioned in Cor. 1, we obtain that for any $\boldsymbol{x}$ and $\boldsymbol{Q} \in \mathcal{C}(\ell^{\mathcal{X}}, \ell^{\mathcal{U}})$, the following holds:

$$\|\boldsymbol{Q}(\boldsymbol{x})\|_p^p = \|\overline{\boldsymbol{P}_m^t}\boldsymbol{Q}(\boldsymbol{x})\|_p^p + \|\boldsymbol{P}_m^t \boldsymbol{Q}(\boldsymbol{x})\|_p^p$$
$$\|\boldsymbol{Q}(\boldsymbol{x})\|_\infty = \|\overline{\boldsymbol{P}_m^t}\boldsymbol{Q}(\boldsymbol{x})\|_\infty \vee \|\boldsymbol{P}_m^t \boldsymbol{Q}(\boldsymbol{x})\|_\infty.$$

Remark. *The above equations are consistent in the case $\boldsymbol{Q}(\boldsymbol{x}) \notin \ell_p^{\mathcal{X}}$ or $\boldsymbol{Q}(\boldsymbol{x}) \notin \ell_p^{\mathcal{X}}$, if we define $\|\boldsymbol{Q}(\boldsymbol{x})\|_p$ and $\|\boldsymbol{Q}(\boldsymbol{x})\|_\infty$ to take on the value $+\infty$.*

Using the definitions of $U^{*[t]}$ and U^t, we can also concisely define the component functions of an operator Q as $Q_t = U_m^{*[t]}QU_n^t$. Similarly, we can factor $P_m^t Q$ as

$$P_m^t Q = \sum_{k=0}^{t} U_m^{[k]} Q_k U_n^{*k}.$$

Another important consequence of this decomposition is that appending a truncation P_m^t to a causal operator Q always results in smaller $\ell_p^{\mathcal{U}}$ norms than prepending a truncation P_n^t. To see this, we prepend P_n^t to Q and use right-distributiveness and causality to obtain the equivalence:

$$QP_n^t = ((I_m - P_m^t)Q + P_m^t Q)P_n^t = \overline{P_m^t}QP_n^t + P_m^t QP_n^t$$
$$= \overline{P_m^t}QP_n^t + P_m^t Q.$$

Then, for any $x \in \ell^{\mathcal{X}}$, $Q \in \mathcal{C}(\ell^{\mathcal{X}}, \ell^{\mathcal{U}})$, and $t \in \mathbb{N}$, the following hold:

$$\|QP_n^t(x)\|_p^p = \|\overline{P_m^t}QP_n^t(x)\|_p^p + \|P_m^t Q\|_p^p \geq \|P_m^t Q\|_p^p$$
$$\|QP_n^t(x)\|_\infty = \|\overline{P_m^t}QP_n^t(x)\|_\infty \vee \|P_m^t Q\|_\infty \geq \|P_m^t Q\|_\infty.$$

We summarize our findings in the lemma below:

Lemma 3. *Let $\{\mathcal{T}_k = [\underline{t}_k, \bar{t}_k]\}$, $(\underline{t}_k \leq \bar{t}_k)$, be a collection of pairwise disjoint intervals of $\mathbb{N}$ over some index set $k \in \mathcal{I}$, and let $Q \in \mathcal{C}(\ell^{\mathcal{X}}, \ell^{\mathcal{U}})$ be a causal operator. Then, for any $t \in \mathbb{N}$, $p \in \{1, \ldots, \infty\}$, and $x \in \ell^{\mathcal{U}}$, the following inequality holds:*

$$\left\|\left(\sum_{k\in\mathcal{I}} P_n^{\mathcal{T}_k} Q\right)x\right\|_p \leq \sum_{k\in\mathcal{I}} \|QP_m^{\bar{t}_k}x\|_p. \tag{2.2}$$

The above lemma will be used in the derivation of the small gain theorem, which we derive in the next section.

Causally Invertible Operators

A square operator $A : \ell^{\mathcal{X}} \to \ell^{\mathcal{X}}$ is invertible if there exists another operator $B : \ell^{\mathcal{X}} \to \ell^{\mathcal{X}}$ such that $AB = BA = I_n$. If such a B exists, it is unique and is therefore called the inverse of A and is denoted as A^{-1}. We call $A \in \mathcal{C}(\ell^{\mathcal{X}})$ causally invertible if it is invertible and its inverse is causal, that is, $A^{-1} \in \mathcal{C}(\ell^{\mathcal{X}})$.

Definition 2.2. *$A \in \mathcal{C}(\ell^{\mathcal{X}})$ is causally invertible if there exists $A^{-1} \in \mathcal{C}(\ell^{\mathcal{X}})$ such that $AA^{-1} = A^{-1}A = I_n$.*

In general, the inverse of a causal operator is not necessarily a causal operator [110]. In fact, it is possible to find counterexamples even for operators $A : \ell_{[0,t]} \to \ell_{[0,t]}$ that map between finite-dimensional spaces. However, if causality of the inverse can be proven, it also provides instructions on how to implement the inverse operator.

If we are given a causally invertible operator $A \in \mathcal{C}(\ell^{\mathcal{X}})$, then by definition, for any $t \in \mathbb{N}$, $P^t A^{-1} = P^t A^{-1} P^t$, $\forall t \in \mathbb{N}$. Combining this with causality of A shows that we obtain an invertible function $\underline{A}_t$ by concatenating the first t component functions:

$$\underline{A}_t : (x_t, \ldots, x_0) \mapsto (A_t(x_t, \ldots, x_0), \ldots, A_0(x_0)). \tag{2.3}$$

The map $\underline{A}_t$ can be written more compactly as $\underline{A}_t = U^{*t} A U^t$. For the next derivation, let $B = A^{-1} \in \mathcal{C}(\ell^{\mathcal{X}})$ denote the inverse of A and similarly, let the concatenation of the components of B be denoted as $\underline{B}_t = U^{*t} B U^t$. Below, we derive that $\underline{B}_t$ is in fact the inverse of $\underline{A}_t$ and therefore invertibility of $\underline{A}_t$, $\forall t \in \mathbb{N}$ is a necessity for causal invertibility:

$$P^t = P^t A B = P^t B A$$

$$\implies \quad U^t U^{*t} = P^t A P^t B P^t = P^t B P^t A P^t$$

$$\overset{X \mapsto U^{*t} X U^t}{\implies} \quad I_{N_t} = \underbrace{(U^{*t} B U^t)}_{\underline{B}_t} \underbrace{(U^{*t} A U^{*t})}_{\underline{A}_t} = (U^{*t} A U^t)(U^{*t} B U^{*t})$$

$$\implies \quad \underline{B}_t = (\underline{A}_t)^{-1} .$$

Having established the equivalence $\underline{B}_t = (\underline{A}_t)^{-1}$, going forward, we can refer to $\underline{B}_t$ as $\underline{A}_t^{-1}$ without causing ambiguity, since $U^{*t} A^{-1} U^t$ and $(\underline{A}_t)^{-1}$ are indeed the same function. Furthermore, the above equivalence aids us in constructing a realization of the inverse operator.

Setting $t = 0$, the above implies that A_0 is invertible. Moving to $t = 1$, it is easy to see that $\underline{A}_1 : (u_0, u_1) \mapsto (A_0(u_0), A_1(u_1, u_0))$ is invertible if and only if the map $u' \mapsto A_1(u', u_0)$ is invertible for any $u_0 \in \mathbb{R}^n$. This observation motivates us to investigate whether this condition extends to all $\mathbb{N}$. To this end, let b and a be such that $b = A a$, and let $\underline{a}_t = (a_t, \ldots, a_0)$ and $\underline{b}_t = (b_t, \ldots, b_0)$. Then, for some $t \in \mathbb{N}$, we can express a_t as

$$a_t = U^{[*t]} A^{-1}(b) = U^{[*t]} A^{-1} P^t(b) = A_t^{-1}(b_t, \underline{b}_{t-1}) = A_t^{-1}(b_t, \underline{A}_{t-1}(\underline{a}_{t-1}))$$

$$= A_t^{-1}(\underline{A}_t(a_t, \underline{a}_{t-1}), \underline{A}_{t-1}(\underline{a}_{t-1})) \tag{2.4}$$

where $A_t^{-1} = U^{[*t]} A^{-1} U^t$ denotes the t-th component function of A^{-1}. For a fixed choice of $\underline{a}_{t-1}$, denote $A_{t|t-1}[\underline{a}_{t-1}] : \mathbb{R}^n \to \mathbb{R}^n$ as the restriction of the component function A_t such that $A_{t|t-1}[\underline{a}_{t-1}](a) = A_t(a, \underline{a}_{t-1})$. Similarly, let $B_{t|t-1}[\underline{a}_{t-1}] : \mathbb{R}^n \to \mathbb{R}^n$ be a function defined for each $b \in \mathbb{R}^n$ as $B_{t|t-1}[\underline{a}_{t-1}](b) = A_t^{-1}(b, \underline{A}_{t-1}(a_{t-1}))$. With these two function definitions, (2.4) can be written as $a_t = B_{t|t-1}[\underline{a}_{t-1}] \circ A_{t|t-1}[\underline{a}_{t-1}](a_t)$. Since this equation has to hold for all $\underline{a}_{t-1}$ and $a_t \in \mathbb{R}^n$, we can conclude that for any fixed $\underline{a}_{t-1}$, the function $B_{t|t-1}[\underline{a}_{t-1}]$ is the left-inverse of $A_{t|t-1}[\underline{a}_{t-1}]$. Hence, equation (2.4) can be written as $a_t = B_{t|t-1}[\underline{a}_{t-1}](b_t)$ and provides us with a realization of $\boldsymbol{A}^{-1}$. For a given b such that $b = \boldsymbol{A}(a)$ with causally invertible $\boldsymbol{A}$, the sequence a is a trajectory of the dynamic system described by the equations

$$a_t = B_t[\underline{a}_{t-1}](b_t). \tag{2.5}$$

Our discussion has so far shown that if $\boldsymbol{A}$ is causally invertible, then the functions $\{\underline{A}_t\}_{t\in\mathbb{N}}$ have to be invertible. Moreover, the latter implies that any function in the set $\mathcal{A} = \{A_{t|t-1}[z_{t-1}] : \mathbb{R}^n \to \mathbb{R}^n \mid t \in \mathbb{N},\ z_{t-1} \in \mathbb{R}^{N_t}\}$ has to be invertible as well and that using the inverse functions $B_t[\underline{a}_{t-1}]$ we can realize the operator $\boldsymbol{A}^{-1}$ as the dynamical system (2.5). Clearly, this realization is also evidence that invertibility of the functions in $\mathcal{A}$ is sufficient for the existence of $\boldsymbol{A}^{-1}$. This conclusion closes a chain of implications and leads to the following Lemma, which characterizes causally invertible operators in terms of their component functions:

Lemma 4. *For a causal operator $\boldsymbol{A} : \ell^{\mathcal{X}} \to \ell^{\mathcal{X}}$, the following statements are all equivalent:*

(i) *$\boldsymbol{A}$ is causally invertible.*
(ii) *$\{\underline{A}_t : z \mapsto \boldsymbol{U}^{*t} \boldsymbol{A} \boldsymbol{U}^t(z) \mid t \in \mathbb{N}\}$ is a family of invertible functions.*
(iii) *$\{A_{t|t-1}[z_{t-1}] : a \mapsto A_t(a, z_{t-1}) \mid t \in \mathbb{N},\ z_{t-1} \in \underline{A}_{t-1}^{-1}(\mathbb{R}^{N_{t-1}})\}$ is a family of invertible functions.*

Proof. We recap the ring implications. $(i) \implies (ii)$: The causal invertibility of $\boldsymbol{A}$ means that there is some causal $\boldsymbol{B}$ such that $\boldsymbol{A}\boldsymbol{B} = \boldsymbol{I}$, which leads to $\boldsymbol{P}^t \boldsymbol{A} \boldsymbol{P}^t \boldsymbol{B} \boldsymbol{P}^t = \boldsymbol{P}^t, \forall t \in \mathbb{N}$ and proves that $\underline{B}_t : \boldsymbol{U}^{*t} \boldsymbol{B} \boldsymbol{U}^t$ is the right inverse of $\underline{A}_t$. $(ii) \implies (iii)$: Pick an arbitrary $t \in \mathbb{N}$ and apply the operator $\boldsymbol{U}^t$ to both sides (multiplication from the left) of the identity $\underline{A}_t \circ \underline{B}_t = I_{N_t}$. The new equation can be restated as:

$$\forall y \in \mathbb{R}^n, \forall y_{t-1} \in \mathbb{R}^{N_{t-1}} : \quad A_t(B_t(y, y_{t-1}), \underline{B}_{t-1}(y_{t-1})) = y.$$

This states that for a fixed $z_{t-1} = \underline{B}_{t-1}(y_{t-1})$, $B_{t|t-1}[y_{t-1}] : y \mapsto B_t(y, y_{t-1})$ is the right inverse of $A_{t|t-1}[z_{t-1}]$. Now, since $\underline{B}_{t-1}$ is the inverse of $\underline{A}_{t-1}$, z_{t-1} ranges over all $\mathbb{R}^{N_t}$. We also showed previously left-invertibility in the same way, which establishes (iii). Lastly, $(iii) \implies (i)$ follows through the realization (2.5). $\qquad\square$

A corollary of this lemma is that operators $A = I_n + A^-$, which are sums of some strictly causal $A^- \in C_s(\ell^{\mathcal{X}})$ and the identity operator I, are always causally invertible, and the inverse operator A^{-1} can be easily realized. This result is derived below:

Corollary 5. *If* $(A - I) \in C_s(\ell^{\mathcal{X}})$, *then* $A^{-1} \in C(\ell^{\mathcal{X}})$ *exists and* $b = A^{-1}(a)$ *satisfies*

$$b_t = a_t - A_t(0, b_{t-1:0}).$$

Proof. Assume given a, we want to find b such that $A(b) = a$. Equivalently, we can write $b = a - (A - I)(b)$ Now, since $A - I$ is strictly causal, the component function A_t satisfies $A_t(x_t, x_{t-1:0}) = A_t(0, x_{t-1:0}) + x_t$. Using this factorization, the component form of $b = a - (A - I)(b)$ becomes

$$b_t = a_t - A_t(0, b_{t-1:0}), \tag{2.6}$$

which proves the existence and uniqueness of b as it describes a concrete recursive procedure for its computation. $\qquad\square$

2.3　Global and Local Stability of Operators

We review some standard stability results from the control literature, which are needed for later analysis. In the following, we derive various versions of the small-gain theorem, which are used in our main results.

If we view a dynamical system as a causal map T between input sequences w and output sequences $y = Tw$, then many common notions of system stability, such as ℓ_p stability, can be mapped to notions of boundedness and continuity of the operator T. Sometimes it is more natural to define the relationship between input and output in an implicit fashion, where for an input w, the output y is a solution to equation $Qy = w$, for some fixed operator Q. In that scenario, input-output stability means that for a bounded w, any corresponding solution y is also bounded.

Approaching stability analysis from this operator-theoretic perspective was first studied in the late 1960s, most notably pioneered by George Zames [139], and led to a large class of important stability analysis tools often grouped together under the name of *small gain* theorems.

In this section, we derive some small-gain theorem conditions that match our problem setup and that we use frequently in our analysis.

We use the following notions of ℓ_p-stability for operators in the statement of our results:

Definition 2.3. *An operator $A \in \mathcal{C}(\ell^{\mathcal{X}}, \ell^{\mathcal{U}})$ is called:*

- *ℓ_p-stable if $A(a) \in \ell_p^{\mathcal{U}}$ for all $a \in \ell_p^{\mathcal{X}}$.*

- *finite gain (f.g.) ℓ_p-stable [1] at $a_0 \in \ell_p^{\mathcal{X}}$, if there exists $\gamma, \beta \geqslant 0$ such that for all $a \in \ell_p^{\mathcal{X}}$:*

$$\|A(a) - A(a_0)\|_p \leqslant \gamma \|a - a_0\|_p + \beta.$$

- *incrementally finite gain [2] (i.f.g.) ℓ_p-stable if there exists $\gamma, \beta \geqslant 0$ such that for all $a, a' \in \ell_p^{\mathcal{X}}$, it holds:*

$$\|A(a) - A(a')\|_p \leqslant \gamma \|a - a'\|_p + \beta.$$

Remark 1. *Finite gain ℓ_p stability and operator continuity in ℓ_p-norm are closely related. In fact, if we set $\beta = 0$ in the above definitions, then, i.f.g. ℓ_p-stability of A with gain γ is equivalent to Lipschitz continuity in ℓ_p-norm of A over $\ell_p^{\mathcal{X}}$. On the other hand, f.g.-ℓ_p-stability at a_0 implies that A is ℓ_p-continuous at a_0 and that $\delta(\varepsilon)$ can be chosen, in the context of the standard continuity definition [109], as a linear function $\delta : \varepsilon \mapsto \gamma^{-1}\varepsilon$. Allowing β to be nonzero can be thought of as "almost" (Lipschitz) continuity, as possible discontinuities are deemed irrelevant for the purpose of stability analysis.*

Notice that while the above definition might not be standard, it allows for the stability analysis of equilibria, trajectories, and limit cycles all within the same definition.

In the following discussion, we consider a fixed causal operator $Q \in \mathcal{C}(\ell^{\mathcal{X}}, \ell^{\mathcal{X}})$ and investigate pairs (w, y) of sequences $w \in \ell^{\mathcal{X}}$ and $y \in \ell^{\mathcal{X}}$ that are solutions to the nonlinear operator equation:

$$Q(y) = w \qquad \Leftrightarrow \qquad y = (I - Q)(y) + w. \qquad (2.7)$$

---

[1] If we say A is f.g. ℓ_p-stable without specifying a_0, it is assumed that a_0 is to be taken as 0.
[2] We write (γ, β)-f.g. and (γ, β)-i.f.g. if we want to specify the constants of the ℓ_p-stability property.

For the purpose of stability analysis, we would ideally like to characterize the space of solutions (w, y) that satisfy relations of the following types:

$$(w \rightarrow y)\ \ell_p - stab. : \quad Q(y) = w,\ w \in \ell_p^{\mathcal{X}} \implies y \in \ell_p^{\mathcal{X}}, \tag{2.8a}$$

$$(y \rightarrow w)\ \ell_p - stab. : \quad Q(y) = w,\ y \in \ell_p^{\mathcal{X}} \implies w \in \ell_p^{\mathcal{X}}, \tag{2.8b}$$

$$bi\text{-}stab.\ \ell_p : \quad Q(y) = w \implies y, w \in \ell_p^{\mathcal{X}}. \tag{2.8c}$$

If we identify w and y with some inputs and outputs of a dynamical system that has dynamic equations of the form (2.7), then condition (2.8a) represents the standard notion of *bounded-input-bounded-output* (BIBO) [75] with respect to the norm ℓ_p. The statement (2.8b) represents the reverse condition, which is more commonly seen as a notion of system observability. Similarly to the definitions in (2.8), one can examine the space of solutions (y, w) for more refined notions of stability, such as (γ, β)-finite gain stability (fgs) and (γ, β)-incremental finite gain stability (ifgs). With respect to the input-output mapping $(w \rightarrow y)$, these are defined as follows:

$$(\gamma, \beta) - \ell_p - fgs : \quad Q(y) = w,\ w \in \ell_p^{\mathcal{X}} \implies \|y\|_p \leqslant \gamma \|w\|_p + \beta,$$

$$(\gamma, \beta) - \ell_p - ifgs \text{ at } w_0 : \quad Q(y) = w,\ w \in \ell_p^{\mathcal{X}} \implies \|\delta y\|_p \leqslant \gamma \|\delta w\|_p + \beta.$$

$$Q(y_0) = w_0,\ w_0 \in \ell_p^{\mathcal{X}}$$

$$\delta y = y - y_0$$

$$\delta w = w - w_0.$$

Next, we focus on deriving sufficient conditions, in the form of small-gain theorems, for finite gain ℓ_p-stability as defined above.

Conditions for Finite-Gain Stability

For the following derivations, we investigate pairs (y, w) which are solutions to (2.7) and from y we define the scalar sequence $s \in \mathbb{R}^\infty$ such that $s_t := \|P^t(y)\|_p$, i.e.,

$$s_t := \begin{cases} \sqrt[p]{\sum_{k=0}^{t} |y_k|^p} & \text{for } p < \infty \\ \sup_{k \leqslant t} |y_k| & \text{for } p = \infty. \end{cases} \tag{2.10}$$

Furthermore, for convenience we abbreviate the causal operator $(I - Q)$ as $\Delta := I - Q$ to rewrite the equation (2.8a) as

$$y = \Delta(y) + w.$$

Next, we bound s_t from above, which results in a key inequality used to prove the later small-gain stability conditions. To this end, substitute $y = \Delta(y) + w$, into the definition (2.10) and by causality of Δ we get an equivalent expression for s_t:

$$s_t = \|\boldsymbol{P}^t(\boldsymbol{y})\|_p = \|\boldsymbol{P}^t(\Delta(\boldsymbol{y}) + \boldsymbol{w})\|_p. \tag{2.11}$$

From (2.11) and with the help of Lem. 3 we obtain the inequalities below:

Lemma 6. *For each term s_t of the sequence $\boldsymbol{s}$ holds:*

$$s_t \leqslant \|\Delta(\boldsymbol{P}^t(\boldsymbol{y}))\|_p + \|\boldsymbol{w}\|_p \tag{2.12}$$

and if $\Delta := (\boldsymbol{I} - \boldsymbol{Q})$ is strictly causal, then the bound below holds:

$$s_t \leqslant \|\Delta(\boldsymbol{P}^{t-1}(\boldsymbol{y}))\|_p + \|\boldsymbol{w}\|_p. \tag{2.13}$$

Proof. The first bound follows by applying Lem. 3 to (2.11) directly:

$$s_t = \|\boldsymbol{P}^t(\Delta(\boldsymbol{y}) + \boldsymbol{w})\|_p \leqslant \|\Delta(\boldsymbol{P}^t(\boldsymbol{y}))\|_p + \|\boldsymbol{w}\|_p.$$

For the second inequality, we have to do a little more work:

$$s_t = \|(\boldsymbol{P}^t\Delta)(\boldsymbol{y}) + \boldsymbol{w}\|_p = \|(\boldsymbol{P}^t\Delta\boldsymbol{P}^{t-1})(\boldsymbol{y}) + \boldsymbol{w}\|_p \tag{2.14a}$$

$$\leqslant \|(\boldsymbol{P}^t\Delta)(\boldsymbol{P}^{t-1}\boldsymbol{y})\|_p + \|\boldsymbol{w}\|_p$$

$$\leqslant \|(\Delta\boldsymbol{P}^t)(\boldsymbol{P}^{t-1}\boldsymbol{y})\|_p + \|\boldsymbol{w}\|_p = \|(\Delta\boldsymbol{P}^t\boldsymbol{P}^{t-1})\boldsymbol{y}\|_p + \|\boldsymbol{w}\|_p$$

$$= \|\Delta(\boldsymbol{P}^{t-1}(\boldsymbol{y}))\|_p + \|\boldsymbol{w}\|_p. \tag{2.14b}$$

$$\square$$

With the above Lemma 6, we present sufficient conditions for the stability (2.8a) with their local/global finite-gain versions.

Theorem 2 (Small Gain Theorem). *Assume that the operator Δ satisfies $\|\Delta(\boldsymbol{x})\|_p \leqslant \gamma\|\boldsymbol{x}\|_p + \beta$ for all $\boldsymbol{x} \in \ell_p^{\mathcal{X}}$ and some small gain $\gamma < 1$. Then, correspondingly for all $\boldsymbol{w} \in \ell_p^{\mathcal{X}}$, the system response $\boldsymbol{y}$ satisfies the bound*

$$\|\boldsymbol{y}\|_p \leqslant \frac{1}{1-\gamma}(\|\boldsymbol{w}\|_p + \beta).$$

Proof. Applying our assumption to the bound (2.12), we obtain:

$$s_t \leqslant \|\Delta(\boldsymbol{P}^t(\boldsymbol{y}))\|_p + \|\boldsymbol{w}\|_p \leqslant \gamma\boldsymbol{P}^t(\boldsymbol{y}) + \beta + \|\boldsymbol{w}\|_p \tag{2.15}$$

$$\leqslant \gamma s_t + \beta + \|\boldsymbol{w}\|_p \tag{2.16}$$

$$\Leftrightarrow \quad s_t \leqslant \frac{1}{1-\gamma}(\beta + \|\boldsymbol{w}\|_p). \tag{2.17}$$

This establishes the boundedness of $\boldsymbol{s}$. Since $\boldsymbol{s}$ is nondecreasing by construction, we know that $\lim_{t\to\infty} s_t = s^* = \|\boldsymbol{y}\|_p$ exists and satisfies $\|\boldsymbol{y}\|_p \leqslant \frac{1}{1-\gamma}(\|\boldsymbol{w}\|_p + \beta)$. $\quad\square$

We can also obtain a local version of that result, provided $\mathbf{\Delta}$ is strictly causal and satisfies the small-gain property locally:

Lemma 7 (Local Small Gain Theorem). *Assume that for some $\rho > 0$ and $0 \leqslant \gamma < 1$, $\beta \geqslant 0$ the operator $\mathbf{I} - \mathbf{Q}$ is strictly causal and satisfies $\|(\mathbf{I} - \mathbf{Q})(\mathbf{x})\|_p \leqslant \gamma \|\mathbf{x}\|_p + \beta$ for all $\|\mathbf{x}\|_p < \rho$ and some $p \in \{1, 2, \ldots, \infty\}$. Then, for any $(\mathbf{y}, \mathbf{w})$ such that $\mathbf{Q}(\mathbf{y}) = \mathbf{w}$ holds:*

$$\|\mathbf{w}\|_p < (1 - \gamma)\rho - \beta \qquad \Longrightarrow \qquad \|\mathbf{y}\|_p \leqslant \tfrac{1}{1-\gamma}(\|\mathbf{w}\|_p + \beta).$$

Proof. Our assumption $\|\mathbf{w}\|_p < (1 - \gamma)\rho - \beta$ can be equivalently stated as:

$$\|\mathbf{w}\|_p < (1 - \gamma)\rho - \beta \Leftrightarrow \frac{\|\mathbf{w}\|_p + \beta}{(1 - \gamma)} < \rho. \tag{2.18}$$

We proceed to show $s_t \leqslant (\|\mathbf{w}\|_p + \beta)/(1 - \gamma)$ for all t per induction: $\underline{t = 0}$: $s_0 = |w_0| \leqslant \|\mathbf{w}\|_p \leqslant (\|\mathbf{w}\|_p + \beta)/(1 - \gamma) < \rho$, since according to assumption $\|\mathbf{w}\|_p < (1-\gamma)\rho-\beta$ which is equivalent to the last inequality. $\underline{t \to t + 1}$: Assume s_t satisfies $s_t \leqslant (\|\mathbf{w}\|_p + \beta)/(1 - \gamma)$. Then, due to (2.18), we have $s_t = \|\mathbf{P}^t(\mathbf{y})\|_p < \rho$ and using the small gain property we know:

$$\|\mathbf{\Delta}(\mathbf{P}^t(\mathbf{y}))\|_p \leqslant \gamma \|\mathbf{P}^t(\mathbf{y})\|_p + \beta. \tag{2.19}$$

Substituting the above into (2.13) and using our induction assumption $s_t \leqslant (\|\mathbf{w}\|_p + \beta)/(1 - \gamma)$, we obtain:

$$s_{t+1} \leqslant \gamma \|\mathbf{P}^t(\mathbf{y})\|_p + \beta + \|\mathbf{w}\|_p = \gamma s_t + (1 - \gamma)\frac{\|\mathbf{w}\|_p + \beta}{1 - \gamma} \tag{2.20a}$$

$$\leqslant \gamma \frac{\|\mathbf{w}\|_p + \beta}{1 - \gamma} + (1 - \gamma)\frac{\|\mathbf{w}\|_p + \beta}{1 - \gamma} = \frac{\|\mathbf{w}\|_p + \beta}{1 - \gamma}, \tag{2.20b}$$

hence, (2.20b) completes the induction step and we can conclude that s_t is bounded above by $(\|\mathbf{w}\|_p + \beta)/(1 - \gamma) < \rho$ for all t.

Finally, since s_t is non-decreasing per construction, we know that $\lim_{t \to \infty} s_t = s^* = \|\mathbf{y}\|_p$ exists and satisfies

$$\|\mathbf{y}\|_p \leqslant \frac{1}{1 - \gamma}(\|\mathbf{w}\|_p + \beta).$$

$\square$

If we associate w_0 as the initial condition y_0 and we can rewrite the bounds of Theorem 7 and Theorem 2 as:

$$\|\boldsymbol{y}\|_p \leqslant \frac{1}{1-\gamma}\left(\sqrt[p]{|y_0|_p^p + \|\boldsymbol{w}\|_p^p} + \beta\right) \qquad 1 \leqslant p < \infty \qquad (2.21\text{a})$$

$$\|\boldsymbol{y}\|_\infty \leqslant \frac{1}{1-\gamma}\left(\max\{|y_0|, \|\boldsymbol{w}\|_\infty\} + \beta\right) \qquad p = \infty. \qquad (2.21\text{b})$$

Eventually Small Gain Condition

As so far discussed, the small gain condition $(\boldsymbol{I} - \boldsymbol{Q})(\boldsymbol{y}) < \boldsymbol{y} + \beta$ is sufficient to prove stability of systems governed by the equation $\boldsymbol{Q}(\boldsymbol{y}) = \boldsymbol{w}$. However, it is not a necessary requirement for stability. The next Lemma shows that we can relax the small gain condition and instead demand, that the gain of the operator $\boldsymbol{\Delta} := \boldsymbol{I} - \boldsymbol{Q}$ is *eventually* smaller than 1: There exists some t_0 such that the operators $\overline{\boldsymbol{P}^{t_0}}\boldsymbol{x} \mapsto \overline{\boldsymbol{P}^{t_0}}\boldsymbol{\Delta}(\boldsymbol{U}^{t_0}z + \overline{\boldsymbol{P}^{t_0}}(\boldsymbol{x}))$ satisfy the small gain property for any fixed bounded $z \in \mathbb{R}^{N_{t_0}}$.

Lemma 8. *Let $(\boldsymbol{y}, \boldsymbol{w})$ be solutions to the operator equation $\boldsymbol{Q}\boldsymbol{y} = \boldsymbol{w}$ for some causal $\boldsymbol{Q} \in \mathcal{C}(\ell^{\mathcal{X}})$. Then, the implication $\boldsymbol{w} \in \ell_p^{\mathcal{X}} \implies \boldsymbol{y} \in \ell_p^{\mathcal{X}}$ holds true, if there exists $t_0 \in \mathbb{N}$ for which the operator $\boldsymbol{\Delta} = \boldsymbol{I} - \boldsymbol{Q}$ meets the following conditions:*

(i) There exist $C_1, C_2 > 0$ such that for all solutions $(\boldsymbol{y}, \boldsymbol{w})$, the following ℓ_p-f.g.s condition holds true:

$$\boldsymbol{P}^{t_0}\boldsymbol{w} \in \ell_p^{\mathcal{X}} \implies \|\boldsymbol{P}^{t_0}\boldsymbol{y}\|_p \leqslant C_1\|\boldsymbol{P}^{t_0}\boldsymbol{w}\|_p + C_2. \qquad (2.22)$$

(ii) There exist some $\gamma < 1$ and non-decreasing function $\beta : \mathbb{R}^+ \to \mathbb{R}^+$ such that for all $\boldsymbol{x} \in \ell_p^{\mathcal{X}}$:

$$\|\overline{\boldsymbol{P}^{t_0}}\boldsymbol{\Delta}(\boldsymbol{x})\|_p \leqslant \gamma\|\overline{\boldsymbol{P}^{t_0}}(\boldsymbol{x})\|_p + \beta(\|\boldsymbol{P}^{t_0}(\boldsymbol{x})\|_p).$$

Proof. Let $(\boldsymbol{y}, \boldsymbol{w})$ be an arbitrary pair such that $\boldsymbol{Q}\boldsymbol{y} = \boldsymbol{w}$ or equivalently $\boldsymbol{y} = \boldsymbol{\Delta}\boldsymbol{y} + \boldsymbol{w}$, and assume that $\boldsymbol{w} \in \ell_p^{\mathcal{X}}$. Split $\boldsymbol{y}$ into the sum $\boldsymbol{P}^{t_0}\boldsymbol{y} + \overline{\boldsymbol{P}^{t_0}}\boldsymbol{y}$ and recall from (2.22) that $\|\boldsymbol{P}^{t_0}\boldsymbol{y}\|_p \leqslant C_1\|\boldsymbol{P}^{t_0}\boldsymbol{w}\|_p + C_2$. Hence, to show $\boldsymbol{y} \in \ell_p^{\mathcal{X}}$, we need to show that $\overline{\boldsymbol{P}^{t_0}}\boldsymbol{y} \in \ell_p^{\mathcal{X}}$. Let $s_{t_0} = \|\boldsymbol{P}^{t_0}\boldsymbol{y}\|$, $s_{k|t_0} = \|\boldsymbol{P}_n^{\mathcal{I}_k}(\boldsymbol{y})\|_p$, where $\mathcal{I}_k$ denote the intervals $[t_0, t_0 + k]$ for $k \in \mathbb{N}$, and notice the following chain of inequalities:

$$s_{k|t_0} = \|\boldsymbol{P}^{\mathcal{I}_k}\boldsymbol{y}\|_p \leqslant \|\boldsymbol{P}^{\mathcal{I}_k}\boldsymbol{\Delta}\boldsymbol{P}^{t_0+k}\boldsymbol{y}\|_p + \|\boldsymbol{P}^{\mathcal{I}_k}\boldsymbol{w}\|_p \leqslant \|\boldsymbol{\Delta}\boldsymbol{P}^{t_0+k}\boldsymbol{y}\|_p + \|\boldsymbol{P}^{\mathcal{I}_k}\boldsymbol{w}\|_p$$

$$\leqslant \gamma\|\boldsymbol{P}^{\mathcal{I}_k}(\boldsymbol{y})\|_p + \beta(\|\boldsymbol{P}^{t_0}\boldsymbol{y}\|_p) + \|\boldsymbol{P}^{\mathcal{I}_k}\boldsymbol{w}\|_p$$

$$\Leftrightarrow \qquad (1 - \gamma)s_{k|t_0} \leqslant \beta(s_{t_0}) + \|\boldsymbol{P}^{\mathcal{I}_k}\boldsymbol{w}\|_p$$

$$\Leftrightarrow \qquad s_{k|t_0} \leqslant \frac{1}{1-\gamma}\left(\beta(s_{t_0}) + \|\overline{\boldsymbol{P}^{t_0}}\boldsymbol{w}\|_p\right).$$

This shows that $s_{k|t_0}$ is bounded for any $k \in \mathbb{N}$ and together with the fact that $s_{k|t_0}$ is non-decreasing in k, it also proves convergence of the sequence $(s_{[0,t]_0}, s_{1|t_0}, \dots)$ to $\lim_{k \to \infty} s_{k|t_0} = s_{\infty|t_0} < \infty$. Hence, we showed $\overline{P^{t_0}}y \in \ell_p^{\mathcal{X}}$ and $\|y\|_p \leqslant s_{t_0} + s_{\infty|t_0}$ is bounded above as

$$\|y\|_p \leqslant \tfrac{1}{1-\gamma}\eta(\|w\|_p) + C_1\|w\|_p + C_2,$$

where $\eta(x) := \beta(C_1 x + C_2) + x$. $\qquad\qquad\square$

2.4 Representing Dynamical Systems as Sets, Relations, and Maps

A discrete-time dynamical system S is a system whose behavior over time can be described by a function $S : \mathbb{N} \to \mathcal{S}$ over the index set $\mathbb{N}$, where each $S(t)$ represents the state of the system S at time-step t. Correspondingly, $\mathcal{S}$ is usually called the state space of the dynamical system and represents the granularity of our system description. Depending on the application, $\mathcal{S}$ can be a finite set, a finite- or infinite-dimensional space of vectors, or even functions. A sequence $s = (s_0, s_1, \dots) \in \mathcal{S}^{\mathbb{N}}$, $S(t) = s_t$, describing a specific realization of the system's behavior is called a state trajectory of S.

Models are used to characterize the behavior of a dynamical system and ideally provide a minimal, yet complete representation of the behavior of S over time, capable of describing all possible state trajectories s of S. There are many ways to formulate such a representation. A natural approach, often easiest when S is known to obey certain laws (e.g., the laws of physics), is to describe the system behavior implicitly as a set of solutions to a set of difference equations. If the behavior of S is random, another suitable representation is to model S as a stochastic process over $\mathbb{N}$, where the system state S_t at time t is represented as a random variable with support $\mathcal{S}$.

However, as our starting point, we choose a more explicit (and deterministic) formulation by identifying the behavior of a dynamical system S with its set $\mathsf{M}_S \subset \mathcal{S}^{\mathbb{N}}$ of all possible state trajectories τ^{x}. We call the set M_S the *dynamic model* or model of the system S.

Remark 3. *Many fundamental concepts of systems theory such as controllability, observability, stability, etc., have an equivalent formulation in this representation. This alternative framework of control and systems theory was first developed by Jan C. Willems [131] and is known as the Behavioral Approach to Control.*

Input-Output Maps of Dynamical Models Often the state space $\mathcal{S}$ is represented by a family $\{\mathcal{Z}_i\}_i$, $i \in \mathcal{I}$, of smaller spaces indexed over some index set $\mathcal{I}$; often $\mathcal{I}$ is used to further specify the internal structure of the system S (see example below).

Example 1. *Assume that S represents the behavior of a large-scale system composed of a family of interconnected dynamical systems $\{X_i\}_{i \in \mathcal{I}_1}$ driven by a distributed family of actuators $\{U_j\}_{j \in \mathcal{I}_2}$ and a collection of external disturbance sources $\{W_j\}_{j \in \mathcal{I}_3}$. Then a natural choice to define $\mathcal{S}$ is as the Cartesian product $\mathcal{X} \times \mathcal{U} \times \mathcal{W}$ of the spaces $\mathcal{X} = \times_{i \in \mathcal{I}_1} \mathcal{X}_i$, $\mathcal{U} = \times_{j \in \mathcal{I}_2} \mathcal{U}_j$, and $\mathcal{Z} = \times_{k \in \mathcal{I}_3} \mathcal{Z}_k$ corresponding to the families of subsystems, actuators, and disturbance sources.*

Sometimes a dynamical system offers a natural way to divide $\mathcal{I}$ into a union $\mathcal{I}^{in} \cup \mathcal{I}^{out}$ of inputs $\mathcal{I}^{in}$ representing independent variables $Z_i, i \in \mathcal{I}^{in}$ and outputs $\mathcal{I}^{out}$ such that the variables $Z_i, i \in \mathcal{I}^{out}$ depend causally on the variables $Z_i, i \in \mathcal{I}$. A dynamical model has a viable assignment of inputs $\mathcal{I}^{in}$ and outputs $\mathcal{I}^{out}$ if and only if the binary relation $\{(\{z_i\}_{i \in \mathcal{I}^{in}}, \{z_j\}_{j \in \mathcal{I}^{out}}) \,|\, \{z_i\}_{i \in \mathcal{I}} \in \mathsf{M}_S\}$ is causal according to the following definition:

Definition 2.4. *A binary relation $\mathcal{R} \subset \mathcal{X}^{\mathbb{N}} \times \mathcal{Y}^{\mathbb{N}}$ is said to be causal if for any $(x, y) \in \mathcal{R}$ and $(x', y') \in \mathcal{R}$, and any $t \in \mathbb{N}$, the following implication holds true:*

$$P^t x = P^t x' \quad \Longrightarrow \quad P^t y = P^t y'.$$

Remark 4. *The above condition is equivalent to requiring that for any t, the relation $\{(P^t x, P^t y) \,|\, (x, y) \in \mathcal{R}\}$ is functional. Hence, if true, it defines a unique causal operator $\Psi : \mathrm{do}(\mathcal{R}) \mapsto \mathcal{Y}^{\mathbb{N}}$ such that $\Psi(x) = y \Leftrightarrow (x, y) \in \mathcal{R}$.*

From the above definition, we see that a viable assignment of inputs $\mathcal{I}^{in}$ and outputs $\mathcal{I}^{out}$ implicitly defines a unique causal mapping between inputs $s^{in} = \{z_i\}_{i \in \mathcal{I}^{in}}$ and outputs $s^{out} = \{z_i\}_{i \in \mathcal{I}^{out}}$ of trajectories $s = \{z\}_{i \in \mathcal{I}} \in \mathsf{M}_S$. We call the operator representing the mapping $\Psi : s^{in} \mapsto s^{out}$ the $\mathcal{I}^{in} \mapsto \mathcal{I}^{out}$-map of the dynamical model M_S. We formulate this definition and summarize our discussion below:

Definition 2.5 (Input-Output Maps of Dynamical Models). *Let $\mathsf{M}_S \subset \mathcal{S}^{\mathbb{N}}$ be a dynamical model, where $\mathcal{S} = \{\mathcal{Z}_i\}_{i \in \mathcal{I}}$. Consider two subsets $\mathcal{I}^{in}, \mathcal{I}^{out} \subset \mathcal{I}$ such that $\mathcal{I}^{in} \cup \mathcal{I}^{out} = \mathcal{I}$, and for a trajectory $s = \{z\}_{i \in \mathcal{I}} \in \mathsf{M}_S$, let $s^{in} = \{z\}_{i \in \mathcal{I}^{in}}$ and $s^{out} = \{z\}_{i \in \mathcal{I}^{out}}$.*

We say that the pair $\mathcal{I}^{in}$ and $\mathcal{I}^{out}$ is an input-output assignment for M_S if the binary relation $\mathcal{R}^{\mathcal{I}^{in} \mapsto \mathcal{I}^{out}}$ defined below is causal:

$$\mathcal{R}^{\mathcal{I}^{in} \mapsto \mathcal{I}^{out}} \triangleq \{(s^{in}, s^{out}) \mid s \in \mathsf{M}_S\}.$$

Furthermore, under this condition, there exists a unique causal operator Ψ : $\mathrm{do}(\mathcal{R}^{\mathcal{I}^{in} \mapsto \mathcal{I}^{out}}) \to \mathcal{S}^{out}$ such that

$$s^{out} = \Psi(s^{in}) \Leftrightarrow s \in \mathsf{M}_S.$$

This causal operator is referred to as the $\mathcal{I}^{in} \mapsto \mathcal{I}^{out}$-map of M_S.

If an input-output map exists, it provides a complete description of the dynamical system S, similar to M_S. However, in general, a viable input-output assignment is not always possible, and if one exists, it is not always unique. Hence, while M_S is uniquely tied to the dynamical system S, there might be different input-output maps of S depending on how many viable assignments of input-output pairs there are. Therefore, input-output maps of dynamic model M_S have to be distinguished by the respective choice of inputs $\mathcal{I}^{in}$ and outputs $\mathcal{I}^{out}$.

Closed Loop Maps of Dynamical Models   In a standard control problem setup, we are given a dynamical system, called the plant, with state X, which we control with some input U using measurements Y, and which is subjected to external (non-measurable) disturbances W. We represent the time behavior of this control system as a dynamical system with state $S = (X, U, W, Y)$ and state space $\mathcal{S} = (\mathcal{X}, \mathcal{U}, \mathcal{W}, \mathcal{Y})$ and denote trajectories of S as a tuple $\tau = (\tau^x, \tau^u, \tau^y, \tau^w)$ composed of the individual state, input, output, and disturbance sequences $\tau^x \in \mathcal{X}^{\mathrm{N}}, \tau^u \in \mathcal{U}^{\mathrm{N}}$, $\tau^y \in \mathcal{Y}^{\mathrm{N}}$, and $\tau^w \in \mathcal{W}^{\mathrm{N}}$ corresponding to a realization of the dynamics of the control system S. Correspondingly, the set of all such realizations forms the dynamical model $\mathsf{M}_S \subset \mathcal{S}^{\mathrm{N}}$, which describes the behavior of the control system S in open-loop.

On the other hand, introducing a feedback controller $K \in \mathcal{C}(\mathcal{Y}^{\mathrm{N}}, \mathcal{U}^{\mathrm{N}})$ leads to a different dynamical system S_K, describing the closed-loop dynamics of the interconnection of the open-loop system S and controller K. We say that the controller K realizes the closed-loop system S_K or equivalently, realizes the closed-loop model M_{S_K}. The relation between open and closed-loop system can be simply stated in terms of their dynamic models:

$$\mathsf{M}_{S_K} = \{\tau \in \mathsf{M}_S \mid \tau^u = K(\tau^y)\} = \mathsf{M}_S \cap \mathsf{M}_K.$$

Thus, the model of the closed-loop system S_K is a subset of the model of the open-loop system S, and more specifically, it is obtained from M_S by imposing the additional constraint $\tau^{\mathrm{u}} = \boldsymbol{K}(\tau^y)$ on the trajectories of S. If we represent the controller $\boldsymbol{K}$ as a dynamical model $\mathsf{M}_K = \{\tau \in \mathcal{S}^{\mathbb{N}} \mid \tau^u = \boldsymbol{K}(\tau^y)\}$ embedded in the space $\mathcal{S}^{\mathbb{N}}$, then we can also simply view the closed-loop model as an intersection of the open-loop and controller models. From this point of view, we can characterize the space of closed-loop systems realizable by some causal operator $\boldsymbol{K} : \mathcal{Y}^{\mathbb{N}} \to \mathcal{U}^{\mathbb{N}}$ as the following set of dynamical models, which can be referred to as the space of realizable closed-loop models of the system S:

$$\mathbb{M}_S = \left\{ \mathsf{M}_S \cap \mathsf{M}_K \mid \boldsymbol{K} \in \mathcal{C}(\mathcal{Y}^{\mathbb{N}}, \mathcal{U}^{\mathbb{N}}) \right\}.$$

Maps of closed-loop models $\mathsf{M}_{S_{K_\star}} \in \mathbb{M}_S$ are usually called closed-loop maps of S, which will be the focus of the discussion in the next section.

Remark 5. *In later discussion, we consider dynamical systems where for all* $\mathsf{M}^\star \in \mathbb{M}_S$, *the pair* $\mathcal{I}_{in} = \{\mathrm{w}\}$ *and* $\mathcal{I}_{out} = \{\mathrm{xuy}\}$ *is a valid input-output pairing. This permits us to represent the space of realizable closed-loop system behavior as a space of operators, which we later denote by* $\Phi_S^{\mathrm{w} \mapsto \{uxy\}}$ *and refer to as the space of closed-loop maps of* S *or* M_S.

In this section, we explored the concept of general exterior representations of dynamical systems, which can be understood as a collection of sequences that encompass all possible manifestations of system behavior. This discussion led us to examine dynamic models and the causality conditions that enable us to establish correspondences between dynamic models and maps linking inputs to outputs. Additionally, we delved into the relationship between the model of an open-loop system and the space of realizable closed-loop models through feedback control.

Building upon these ideas, in the upcoming section, we will shift our focus to an interior representation of dynamical systems. This representation takes the form of nonlinear difference and operator equations, allowing us to establish connections and equivalences with the exterior representations discussed earlier, and leads to the main result of this chapter: the characterization and realization of closed-loop maps via solutions of the CLM operator equation.

2.5 Dynamic Equations of the Open-Loop System

Here we discuss the class of discrete-time systems of interest and define our setup of the open-loop system S and the governing dynamic equations.

The open-loop system S is a dynamical system with state space $\mathcal{S} = \mathcal{S}_x \times \mathcal{S}_u \times \mathcal{S}_w = \times_{i \in \mathcal{I}} \mathcal{S}_i, \mathcal{I} = \{x, u, w\}$, where $\mathcal{S}_x, \mathcal{S}_w, \mathcal{S}_u$ denote the spaces corresponding to the state of the plant, control action, and external disturbance. Moreover, let $\mathcal{X} = \mathcal{S}_x = \mathcal{S}_w$ and $\mathcal{U} = \mathcal{S}_u$, where $\mathcal{X}$ and $\mathcal{U}$ are some fixed finite-dimensional vector spaces, e.g., $\mathcal{X} = \mathbb{R}^n, \mathcal{U} = \mathbb{R}^m$. We define the open-loop system S in terms of its trajectories $\tau = (x, u, w)^\top$ as follows:

Let $x \in \ell^{\mathcal{X}}$ be the state trajectory of a nonlinear discrete-time dynamical system, and let $u \in \ell^{\mathcal{U}}$ and $w \in \ell^{\mathcal{X}}$ be the corresponding sequences of control inputs and disturbances that generate the trajectory according to the following set of equations:

$$\forall t \geq 1 : \quad x_t = f_{t-1}(x_{t-1:0}, u_{t-1:0}) + w_t, \quad x_0 = w_0, \tag{2.23}$$

where $f := \{f_t\}_{t=0}^\infty$, $f_t : \mathcal{X}^{[0,t]} \times \mathcal{U}^{[0,t]} \to \mathcal{X}^{[0,t]}$ is a fixed sequence of functions representing the (open-loop) dynamics of the system. We make *no* further assumptions on f, i.e., each f_t can be an arbitrary nonlinear function. We can obtain a more compact description of the dynamics by embedding the function sequence f as the component functions of a causal operator $F : \ell^{\mathcal{X} \times \mathcal{U}} \to \ell^{\mathcal{X}}$. Thus, let $F \in \mathcal{C}(\ell^{\mathcal{X} \times \mathcal{U}}, \ell^{\mathcal{X}})$, and its strictly causal right-shifted version $F^+ := \mathcal{S}^+ F \in \mathcal{C}_s(\ell^{\mathcal{X} \times \mathcal{U}}, \ell^{\mathcal{X}})$ be defined in terms of f as:

$$F(x, u) := (f_0(x_0, u_0), f_1(x_1, x_0, u_1, u_0), \ldots, f_t(x_{t-1:0}, u_{t-1:0}), \ldots) \tag{2.24}$$

$$F^+(x, u) := \mathcal{S}^+ F(x, u) = (0, f_0(x_0, u_0), f_1(x_{1:0}, u_{1:0}), \ldots). \tag{2.25}$$

We refer to the operator F as the open-loop *dynamics* of the system, since the dynamic equations (2.23) can be equivalently defined in sequence space as

$$x = \mathcal{S}^+ F(x, u) + w =: F^+(x, u) + w. \tag{2.26}$$

The above equation is a compact description of the state transitions of the dynamical system; hence, one can view this as an *internal* characterization of the system dynamics. On the other hand, the corresponding external representation of the open-loop system S is given by the dynamical model $\mathrm{M}_F \subset \ell^{\mathcal{X}} \times \ell^{\mathcal{U}} \times \ell^{\mathcal{X}}$, which is defined for a fixed dynamics operator F as:

$$\mathrm{M}_F := \left\{ \tau = (\tau^x, \tau^u, \tau^w)^\top \mid \tau^x = F^+(\tau^x, \tau^u) + \tau^w \right\}. \tag{2.27}$$

Hence, as discussed in the previous section, M_F is the set of all possible trajectories $\tau = (x, u, w)^\top$ of S, represented as a tuple of state x, input u, and disturbance w sequences corresponding to realizations of the open-loop system.

From the dynamic equations, it is also easy to see that w, u can be treated as input signals of the open-loop system and the resulting state trajectory x as the corresponding output. In other words, $\{u, w\}$ and $\{x\}$ is a valid input-output assignment of the open-loop model M_F, and it has, therefore, a $\{u, w\} \mapsto x$-map, which we refer to as the open-loop map G_F. We derive the closed-form of G_F next.

Remark. *Aside from G_F, M_F also has other valid input-output map representations. For example, $\mathcal{I}_{in} = \{x, u\}$, $\mathcal{I}_{out} = \{w\}$ is a valid input-output assignment as well, and therefore M_F can be equivalently represented as a $\{x, u\} \mapsto w$-map.*

Open-Loop Maps

Since a system state trajectory x is always a causal function of the input u and the disturbance w, it is clear that $G_F : (u, w) \mapsto x$, st.: $(x, u, w)^\top \in \mathsf{M}_F$, represents a well-defined mapping. We call $G_F \in \mathcal{C}(\ell^{\mathcal{X} \times \mathcal{U}}, \ell^{\mathcal{X}})$ the open-loop map of dynamics F, and notice that the model set M_F is the graph of G_F. For a fixed u, denote $F^+|_u : \ell^{\mathcal{X}} \mapsto \ell^{\mathcal{X}}$ as the restriction of F^+ such that $F^+|_u(x) = F^+(x, u)$. Then, for each trajectory $(x, u, w)^\top \in \mathsf{M}_F$, holds $x = (I - F^+|_u)^{-1}w$ (inverse exists due to Proposition 5), and therefore G_F is the mapping

$$G_F : (u, w) \mapsto (I - F^+|_u)^{-1}w.$$

In this section, we discussed the relationships between different representations of the open-loop system S: as a set of dynamic equations $x = F^+(x, u) + w$, as a dynamic model M_F, and as a causal input-output map $G_F : (u, w) \mapsto (I - F^+|_u)^{-1}w$. In the next section, we discuss this trinity of representations for the closed-loop system.

2.6 Closed-Loop Maps and Realizing Controllers

In this section, we introduce feedback control and characterize the dynamics of the closed-loop, by establishing a correspondence between the closed-loop system's exterior representation, as closed-loop maps, and the internal representation, as dynamics operators and system equations. In short, the key theoretical results discuss a characterization of the space of realizable CLMs as the solution space of an operator equation and the one-to-one correspondence between CLMs and their realizing controllers.

The Closed-Loop System

Let $K \in \mathcal{C}(\ell^{\mathcal{X}}, \ell^{\mathcal{U}})$ represent a causal feedback control law, and let the component functions K_t denote the decision rule for u_t based on the state observations $x_0, \ldots, x_t$:

$$K(x) := (K_0(x_0), \ldots, K_t(x_{t:0}), \ldots).\tag{2.28}$$

Correspondingly, let $\mathsf{M}_K = \{\tau^{\mathrm{xu}} = (\tau^{\mathrm{x}}, \tau^{\mathrm{u}}) \mid \tau^{\mathrm{u}} = K\tau^{\mathrm{x}}\}$ denote the dynamic model of K. The interconnection of (2.23) with the open-loop system S defines a dynamical system CL, which we refer to as the closed-loop, defined by the following dynamic equations:

$$\text{CL}: \qquad x_t = f_{t-1}(x_{t-1:0}, u_{t-1:0}) + w_t, \quad x_0 = w_0 \tag{2.29a}$$

$$u_t = K_t(x_{t:0}). \tag{2.29b}$$

Equivalently, the dynamic system equations of the closed-loop CL can be written as:

$$\text{CL}: \qquad x = F^+(x, K(x)) + w, \quad u = K(x) \tag{2.30}$$

$$= F_K^+(x) + w,$$

where F_K denotes the mapping $x \mapsto F(x, K(x))$, also referred to as the closed-loop dynamics operator.

Definition (Operator of Closed-Loop Dynamics). *For $F \in \mathcal{C}(\ell^{\mathcal{X} \times \mathcal{U}}, \ell^{\mathcal{X}})$ and $K \in \mathcal{C}(\ell^{\mathcal{X}}, \ell^{\mathcal{U}})$, define the operator $F_K \in \mathcal{C}(\ell^{\mathcal{X}})$ for all $x \in \ell^{\mathcal{X}}$ as $F_K(x) := F(x, K(x))$.*

As formulated in Section 2.4, intersecting the open-loop model M_F with the controller model M_K yields $\text{CL}_{[F,K]} \triangleq \mathsf{M}_F \cap \mathsf{M}_K$, the dynamic model of the closed-loop system CL. $\text{CL}_{[F,K]}$ is the set of all trajectories $\tau_{\mathrm{xuw}} = (x, u, w)$ that satisfy the equations 2.30 for some fixed pair of F and K and hence admits the equivalent definition: and define the set of all trajectories $\tau = (x, u, w)^\top$, $x \in \ell^{\mathcal{X}}$, $u \in \ell^{\mathcal{U}}$, and $w \in \ell^{\mathcal{X}}$ which are solutions to the above equation as the *closed-loop* model $\text{CL}_{[F,K]}$:

$$\text{CL}_{[F,K]} = \left\{ (x, u, w)^\top \mid x = F^+(x, u) + w, \quad u = K(x) \right\}. \tag{2.31}$$

Recall that $F^+ \in \mathcal{C}_s(\ell^{\mathcal{X} \times \mathcal{U}}, \ell^{\mathcal{X}})$ and therefore the map $F_K^+ \in \mathcal{C}_s(\ell^{\mathcal{X}})$ is also strictly causal. As shown in (5), this observation certifies that the operator $I - F_K^+$ has a causal inverse $(I - F_K^+)^{-1} \in \mathcal{C}(\ell^{\mathcal{X}})$. Hence, we can equivalently rewrite the equation $x = F_K(x) + w$ as $x = (I - F_K^+)^{-1}w$ and see that $(x, u, w)^\top$ is a trajectory of $\text{CL}_{[F,K]}$ if and only if it satisfies the equations:

$$x = (I - F_K^+)^{-1}w \tag{2.32a}$$

$$u = K(I - F_K^+)^{-1}w. \tag{2.32b}$$

In other words, for any F and K, the operator $\left[\begin{smallmatrix} I \\ K \end{smallmatrix}\right](I - F_K^+)^{-1} \in \mathcal{C}(\ell^{\mathcal{X}}, \ell^{\mathcal{X} \times \mathcal{U}})$ represents the $w \mapsto \left[\begin{smallmatrix} x \\ u \end{smallmatrix}\right]$-map of the closed-loop model $\mathsf{CL}_{[F,K]}$.

Remark. *In terms of the discussion in Section 2.4, 2.32 proves the causality of the relation $\mathcal{R}^{w \mapsto x,u} = \{(w, \left[\begin{smallmatrix} x \\ u \end{smallmatrix}\right]) \mid (x, u, w)^\top \in \mathsf{CL}_{[F,K]}\}$ and provides a formula for the $w \mapsto x, u$-map of the closed-loop model $\mathsf{CL}_{[F,K]}$.*

The input-output representation of a closed loop provides a natural way to define closed-loop stability [139]: The closed-loop trajectories $\tau \in \mathsf{CL}_{[F,K]}$ have ℓ_p-bounded state and input trajectory τ^{xu} for any ℓ_p bounded disturbances τ^{w} if and only if the $w \mapsto \{x, u\}$-map of $\mathsf{CL}_{[F,K]}$ is ℓ_p-stable, and correspondingly, a controller K is ℓ_p-stabilizing if and only if the corresponding $w \mapsto \{x, u\}$-map, i.e., the operator $\left[\begin{smallmatrix} I \\ K \end{smallmatrix}\right](I - F_K^+)^{-1}$ is ℓ_p-stable. The next theorem summarizes our discussion so far, and states important properties of the $w \mapsto \{x, u\}$-map representation of $\mathsf{CL}_{[F,K]}$.

Theorem 6 (Input-Output Maps of closed loops). *For any fixed pair $F \in \mathcal{C}(\ell^{\mathcal{X} \times \mathcal{U}}, \ell^{\mathcal{X}})$ and $K \in \mathcal{C}(\ell^{\mathcal{X}}, \ell^{\mathcal{U}})$, the closed-loop model $\mathsf{CL}_{[F,K]}$ has a $w \mapsto \{x, u\}$-map $\Psi \in \mathcal{C}(\ell^{\mathcal{X}}, \ell^{\mathcal{X} \times \mathcal{U}})$, and it holds:*

i) $\Psi = \left[\begin{smallmatrix} \Psi^{\mathrm{x}} \\ \Psi^{\mathrm{u}} \end{smallmatrix}\right]$*, where* $\Psi^{\mathrm{x}} = (I - F_K^+)^{-1}$ *and* $\Psi^{\mathrm{u}} = K(I - F_K^+)^{-1}$.

ii) $\tau \in \mathsf{CL}_{[F,K]}$ *if and only if* $\tau^{\mathrm{xu}} = \Psi(\tau^{\mathrm{w}})$.

iii) $K = \Psi^{\mathrm{u}}(\Psi^{\mathrm{x}})^{-1}$.

Proof. We have already established that $\mathsf{CL}_{[F,K]}$ always has a $w \mapsto \{x, u\}$-*map*; according to the setup of the theorem we denote this map Ψ.
<u>i)</u>: As discussed previously, i) follows since for any $(x, u, w)^\top \in \mathsf{CL}_{[F,K]}$ holds 2.32, which provides a closed-form expression for Ψ. <u>ii)</u>: This is just the restatement of the general definition Def. 2.5. <u>(iii)</u>: From i), which is already proven, we have $\Psi^{\mathrm{x}} = (I - F_K^+)^{-1}$ and $(\Psi^{\mathrm{x}})^{-1} = (I - F_K^+)$, which implicitly states that Ψ^{x} is always causally invertible. Now since $\Psi^{\mathrm{u}} = K(I - F_K^+)^{-1}$, multiplying both sides from the right by $(\Psi^{\mathrm{x}})^{-1}$ yields $K = \Psi^{\mathrm{u}}(\Psi^{\mathrm{x}})^{-1}$. $\qquad\square$

The above result leads us to a very important conclusion: For fixed F, there is a one-to-one correspondence, via the transformation $\mathbb{H} : K \mapsto \left[\begin{smallmatrix} I \\ K \end{smallmatrix}\right](I - F_K^+)^{-1}$ and its inverse $\mathbb{H}^{-1} : \left[\begin{smallmatrix} \Psi^{\mathrm{x}} \\ \Psi^{\mathrm{u}} \end{smallmatrix}\right] \mapsto \Psi^{\mathrm{u}}(\Psi^{\mathrm{x}})^{-1}$, between a controller K, later called the *realizing controller*, and the corresponding $w \mapsto \{x, u\}$-map of $\mathsf{CL}_{[F,K]}$.

Remark 7. *In fact, there is a one-to-one correspondence between* $w \mapsto \{x, u\}$*-maps of* $\mathsf{CL}_{[F,K]}$ *and the pairs* (K, F_K) *of controller* K *and closed-loop dynamics* F_K*: If* Ψ *is the* $w \mapsto \{x, u\}$*-map of* $\mathsf{CL}_{[F,K]}$*, then* $I - (\Psi^x)^{-1} = F_K^+$ *and* $\Psi^u(\Psi^x)^{-1} = K$*; the reverse direction is clear, since* $w \mapsto \{x, u\}$*-maps of* $\mathsf{CL}_{[F,K]}$ *always have the form* $\Psi = \left[\begin{smallmatrix} I \\ K \end{smallmatrix}\right](I - F_K^+)^{-1}$*.*

This observation motivates us to characterize the space of all $w \mapsto \{x, u\}$-maps of $\mathsf{CL}_{[F,K]}$. We define this as the space $\Phi_{\mathrm{CL}}^{w \mapsto \{x,u\}}[F]$ of *closed-loop maps* (CLMs) of F and call an operator Ψ in the set $\Phi_{\mathrm{CL}}^{w \mapsto xu}[F]$ a closed-loop map of F:

Definition 2.6 (Closed Loop Maps). *An operator* $\Psi \in \mathcal{C}(\ell^{\mathcal{X}}, \ell^{\mathcal{X} \times \mathcal{U}})$ *is called a closed-loop map (CLM) of* F*, if for some* $\tilde{K} \in \mathcal{C}(\ell^{\mathcal{X}}, \ell^{\mathcal{U}})$*, called realizing controller of* Ψ*, the operator* Ψ *coincides with the* $w \mapsto (x, u)$*-map of the closed-loop model* $\mathsf{CL}_{[F,\tilde{K}]}$*. We denote the set of all such* Ψ*, with* $\Phi_{\mathrm{CL}}^{w \mapsto xu}[F]$ *and call it the space of CLMs of* F *or the CLM-space of* F*.*

Remark. *In terms of the discussion of Section 2.4, the above considers the special case where* $y = x$*, i.e., we can measure the state* x *and used as measurement for feedback control.*

The next theorem presents different characterizations of the CLM-space $\Phi_{\mathrm{CL}}^{w \mapsto xu}[F]$ and represents a main result of this chapter.

Theorem 8 (Characterization of CLMs). *For a fixed causal operator* $F \in \mathcal{C}(\ell^{\mathcal{X} \times \mathcal{U}}, \ell^{\mathcal{X}})$*, the following are all equivalent definitions of* $\Phi_{\mathrm{CL}}^{w \mapsto xu}[F]$*:*

    i) $\Phi_{\mathrm{CL}}^{w \mapsto xu}[F] = \{\Psi \in \mathcal{C}(\ell^{\mathcal{X}}, \ell^{\mathcal{X} \times \mathcal{U}}) \mid \Psi \text{ is a CLM of } F\}$.

    ii) $\Phi_{\mathrm{CL}}^{w \mapsto xu}[F] = \{\Psi \mid \forall \left[\begin{smallmatrix} \tau^{xu} \\ \tau^{w} \end{smallmatrix}\right] \in \mathrm{CL}_{[F,\Psi^u(\Psi^x)^{-1}]} \text{ s.t. } \tau^{xu} = \Psi(\tau^w)\}$.

    iii) $\Phi_{\mathrm{CL}}^{w \mapsto xu}[F] = \{\Psi = \left[\begin{smallmatrix} \Psi^x \\ \Psi^u \end{smallmatrix}\right] \in \mathcal{C}(\ell^{\mathcal{X}}, \ell^{\mathcal{X} \times \mathcal{U}}) \mid \Psi^x = F^+(\Psi) + I\}$.

    iv) $\Phi_{\mathrm{CL}}^{w \mapsto xu}[F] = \{\left[\begin{smallmatrix} I \\ K \end{smallmatrix}\right](I - F_K^+)^{-1} \mid K \in \mathcal{C}(\ell^{\mathcal{X}}, \ell^{\mathcal{U}})\}$.

Proof. <u>i)</u> is just reiterating the original definition Def. 2.6. We proceed by proving that the other statements are all equivalent to i). <u>ii)</u>: This statement follows from the fact that there is a one-to-one correspondence between CLMs $\Psi \in \Phi_{\mathrm{CL}}^{w \mapsto xu}[F]$ and their realizing controllers $K = \Psi^u(\Psi^x)^{-1}$ and recalling Theorem 6 ii). Part <u>iv)</u>: This statement follows directly from ii) of Theorem 6. It remains to establish the equivalence between iii) and i).

i) $\implies$ iii): Assume Ψ is a CLM of F. Then, per definition, there exists some $\tilde{K} \in \mathcal{C}(\ell^{\mathcal{X}}, \ell^{\mathcal{U}})$ such that Ψ is an w $\mapsto$ (x, u)-map of $\mathrm{CL}_{[F,\tilde{K}]}$, which is equivalent, by the characterization Theorem 6, to stating that $(x, u, w)^{\top} \in \mathrm{CL}_{[F,\tilde{K}]}$ if and only if $x = \Psi^{\mathrm{x}}(w)$ and $u = \Psi^{\mathrm{u}}(w)$. Since $(x, u, w)^{\top} \in \mathrm{CL}_{[F,\tilde{K}]} \Leftrightarrow x = F^{+}(x, u) + w$, we conclude by substitution that for all $w \in \ell^{\mathcal{X}}$, $\Psi^{\mathrm{x}}(w) = F^{+}(\Psi(w)) + w$, i.e., the operator Ψ^{x} and $F^{+}(\Psi) + I^{\mathrm{x}}$ are the same.

iii) $\implies$ i): Assume Ψ is a solution of (2.30). Then, $\Psi^{\mathrm{x}} - I = F^{+}(\Psi) \in \mathcal{C}_s(\ell^{\mathcal{X}})$, since $F^{+} \in \mathcal{C}_s(\ell^{\mathcal{X} \times \mathcal{U}}, \ell^{\mathcal{X}})$. As discussed in Lem. 5, any operator of the form $I_n + A^{+}$, $A^{+} \in \mathcal{C}_s(\ell^{\mathcal{X}})$ is causally invertible. Therefore, $(\Psi^{\mathrm{x}})^{-1}$ exists and $(\Psi^{\mathrm{x}})^{-1} \in \mathcal{C}(\ell^{\mathcal{X}})$. Now, take $K' = \Psi^{\mathrm{u}}(\Psi^{\mathrm{x}})^{-1}$ and let $(x, u, w)^{\top} \in \mathrm{CL}_{[F,K']}$ be an arbitrary trajectory of the closed-loop model $\mathrm{CL}_{[F,K']}$. Then, $(x, u) = \Phi_{\mathrm{CL}}[F, K'](w)$, which implies that $x = F(x, \Psi^{\mathrm{u}}(\Psi^{\mathrm{x}})^{-1}x) + w$. We apply the identity $(\Psi^{\mathrm{x}} - I)(\Psi^{\mathrm{x}})^{-1} + (\Psi^{\mathrm{x}})^{-1} = I$ to the left side of the equation and obtain

$$(\Psi^{\mathrm{x}} - I)(\Psi^{\mathrm{x}})^{-1}x + (\Psi^{\mathrm{x}})^{-1}x = F^{+}\Psi(\Psi^{\mathrm{x}})^{-1}x + w$$
$$\Leftrightarrow (\Psi^{\mathrm{x}} - I - F^{+}\Psi)(\Psi^{\mathrm{x}})^{-1} + (\Psi^{\mathrm{x}})^{-1}x = w,$$

which, due to $\Psi^{\mathrm{x}} - I = F^{+}(\Psi)$, implies $x = \Psi^{\mathrm{x}}w$ and $u = \Psi^{\mathrm{u}}(\Psi^{\mathrm{x}})^{-1}\Psi^{\mathrm{x}}w = \Psi^{\mathrm{u}}w$. $\qquad\square$

As it turns out, the space of realizable CLMs $\Phi_{\mathrm{CL}}^{\mathrm{w} \mapsto \mathrm{xu}}[F]$ can be characterized as solutions to the nonlinear operator equation:

$$\Psi^{\mathrm{x}} = F^{+}(\Psi) + I. \tag{2.33}$$

We therefore also refer to (2.33) as the *CLM equation*. Writing out the CLM equation (2.33) in terms of component functions gives the more explicit condition on the functions Ψ_t^x, Ψ_t^u: The map $\Psi = (\Psi^{\mathrm{x}}, \Psi^{\mathrm{u}})$ satisfies (2.33) if and only if its component functions satisfy the following infinite set of function equations for all inputs $w_{t:0}$:

$$\Psi_t^x(w_{t:0}) = F_{t-1}(\Psi_{t-1:0}^x(w_{t-1:0}), \Psi_{t-1:0}^u(w_{t-1:0})) + w_t. \tag{2.34}$$

As proven in Theorem 6, the mapping between CLMs $(\Psi^{\mathrm{x}}, \Psi^{\mathrm{u}}) \in \Phi_{\mathrm{CL}}^{\mathrm{w} \mapsto \mathrm{xu}}[F]$ and the corresponding realizing controllers K' is one-to-one, via the relationship $K' = \Psi^{\mathrm{u}}(\Psi^{\mathrm{x}})^{-1}$. While the realizing controller K of a CLM Ψ is a unique operator, we have degrees of freedom regarding the implementation – usually in the form of a dynamical system – of K. Next, we discuss a particularly simple approach enabled by the fact that Ψ^{x} is causally invertible for any $\Psi \in \Phi_{\mathrm{CL}}^{\mathrm{w} \mapsto \mathrm{xu}}[F]$.

System Level Implementations of Realizing Controllers

A crucial step of establishing Theorem 8, is to show that $(\Psi^x)^{-1}$ always exists. We briefly rederive this partial result as a consequence of Lem. 5, restated below, which highlights the main idea behind the system level implementation.

Lemma. *The inverse A^{-1} of a square operator $A \in \mathcal{C}(\ell^{\mathcal{X}})$ exists if $A - I \in \mathcal{C}_s(\ell^{\mathcal{X}})$, and $b = A^{-1}(a)$ is calculated by evaluating equations $b_t = a_t - A_t(0, b_{t-1:0})$ in the order $t = 0, 1, 2, \ldots$.*

Due to Theorem 8, we know that for any CLM $\Psi \in \Phi_{\mathrm{CL}}^{\mathrm{w \mapsto xu}}[F]$ holds $\Psi^x - I = F^+(\Psi)$. Furthermore, since $F^+ \in \mathcal{C}_s(\ell^{\mathcal{X} \times \mathcal{U}}, \ell^{\mathcal{X}})$ holds, we are also ensured that $\Psi^x - I \in \mathcal{C}_s(\ell^{\mathcal{X}})$. This allows us to apply the previous lemma and proves that $(\Psi^x)^{-1}$ always exists and that it is a causal operator in $\mathcal{C}(\ell^{\mathcal{X}})$.

Furthermore, as shown in Lem. 5, the condition $\Psi^x - I \in \mathcal{C}_s(\ell^{\mathcal{X}})$ allows a particularly simple method of implementing $K' = \Psi^u(\Psi^x)^{-1}$ of Theorem 8: Given an input a, the output $b = K'(a)$ can be computed recursively through the equations

$$c_t = a_t - \Psi_t^x(0, c_{t-1:0}), \tag{2.35a}$$

$$b_t = \Psi_t^u(c_{t:0}). \tag{2.35b}$$

The above implementation represents a dynamical system with input a, output b and internal state c and will be referred to as the *system level implementation* of K'.

Remark. *It is more common in control literature [49, 142] to refer to (2.35) as the "realization" of the controller, however to prevent overloading and potential confusion with the "realizing controller" (see Def. 2.6), we instead choose to call (2.35) the implementation of K.*

Moreover, in later sections, we make use of this implementation to define controllers K that are parameterized by operators Ψ^x, Ψ^u that are not necessarily CLMs. In particular, the next section will show that such an implementation can yield closed-loop stability if Ψ approximately satisfies (2.33). Therefore, we define controllers with (or which permit) the above implementation separately as *System Level* (SL)-controllers:

Definition 2.7. *Assume given operators $A \in \mathcal{C}(\ell^{\mathcal{X}})$, $B \in \mathcal{C}(\ell^{\mathcal{X}}, \ell^{\mathcal{U}})$, where $A - I \in \mathcal{C}_s(\ell^{\mathcal{X}})$. The above dynamical system will be referred to as the system level controller*

SL$[A, B]$. *Consider the dynamical system with input a, output b, and internal state c according to the equations*

$$c_t = a_t - A_t(0, c_{t-1:0}), \tag{2.36}$$

$$b_t = B_t(c_{t:0}). \tag{2.37}$$

The above dynamical system will be referred to as the system level controller SL$[A, B]$.

Remark. *A trivial yet important consequence of the definition Def. 2.7 is that both input a and output b can always be expressed through the internal state c as $a = Ac$, $b = Bc$.*

The only requirements for the above implementation are that the operator $A \in \mathcal{C}(\ell^{\mathcal{X}})$ is square and $A - I$ strictly causal, and that $B \in \mathcal{C}(\ell^{\mathcal{X}}, \ell^{\mathcal{U}})$ is of compatible dimensions. We will abbreviate these conditions with the term *candidate closed-loop map* (cCLM):

Definition 2.8 (Candidate Closed-Loop Map). *A causal operator $\Psi = \left[\begin{smallmatrix} A \\ B \end{smallmatrix}\right]$ is called a candidate closed-loop map (cCLM), if A is square and such that $A - I$ is strictly causal. More specifically, we say that Ψ is a candidate CLM of some $F \in \mathcal{C}(\ell^{\mathcal{X} \times \mathcal{U}}, \ell^{\mathcal{X}})$ if $A \in \mathcal{C}(\ell^{\mathcal{X}})$ and $B \in \mathcal{C}(\ell^{\mathcal{X}}, \ell^{\mathcal{U}})$, i.e., F and Ψ are of compatible domain and co-domain.*

The dynamic model SL$[\Psi]$ of the implementation can be defined as a subset SL$[\Psi] \subset \ell^{\mathcal{S}}$, where $\mathcal{S} = \{\mathcal{S}_i\}_{i \in \mathcal{I}}$, $\mathcal{I} = \{x, u, \hat{w}\}$ s.t.: $\mathcal{S}_{\hat{w}} = \mathcal{S}_x = \mathcal{X}$. We can define SL$[\Psi]$ in terms of its trajectories $\tau = (\tau^{\hat{w}}, \tau^{x}, \tau^{u})^{\top} \in$ SL$[\Psi]$ as

$$\mathrm{SL}[\Psi] = \{\tau \in \ell^{\mathcal{X} \times \mathcal{X} \times \mathcal{U}} \mid \tau^{\hat{w}} = (\Psi^{x})^{-1}\tau^{x}, \ \tau^{u} = \Psi^{u}\tau^{\hat{w}}\}, \tag{2.38}$$

and notice that it can be equivalently represented by its $x \mapsto \{\hat{w}, u\}$-map, the operator $\left[\begin{smallmatrix} I \\ \Psi^{u} \end{smallmatrix}\right](\Psi^{x})^{-1}$. We denote the former as $K_{SL} = \left[\begin{smallmatrix} K_{SL}^{\hat{w}} \\ K_{SL}^{u} \end{smallmatrix}\right] \in \mathcal{C}(\ell^{\mathcal{X}}, \ell^{\mathcal{X} \times \mathcal{U}})$ for a fixed $\Psi \in \Phi_{\mathrm{CL}}^{w \mapsto xu}[F]$. The partial map K_{SL}^{u} corresponds to the realizing control law $\Psi^{u}(\Psi^{x})^{-1}$, while $K_{SL}^{\hat{w}}$ represents the dynamics of the internal state of the implementation used to compute the control action u from the internal state $\hat{w}$ via $u = \Psi^{u}\hat{w}$.

For a CLM $\Psi \in \Phi_{\mathrm{CL}}^{w \mapsto xu}[F]$, the dynamic system SL$[\Psi]$ provides a simple and straightforward implementation, of the realizing controller $K = \Psi^{u}(\Psi^{x})^{-1}$, via the dynamic equations (2.35). However, the system level implementation of K is not unique, that is there exist cCLMs $\hat{\Psi} \neq \Psi$ such that $K = \Psi^{\mathrm{u}}(\Psi^{\mathrm{x}})^{-1} = \hat{\Psi}^{\mathrm{u}}(\hat{\Psi}^{\mathrm{x}})^{-1}.;$ a trivial example is $\hat{\Psi} = \left[\begin{smallmatrix} I \\ K \end{smallmatrix}\right]$. The space of all such cCLM $\hat{\Psi}$ defines an equivalence class which we refer to as the *Implementation Space* of a controller K.

Definition 2.9. *The (SL)-Implementation Space* $\mathbb{I}_K$ *of causal controller* $K \in \mathcal{C}(\ell^{\mathcal{X}}, \ell^{\mathcal{U}})$ *is the set of all cCLMs* $\Psi \in \mathcal{C}(\ell^{\mathcal{X}}, \ell^{\mathcal{X} \times \mathcal{U}})$ *such that* $\Psi^{\mathrm{u}}(\Psi^{\mathrm{x}})^{-1} = K$.

Remark. *This was first formulated and studied in [6] for the case of LTI systems. The same work shows that the implementation space can be leveraged to formulate novel robust control synthesis procedures for LTI systems.*

The following result provides a characterization of the implementation space $\mathbb{I}_K$ of a controller K in terms of the CLM it realizes:

Lemma 9 (Space of System Level Implementations). *Let* $\Phi \in \Phi_{\mathrm{CL}}^{\mathrm{w} \mapsto \mathrm{xu}}[F]$ *be a CLM of some fixed* F *and let* $K = \Phi^{\mathrm{u}}(\Phi^{\mathrm{x}})^{-1}$ *be the corresponding unique realizing controller. Then,*

$$\mathbb{I}_K = \{ \left[\begin{smallmatrix} \Psi^{\mathrm{x}} \\ \Psi^{\mathrm{u}} \end{smallmatrix}\right] \in \mathcal{C}(\ell^{\mathcal{X}}, \ell^{\mathcal{X} \times \mathcal{U}}) \mid \Phi(\Psi^{\mathrm{x}} - F^{+}\Psi) = \Psi \}.$$

Despite that for any $\Psi \in \mathbb{I}_K$, the dynamic model $\mathrm{SL}[\Psi]$ realizes the controller K, special care needs to be taken when selecting $\Psi \in \mathbb{I}_K$ in practice, since the choice of the realizing dynamical system $\mathrm{SL}[\Psi]$ can impact the stability of the overal closed-loop: We need to make sure that the dynamical system $\mathrm{SL}[\Psi]$ is stable even if we add small perturbations v to the computation of the internal state as $\hat{w} = (\Psi^{\mathrm{x}})^{-1}x + v$. This is a crucial concern for control applications since even if our evaluation of $\hat{w} = (\Psi^{\mathrm{x}})^{-1}x$ is entirely digital, numerical errors can introduce perturbations v that are capable of causing numerical instability in our implementation. In the worst-case scenario, this can jeopardize the stability of the entire closed-loop $\mathrm{CL}_{[F,K]}$. In control literature, this problem is known as the question of closed-loop internal stability. In Section 2.7, we show that the system-level implementation $\mathrm{SL}[\Psi]$ of a realizing controller $K = \Psi^{\mathrm{u}}(\Psi^{\mathrm{x}})^{-1}$ is internally stable under suitable assumptions and provided Ψ is a stable CLM of F.

2.7 Robust and Internal Stability of Closed-Loop

As shown in the previous section, any closed-loop map $\Psi = (\Psi^{\mathrm{x}}, \Psi^{\mathrm{u}})^{\top} \in \Phi_{\mathrm{CL}}^{\mathrm{w} \mapsto \mathrm{xu}}[F]$ can be realized with the corresponding system level controller $\mathrm{SL}[\Psi^{\mathrm{x}}, \Psi^{\mathrm{u}}]$ as defined in Def. 2.7. Thus, we know that if we choose $K = \mathrm{SL}[\Psi^{\mathrm{x}}, \Psi^{\mathrm{u}}]$, then for any disturbance w, the trajectories (x, u) of the closed-loop CL will be $(x, u)^{\top} = \Psi(w)$. In this section, we show that under mild assumptions, the controller $K = \mathrm{SL}[\Psi^{\mathrm{x}}, \Psi^{\mathrm{u}}]$ guarantees internal stability of the closed-loop. Moreover, we show that closed-loop stability is guaranteed even if Ψ is satisfying the CLM equation (2.33) only approximately.

To setup the stability analysis we take the original closed-loop (2.29a) with K chosen to be $\mathrm{SL}[\Psi^{\mathrm{x}}, \Psi^{\mathrm{u}}]$ and add additional perturbation signals v and d to the internal state of the system level controller and control input. We call the new perturbed closed-loop δCL:

$$\delta\mathrm{CL}: \qquad x_t = F_{t-1}(x_{t-1:0}, u_{t-1:0}) + w_t, \quad x_0 = w_0 \tag{2.39a}$$

$$\hat{w}_t = x_t + v_t - \Psi_t^x(0, \hat{w}_{t-1:0}) \tag{2.39b}$$

$$u_t = \Psi_t^u(\hat{w}_{t:0}) + d_t. \tag{2.39c}$$

As before, w, x and u represent system disturbance, state and input, and the added state $\hat{w}$ represents the internal state of the system level controller. In contrast, to the previous section, we now assume that $\Psi = (\Psi^{\mathrm{x}}, \Psi^{\mathrm{u}})$ are merely *candidate* closed-loop maps (cCLM) of F, and derive conditions sufficient for ℓ_p-stability of δCL with respect to the *residual* operator $\Delta[F, \Psi]$, which is defined for fixed cCLM Ψ and dynamics F as:

Definition 2.10. *Let $\Psi = (\Psi^{\mathrm{x}}, \Psi^{\mathrm{u}})$ be cCLMs of some dynamics F. Then, the (CLM)-residual of the pair $[F, \Psi]$, denoted by $\Delta[F, \Psi]$, is the operator defined as $\Delta[F, \Psi] := F^{+}(\Psi) + I - \Psi^{\mathrm{x}}$.*

As alluded to by its name, the residual $\Delta[F, \Psi]$ of a cCLM Ψ represents the CLM-equation error with respect to the dynamics F. Hence, candidate closed-loop map Ψ which satisfy $\Delta[F, \Psi] = 0$ define the space of closed-loop maps of F, i.e., we can equivalently define the set $\Phi_{\mathrm{CL}}^{\mathrm{w} \mapsto \mathrm{xu}}[F]$ as

$$\Phi_{\mathrm{CL}}^{\mathrm{w} \mapsto \mathrm{xu}}[F] = \{\Psi \mid \Delta[F, \Psi] = 0\}.$$

In sequence space, we can equivalently describe δCL by the equations:

$$\delta\mathrm{CL}: \qquad x = F^{+}(x, u) + w \tag{2.40a}$$

$$\hat{w} = x + v - (\Psi^{\mathrm{x}} - I)(\hat{w}) \tag{2.40b}$$

$$u = \Psi^{\mathrm{u}}(\hat{w}) + d. \tag{2.40c}$$

We can rewrite the equations (2.40) into a form, more suitable for analysis. Denote $\hat{u} := u - d$, $\hat{x} := x + v$ to rewrite (2.40) as

$$\delta\text{CL}' : \qquad \hat{x} = F^+(\hat{x} - v, \hat{u} + d) + w + v \qquad (2.41a)$$

$$\hat{w} = \hat{x} - (\Psi^x - I)(\hat{w}) = \hat{x} + \overline{\Psi^x}(\hat{w}) \qquad (2.41b)$$

$$\hat{u} = \Psi^u(\hat{w}). \qquad (2.41c)$$

For fixed sequences v, d, and denote $T_{-v,d} \in \mathcal{C}(\ell^{\mathcal{X} \times \mathcal{U}}, \ell^{\mathcal{X} \times \mathcal{U}})$ as the affine operator such that $\forall \begin{bmatrix} x' \\ u' \end{bmatrix} \in \ell^{\mathcal{X}} : T_{-v,d}\begin{bmatrix} x' \\ u' \end{bmatrix} = \begin{bmatrix} x'-v \\ u'+d \end{bmatrix}$, i.e., $T_{-v,d}$ represents the identity operator shifted by the constant sequence $\begin{bmatrix} -v \\ d \end{bmatrix}$. With this affine operator, we can formally rewrite (2.41a) as $\hat{x} = F^+ T_{-v,d}(\hat{x}, \hat{u}) + w + v$ and substitute it into (2.41b) to obtain an equation describing the dynamics of the internal state $\hat{w}$:

$$\hat{w} = F^+ T_{-v,d}(\Psi)(\hat{w}) - (\Psi^x - I_n)(\hat{w}) + w + v$$
$$\hat{w} = F^+ T_{-v,d}(\Psi)(\hat{w}) - (T^x_{-v,d}\Psi^x - I_n)(\hat{w}) + w = \Delta[F, T_{-v,d}\Psi](\hat{w}).$$
$$(2.42)$$

Since the residual is a sum of two strictly causal operators $F^+(\Psi)$ and $I - \Psi^x$ (by definition of cCLMs), the operator $I_n - \Delta[F, T_{-v,d}\Psi]$ is causally invertible, and we can solve for the lumped disturbance $\hat{w}$ as

$$\hat{w} = (I_n - \Delta[F, T_{-v,d}\Psi])^{-1} w, \qquad (2.43)$$

and $\hat{x}$, $\hat{u}$ as

$$\hat{x} = \Psi^x (I_n - \Delta[F, T_{-v,d}\Psi])^{-1} w, \quad \hat{u} = \Psi^x (I_n - \Delta[F, T_{-v,d}\Psi])^{-1} w.$$

For $v, d = 0$, the operator $(I_n - \Delta[F, \Psi])^{-1}$ represents the map $w \mapsto \hat{w}$, and we see that ℓ_p-boundedness of $\hat{w}$ is entirely determined by ℓ_p-stability of the former operator. Suitably, we will refer to the dynamic system governed by equation (2.43) as the *effective/lumped disturbance dynamics*. In fact, as shown in our next result, the stability of the overall closed-loop is entirely determined by the stability of the dynamics (2.43).

Sufficient and Necessary Conditions for Internal Stability

Summarizing our derivations so far, we can describe the input-output mapping $\Phi_{\delta\text{CL}} : (w, v, d) \mapsto (\hat{w}, x, u)$ of the closed-loop δCL by the equations:

$$\hat{w} = (I_n - \Delta[F, T_{-v,d}\Psi])^{-1} w \qquad (2.44a)$$

$$x = \Psi^x (I_n - \Delta[F, T_{-v,d}\Psi])^{-1} w - v \qquad (2.44b)$$

$$u = \Psi^u (I_n - \Delta[F, T_{-v,d}\Psi])^{-1} w + d. \qquad (2.44c)$$

The closed-loop δCL is internally stable if its input-output map $\Phi_{\delta\text{CL}}$ is ℓ_p-stable.
The next theorem states sufficient and necessary conditions for that:

Theorem 9. *Consider the closed-loop system δCL described by the equations (2.40)
for a fixed $F \in C(\ell^{\mathcal{X} \times \mathcal{U}}, \ell^{\mathcal{X}})$ and some fixed compatible ℓ_p-stable candidate CLM
$\Psi \in C(\ell^{\mathcal{X}}, \ell^{\mathcal{X} \times \mathcal{U}})$. Then, the map $\Phi_{\delta\text{CL}} : (w, v, d) \mapsto (\hat{w}, x, u)$ is ℓ_p-stable if and
only if the operator $(I_n - \Delta[F, T_{-v,d}\Psi])^{-1} \in C(\ell^{\mathcal{X}})$ is ℓ_p-stable for any $v \in \ell_p^{\mathcal{X}}$,
$d \in \ell_p^{\mathcal{U}}$.*

Proof. Sufficiency: This is clear from equations (2.44). *Necessity:* Assume
$(I_n - \Delta[F, T_{-v',d'}\Psi])^{-1}$ is not ℓ_p-stable for some $v' \in \ell_p^{\mathcal{X}}$, $d' \in \ell_p^{\mathcal{U}}$, then there
exists some $w' \in \ell_p^{\mathcal{X}}$ such that $(I_n - \Delta[F, T_{-v',d'}\Psi])^{-1} w' \notin \ell_p^{\mathcal{X}}$ and therefore
$\Phi_{\delta\text{CL}}^{\hat{w}}(w', v', d') \notin \ell_p^{\mathcal{X}}$. $\square$

Robustness of SL controllers in the LTV case was first discussed in [66]. Here, we
discuss the generalization of these results in the context of linear operators.

If we assume that F is linear, we can split F into the linear operators $A \in \mathcal{LC}(\ell^{\mathcal{X}}, \ell^{\mathcal{X}})$
and $B \in \mathcal{LC}(\ell^{\mathcal{U}}, \ell^{\mathcal{X}})$ such that $F : (x, u) \mapsto A(x) + B(u)$. In that case, the operator
$\Delta[F, T_{-v,d}\Psi]$ represents the mapping

$$\Delta[F, T_{-v,d}\Psi] : z \mapsto \Delta[F, \Psi](z) + (I_n - A^+)v + B^+d,$$

and correspondingly, $(I_n - \Delta[F, T_{-v,d}\Psi])^{-1}$ represents the mapping:

$$z \mapsto (I_n - \Delta[F, \Psi])^{-1}(z + (I_n - A^+)v + B^+d). \tag{2.45}$$

The dynamic equations (2.44) of δCL can be equivalently stated in the form:

$$\hat{w} = (I_n - \Delta[F, \Psi])^{-1}(w + (I - A^+)v + B^+d), \tag{2.46a}$$

$$x = \Psi^{\text{x}}(I_n - \Delta[F, \Psi])^{-1}(w + (I - A^+)v + B^+d) - v, \tag{2.46b}$$

$$u = \Psi^{\text{u}}(I_n - \Delta[F, \Psi])^{-1}(w + (I - A^+)v + B^+d) + d. \tag{2.46c}$$

Remark. *Notice that Ψ is still allowed to be nonlinear, and therefore, the dynamics
of the closed-loop (2.46) are, in general, nonlinear.*

From the above equations, we can see that the stability of $(I_n - \Delta[F, T_{-v,d}\Psi])^{-1}$
is equivalent to the stability of $(I_n - \Delta[F, \Psi])^{-1}$ and the boundedness of A and B.
Thus, as a corollary of Theorem 9, we obtain a simpler characterization of internal
ℓ_p-stability for the case where F is linear:

Corollary 10. *Consider the closed-loop system δCL described by the equations (2.40) for a fixed $F \in \mathcal{LC}(\ell^{\mathcal{X} \times \mathcal{U}}, \ell^{\mathcal{X}})$ and some fixed compatible ℓ_p-stable candidate CLM $\Psi \in \mathcal{C}(\ell^{\mathcal{X}}, \ell^{\mathcal{X} \times \mathcal{U}})$. Then, the map $\Phi_{\delta\mathrm{CL}} : (w, v, d) \mapsto (\hat{w}, x, u)$ is ℓ_p-stable if and only if the operator $(I_n - \Delta[F, \Psi])^{-1} \in \mathcal{C}(\ell^{\mathcal{X}})$ is ℓ_p-stable, and F is ℓ_p-stable.*

Proof. Envoke Theorem 9 and recall (2.45), which makes it clear that the operator $(I_n - \Delta[F, T_{-v,d}\Psi])^{-1}$ is ℓ_p-stable for all $v \in \ell_p^{\mathcal{X}}$, $d \in \ell_p^{\mathcal{U}}$, if and only if the map $(I_n - \Delta[F, \Psi])^{-1}$ is ℓ_p-stable and both A and B are ℓ_p-stable, i.e., F is ℓ_p-stable. $\qquad\square$

Remark. *It is important to note that requiring ℓ_p-stability of F does **not** mean we assume that the dynamics F is open-loop stable; that requirement would be ℓ_p-stability of $(I_n - F^+|_{u=0})^{-1} = (I_n - A^+)^{-1}$.*

Thus, as long as A and B are ℓ_p-bounded linear operators, we see that $(\hat{w}, x, u)$ is ℓ_p-bounded if and only if the operator $(I_n - \Delta[F, \Psi])^{-1}$ is ℓ_p-bounded. If Ψ is a CLM of F, hence $\Delta[F, \Psi] = 0$, then the former is trivially satisfied. Thus, we see that for the linear dynamics case, any (potentially nonlinear) CLM can be realized via the system level controller $\mathrm{SL}[\Psi^{\mathrm{x}}, \Psi^{\mathrm{u}}]$ in an internally stable way.

In contrast to the linear case, verifying the stability of $(I_n - \Delta[F, \Psi])^{-1}$ is not sufficient alone to ensure internal stability for the case of nonlinear F. Instead, one must also verify ℓ_p-stability of the lumped dynamics (2.44a) for all perturbed versions $\tilde{\Psi} = T_{-v,d}\Psi$ of the original CLM Ψ; this is, in general, not a trivial question.

In the next section, we use small-gain theorem techniques to break down the internal stability criterion of Theorem 9 into simpler ones. However, at the expense of losing necessity.

Remark. *It should be noted that for nonlinear systems, it might even be impossible to achieve ℓ_p-internal stability, despite the controllability and observability of the system. For example, consider the simple scalar system $x_{k+1} = x_k^2 + u_k + w_k$ with $\hat{x}_k = x_k + v_k$. It is easy to verify that there does not exist a causal controller that realizes an ℓ_∞-internally stable closed loop.*

Sufficient Small-Gain Conditions for Internal Stability

We use the small-gain theorems and ideas derived in Section 2.3 to provide sufficient conditions for the stability of the map $\Phi_{\delta\mathrm{CL}}$. For the next theorem, recall the following theorems from Section 2.3:

Theorem (Global Small Gain Condition). *Assume the operator $\Delta \in \mathcal{C}(\ell^{\mathcal{X}})$ satisfies $\|\Delta(x)\|_p \leqslant \gamma \|x\|_p + \beta$ for all $x \in \ell_p^{\mathcal{X}}$ and some small gain $\gamma < 1$. Then, for all $w \in \ell_p^{\mathcal{X}}$ and $y = \Delta y + w$ holds:*

$$\|y\|_p \leqslant \frac{1}{1-\gamma}(\|w\|_p + \beta).$$

Theorem (Local Small Gain Condition). *Assume that for some $\rho > 0$ and $0 \leqslant \gamma < 1$, the operator $\Delta \in \mathcal{C}_s(\ell^{\mathcal{X}})$ satisfies $\|\Delta(x)\|_p \leqslant \gamma \|x\|_p$ for all $\|x\|_p < \rho$. Then, for any $w \in \ell_p^{\mathcal{X}}$ and $y = \Delta y + w$ the following statement holds true:*

$$\|w\|_p < (1-\gamma)\rho \qquad \Longrightarrow \qquad \|y\|_p \leqslant \frac{1}{1-\gamma} \|w\|_p.$$

It suits to measure the gain of the residual in terms of the operator (pseudo)-norm $\|\!|\Delta[F, \tilde{\Psi}]|\!\|_p^{\star 3}$, where $\|\!| \cdot |\!\|_p^{\star}$ and the related operator norm $\|\!| \cdot |\!\|_p$ for some operator A are defined as:

$$\|\!|A|\!\|_p^{\star} := \sup_{x \in \mathcal{X}_p^{\mathrm{N}} : x \neq 0} \frac{\|A(x) - A(0)\|_p}{\|x\|_p} \qquad \|\!|A|\!\|_p := \|\!|A|\!\|_p^{\star} + \|A(0)\|_p. \tag{2.47}$$

The residual norm $\|\!|\Delta[F, \Psi]|\!\|_p^{\star}$ measures how well, up to a constant offset, a candidate CLM Ψ fulfills the CLM equation corresponding to F and can be viewed as a degree of approximation of the CLM equation. As shown in the following results, the closer a cCLM Ψ approximates a true CLM of F, the more robust the closed-loop system is. In particular, we require $\|\!|\Delta[F, \Psi]|\!\|_p^{\star}$ and the perturbed residual norm $\|\!|\Delta[F, T_{-v,d}\Psi]|\!\|_p^{\star}$ to be uniformly smaller than 1; if the condition can only be fulfilled for a subset of perturbations v, d, then we also obtain corresponding local [4] stability results.

Theorem 11. *Consider the closed-loop system $\delta\mathrm{CL}$ described by the equations (2.40) and let $\Phi_{\delta\mathrm{CL}} : (w, v, d) \mapsto (\hat{w}, x, u)$ be the corresponding input-output mapping of $\delta\mathrm{CL}$ for a fixed $F \in \mathcal{C}(\ell^{\mathcal{X} \times \mathcal{U}}, \ell^{\mathcal{X}})$ and some fixed compatible ℓ_p-stable candidate CLM $\Psi \in \mathcal{C}(\ell^{\mathcal{X}}, \ell^{\mathcal{X} \times \mathcal{U}})$ such that $\|\!|\Delta[F, \Psi]|\!\|_p^{\star} = \gamma < 1$. Then, the following statements hold:*

> i) *Sufficient Condition for Global ℓ_p-internal-stability: If F is i.f.g. ℓ_p-stable, then $\Phi_{\delta\mathrm{CL}}$ is ℓ_p-stable. If in addition, Ψ is f.g. ℓ_p-stable, then, respectively, $\Phi_{\delta\mathrm{CL}}$ is f.g. ℓ_p-stable.*

---

[3]See Appendix Section 2.A for proof of norm-properties.
[4]Local means for ℓ_p-bounded subsets around 0 of inputs (w, u, d).

ii) Sufficient Condition for Local ℓ_p-internal-stability: Assume $\Delta[F, \Psi](0) = 0$ *and that at $w = 0$, the map Ψ is ℓ_p-continuous. Moreover, assume that there exists a nondecreasing function[5] $L : \mathbb{R}^+ \to \mathbb{R}^+$ such that:*

$$\|z\|_p, \|z'\|_p \leqslant \eta \implies \|F(z) - F(z')\|_p^{\mathcal{X}} \leqslant L(\eta)\|z - z'\|_p^{\mathcal{X}}.$$

Then, there exists a $\delta > 0$ such that $\Phi_{\delta\mathrm{CL}}(w, v, d) \in \ell_p^{\mathcal{X} \times \mathcal{X} \times \mathcal{U}}$ whenever $\|w\|_p^{\mathcal{X}}, \|v\|_p^{\mathcal{X}}, \|d\|_p^{\mathcal{U}} \leqslant \delta$. If, in addition, $\|\Psi\|_p^* < \infty$, then for any $\delta_w > 0$, there exists a $\delta_{vd} > 0$ such that $\Phi_{\delta\mathrm{CL}}(w, v, d) \in \ell_p^{\mathcal{X} \times \mathcal{X} \times \mathcal{U}}$ whenever $\|w\|_p^{\mathcal{X}} \leqslant \delta_w$ and $\|v\|_p^{\mathcal{X}}, \|d\|_p^{\mathcal{U}} \leqslant \delta_{vd}.$

Proof. Part i): Fix some $v \in \ell_p^{\mathcal{X}}$, $d \in \ell_p^{\mathcal{U}}$, let $\Phi^{\hat{w}}|_{v,d} := (I_n - \Delta[F, T_{-v',d'}\Psi])^{-1}$ and decompose $\Delta[F, T_{-v,d}\Psi]$ into the sum $\Delta[F, \Psi] + \delta\Delta$ where the operator $\delta\Delta$ denotes the difference $\Delta[F, \Psi] - \Delta[F, T_{-v,d}\Psi]$. Now, notice that for all $w \in \ell_p^{\mathcal{X}}$:

$$\delta\Delta(w) = F\Psi(w) + \overline{\Psi^{\mathsf{x}}}(w) - F(\Psi(w) + \left[\begin{smallmatrix}-v\\d\end{smallmatrix}\right]) - \overline{\Psi^{\mathsf{x}}}(w) - v$$
$$= F\Psi(w) - F(\Psi(w) + \left[\begin{smallmatrix}-v\\d\end{smallmatrix}\right]) - v.$$

Next, we show that the ℓ_p-norm of $\Delta_1(w)$ is bounded by a constant independent of w:

- Per assumption, Ψ is ℓ_p-stable, and therefore $z = \Psi w \in \ell_p^{\mathcal{X} \times \mathcal{U}}$. Furthermore, since v and d are assumed to be bounded in the ℓ_p-norm, we have $z + \left[\begin{smallmatrix}-v\\d\end{smallmatrix}\right] \in \ell_p^{\mathcal{X} \times \mathcal{U}}$. Invoking i.f.g.-$\ell_p$ stability of F at z, we know there are fixed constants (γ_f, β_f) such that for any z: $\|F(z) - F(z + \left[\begin{smallmatrix}-v\\d\end{smallmatrix}\right])\|_p \leqslant \gamma_f(\|v\|_p + \|d\|_p) + \beta_f$. This implies that $\|\delta\Delta(w)\|_p$ is bounded above for all $w \in \ell^{\mathcal{X}}$ as:

$$\|\delta\Delta(w)\|_p \leqslant (\gamma_f + 1)\|v\|_p + \gamma_f\|d\|_p + \beta_f. \tag{2.48}$$

Denote $e_0 = \Delta[F, \Psi](0)$ and note that since F is i.f.g. ℓ_p-stable and Ψ is assumed ℓ_p-stable, the sequence e_0 belongs to $\ell_p^{\mathcal{X}}$. Now, per our assumption $\|\Delta[F, \Psi]\|_p^* =: \gamma < 1$, for any $w \in \ell_p^{\mathcal{X}}$ holds:

$$\|\Delta[F, \Psi](w)\|_p \leqslant \|\Delta[F, \Psi](w) - e_0\|_p + \|e_0\|_p \leqslant \gamma\|w\|_p + \|e_0\|_p.$$

From the above inequality and (2.48), we can conclude that $\Delta[F, T_{-v,d}\Psi]$ is (γ, β_1)-f.g. ℓ_p-stable for any $v \in \ell_p^{\mathcal{X}}$, $d \in \ell_p^{\mathcal{U}}$. In particular, for all $w \in \ell_p^{\mathcal{X}}$ holds:

$$\|\Delta[F, T_{-v,d}\Psi](w)\|_p \leqslant \gamma\|w\|_p + \underbrace{\|e_0\|_p + (\gamma_f + 1)\|v\|_p + \gamma_f\|d\|_p + \beta_f}_{\beta_1}. \tag{2.49}$$

---

[5]This can be thought of a local Lipshitz-constant.

According to the above, $\mathbf{\Delta}[\boldsymbol{F}, \boldsymbol{T}_{-v,d}\boldsymbol{\Psi}]$ satisfies the ℓ_p small-gain property. Using the small-gain Theorem 2 (presented in Section 2.3), we confirm that $\boldsymbol{\Phi}^{\hat{w}}|_{v,d}$ achieves $(\frac{1}{1-\gamma}, \beta_2)$-$\ell_p$ fg stability, with $\beta_2 = \frac{1}{1-\gamma}\beta_1$. On substituting the constants defined in (2.49), we establish that the partial input-output map of δCL, $\boldsymbol{\Phi}^{\hat{w}}_{\delta\mathrm{CL}} : (\boldsymbol{w}, \boldsymbol{v}, \boldsymbol{d}) \mapsto \hat{\boldsymbol{w}}$, is (γ_3, β_3)-ℓ_p-stable where

$$\gamma_3 = \frac{\gamma_f + 1}{1 - \gamma}, \quad \beta_3 = \frac{\beta_f + \|\boldsymbol{e}_0\|_p}{1 - \gamma}.$$

The other partial input-output map of the closed loop δCL, $\boldsymbol{\Phi}^{\mathrm{xu}}_{\delta\mathrm{CL}} : (\boldsymbol{w}, \boldsymbol{v}, \boldsymbol{d}) \mapsto (\boldsymbol{x}, \boldsymbol{u})$, is written as $\boldsymbol{\Phi}^{\mathrm{xu}}_{\delta\mathrm{CL}} = \boldsymbol{\Psi}\boldsymbol{\Phi}^{\hat{w}}_{\delta\mathrm{CL}}$. This implies that, at a minimum, it is ℓ_p-stable, given that $\boldsymbol{\Psi}$ is ℓ_p-stable and $\boldsymbol{\Phi}^{\hat{w}}_{\delta\mathrm{CL}}$ is (γ_3, β_3)-f.g.-ℓ_p-stable. If $\boldsymbol{\Psi}$ is additionally fg.-ℓ_p-stable, $\boldsymbol{\Phi}^{\mathrm{xu}}_{\delta\mathrm{CL}}$ also shares this property. Given that $\boldsymbol{\Phi}_{\delta\mathrm{CL}} = (\boldsymbol{\Phi}^{\hat{w}}_{\delta\mathrm{CL}}, \boldsymbol{\Phi}^{\mathrm{x}}_{\delta\mathrm{CL}}, \boldsymbol{\Phi}^{\mathrm{u}}_{\delta\mathrm{CL}})$, we affirm the desired statement.

Part ii): Aside from a few modifications, the proof is structured as in the previous part. Pick $\varepsilon > 0$, then, by the ℓ_p-continuity assumption, there exists some $\delta(\varepsilon) > 0$ such that $\|\boldsymbol{\Psi}(\boldsymbol{z})\|_p^{\mathcal{X} \times \mathcal{U}} \leqslant \varepsilon$ for all $\boldsymbol{z} \in \mathrm{B}^x_{\ell_p}[\delta(\varepsilon)]$ where $\mathrm{B}^x_{\ell_p}[r] := \{\boldsymbol{w}' \in \ell_p^{\mathcal{X} \times \mathcal{U}} \mid \|\boldsymbol{w}'\|_p^{\mathcal{X}} \leqslant r\}$. Fix some $\boldsymbol{v}, \boldsymbol{d}$ and denote $\delta_{vd} := \|\boldsymbol{v}\|_p^{\mathcal{X}} \vee \|\boldsymbol{d}\|_p^{\mathcal{U}}$, then for all $\boldsymbol{z} \in \mathrm{B}^x_{\ell_p}[\delta(\varepsilon)]$ holds

$$\|\delta\boldsymbol{\Delta}(\boldsymbol{z})\|_p \leqslant \|\boldsymbol{F}(\boldsymbol{\Psi}(\boldsymbol{z})) - \boldsymbol{F}(\boldsymbol{\Psi}(\boldsymbol{z}) + \begin{bmatrix} -\boldsymbol{v} \\ \boldsymbol{d} \end{bmatrix})\|_p \leqslant L(\varepsilon + \delta_{vd})\delta_{vd}.$$

From $\|\boldsymbol{\Delta}[\boldsymbol{F}, \boldsymbol{\Psi}]\|_p^\star = \gamma < 1$ and since we assumed $\boldsymbol{\Delta}[\boldsymbol{F}, \boldsymbol{\Psi}](0) = 0$, we have $\boldsymbol{e}_0 = 0$ and therefore $\boldsymbol{\Delta}[\boldsymbol{F}, \boldsymbol{\Psi}](\boldsymbol{z}) \leqslant \gamma\|\boldsymbol{z}\|_p$. We proceed to bound $\|\boldsymbol{\Delta}[\boldsymbol{F}, \boldsymbol{T}_{-v,d}\boldsymbol{\Psi}](\boldsymbol{z})\|_p^{\mathcal{X}}$ for $\boldsymbol{z} \in \mathrm{B}^{xu}_{\ell_p}[\delta(\varepsilon)]$:

$$\|\boldsymbol{\Delta}[\boldsymbol{F}, \boldsymbol{T}_{-v,d}\boldsymbol{\Psi}](\boldsymbol{z})\|_p^{\mathcal{X}} \leqslant \gamma\|\boldsymbol{z}\|_p^{\mathcal{X}} + L(\varepsilon + \delta_{vd})\delta_{vd}.$$

We can now use the local small gain theorem Lem. 7, letting $\delta\boldsymbol{\Delta}(\boldsymbol{z})$ play the role of disturbance and setting $\rho = \delta(\varepsilon)$. Notice that since L is not decreasing, the term $L(\varepsilon + \delta_{vd})\delta_{vd}$ is not decreasing and converges to 0 as $\delta_{vd} \to 0$. This implies that there exists $r_{vd} > 0$ such that $L(\varepsilon + r_{vd})r_{vd} \leqslant (1 - \gamma)\delta(\varepsilon)$. Therefore, if $\delta_{vd} \leqslant r_{vd}$, then $\|\delta\boldsymbol{\Delta}(\boldsymbol{z})\|_p \leqslant (1 - \gamma)\delta(\varepsilon)$ and we can conclude that $\boldsymbol{\Phi}^{\hat{w}}|_{v,d}$ maps to $\ell_p^{\mathcal{X}}$ when restricted to $\boldsymbol{w} \in \mathrm{B}^x_{\ell_p}[\delta(\varepsilon)]$ and provided that $\|\boldsymbol{v}\|_p^{\mathcal{X}} \vee \|\boldsymbol{d}\|_p^{\mathcal{U}} \leqslant r_{vd}$. With $\delta^* = \min\{\delta(\varepsilon), r_{vd}\}$, this shows that $\boldsymbol{\Phi}^{\mathrm{xu}}_{\delta\mathrm{CL}}(\boldsymbol{w}, \boldsymbol{v}, \boldsymbol{d}) \in \ell_p^{\mathcal{X}}$ whenever $\|\boldsymbol{w}\|_p^{\mathcal{X}}, \|\boldsymbol{v}\|_p^{\mathcal{X}}, \|\boldsymbol{d}\|_p^{\mathcal{U}} \leqslant \delta^*$ and leads to the desired result, since $\boldsymbol{\Psi}$ is ℓ_p-stable, $\boldsymbol{\Phi}^{\mathrm{xu}}_{\delta\mathrm{CL}} = \boldsymbol{\Psi}\boldsymbol{\Phi}^{\hat{w}}_{\delta\mathrm{CL}}$, and $\boldsymbol{\Phi}_{\delta\mathrm{CL}} = (\boldsymbol{\Phi}^{\hat{w}}_{\delta\mathrm{CL}}, \boldsymbol{\Phi}^{\mathrm{x}}_{\delta\mathrm{CL}}, \boldsymbol{\Phi}^{\mathrm{u}}_{\delta\mathrm{CL}})$. If in addition, $\|\boldsymbol{\Psi}\|_p^\star = L_\Psi < \infty$, then $\delta(\varepsilon) = L_\Psi^{-1}\varepsilon$ is onto $\mathbb{R}_+$ and we can replace $\delta(\varepsilon)$ simply by a variable $\delta_w \geqslant 0$ and construct r_{vd} as a function of δ_w such that $L(L_\Psi\delta_w + r_{vd})r_{vd} \leqslant (1 - \gamma)\delta_w$, which, as argued before, is possible for any $\delta_w \in \mathbb{R}_+$.

$\square$

In the previous theorem, the global stability condition i) required the dynamics operator F to be i.f.g ℓ_p-stability and the local one ii) required the statement

$$\|z\|_p, \|z'\|_p \leqslant \eta \implies \|F(z) - F(z')\|_p^{\mathcal{X}} \leqslant L(\eta)\|z - z'\|_p^{\mathcal{X}} \qquad (2.50)$$

to hold for an appropriate choice of non-decreasing[6] function $L : \mathbb{R}^+ \to \mathbb{R}^+$. It is important to point out that both requirements are continuity-type conditions and are **completely unrelated** to the question of whether the dynamics F are open-loop stable or unstable! As mentioned in our discussion of the different notions of ℓ_p -stability of Def. 2.3, requiring F to be (γ, β)-i.f.g.-ℓ_p-stable, with $\beta = 0$, is the same as requiring F to be Lipschitz continuous over the subspace $\ell_p^{\mathcal{X} \times \mathcal{U}}$ with respect to the corresponding ℓ_p-norms of the domain and codomain. On the other hand, the statement (2.50) is a weaker continuity requirement that is equivalent to uniform continuity of F in the ℓ_p-norm. For most types of dynamical system, these continuity assumptions can be further reduced to the standard continuity properties of functions over finite-dimensional normed spaces. For example, the dynamics operator F corresponding to any of the following dynamical system equations is (γ, β)-i.f.g.-ℓ_p-stable:

- *LTI finite-dim.*[7] *dynamics*: $x_t = \sum_{k=1}^{h_x} A_k x_{t-k} + \sum_{k=1}^{h_u} B_k u_{t-k} + w_t$.

- *LTV finite-dim. dynamics*:
  $x_t = \sum_{k=1}^{h_x} A_{t,k} x_{t-k} + \sum_{k=1}^{h_u} B_{t,k} u_{t-k} + w_t$, for a bounded set of matrices $\{(A_{t,k}, B_{t,k}) \mid t \in \mathbb{N}, 1 \leqslant k \leqslant h\}$.

- *Nonlinear "almost"-Lipshitz TI*[8] *finite-dim. dynamics*:
  $x_t = f(x_{t-1:t-h_x}, u_{t-1:t-h_u}) + w_t$ where $f = f_1 + f_2$ and with f_1 being γ-Lipshitz-continuous and f_2 being a (possibly discontinuous) β-bounded function, that is: $\sup_{x,u} |f_2|(x, u) \leqslant \beta$.

- *Nonlinear "almost"-equi-Lipshitz TV finite-dim. dynamics*:
  $x_t = f_t(x_{t-1:t-h_x}, u_{t-1:t-h_u}) + w_t$ where <u>for each t</u>, the function $f_t = f_{t,1} + f_{t,2}$ can be decomposed by a γ-Lipshitz continuous $f_{1,t}$ and β-bounded (possibly discontinuous) function $f_{2,t}$.

Similarly, F satisfies the condition (2.50) for the above nonlinear systems if we replace Lipschitz continuity by uniform continuity; for the linear system examples,

---

[6]It is clear that the requirement L to be non-decreasing is without loss of generality.
[7]Finite-dimensional.
[8]TI=time-invariant, TV=time-varying.

the conditions are the same. On the other hand, if a dynamical system cannot be realized with a finite-dimensional internal state, then correspondingly the continuity requirements on the dynamics operator F may need to be verified directly in the corresponding infinite-dimensional ℓ_p-normed spaces.

In summary, Theorem 11 showed to us that for any of the dynamical systems listed above, any ℓ_p-stable CLM Ψ of F, i.e., $\Delta[F, \Psi] = 0$, can be realized with an internally ℓ_p-stable system level controller $\mathrm{SL}[\Psi^x, \Psi^u]$; in the context of closed loop of $\delta\mathrm{CL}$, Ψ being realized means that the operator $\Phi^{\mathrm{xu}}_{\delta\mathrm{CL}}|_{v,d=0} : w \mapsto \Phi^{\mathrm{xu}}_{\delta\mathrm{CL}}(w, 0, 0)$ is the same as Ψ. Moreover, if a candidate CLM Ψ is an accurate enough approximate solution to the CLM equation, i.e., the residual operator $\Delta[F, \Psi]$ has a small enough norm $\|\Delta[F, \Psi]\|_p^\star < 1$, then the system level controller $\mathrm{SL}[\Psi^x, \Psi^u]$ still stabilizes the system and guarantees ℓ_p-internal stability of the closed loop dynamics. However, in that scenario, the resulting mapping $\Phi^{\mathrm{xu}}_{\delta\mathrm{CL}}|_{v,d=0}$ is the operator

$$\Phi^{\mathrm{xu}}_{\delta\mathrm{CL}}|_{v,d=0} = \Psi \left(I_n - \Delta[F, \Psi]\right)^{-1}$$

and no longer matches Ψ. Nevertheless, if Ψ has stronger stability properties such as fg-ℓ_p-stability, then in either case, whether Ψ is an exact or approximate CLM of F, the resulting input-output map $\Phi_{\delta\mathrm{CL}} : (w, v, d) \mapsto (\hat{w}, x, u)$ of the closed loop $\delta\mathrm{CL}$ is guaranteed to be fg-ℓ_p-stable. Moreover, if F is only uniformly ℓ_p-continuous but not necessarily i.f.g.-ℓ_p-stable, then as stated in Theorem 11ii), we can ensure ℓ_p-internal closed loop stability of $\Phi_{\delta\mathrm{CL}}$ for a bounded set of small enough perturbations (w, v, d).

2.8 Nonlinear Closed Loop Maps of Linear Systems

Linear techniques, such as loop-shaping, $\mathcal{H}_\infty$, $\mathcal{H}_2$-optimal control, to name a few, provide powerful tools for feedback control design of linear systems. However, there are many applications where, despite the linearity of the system, nonlinear controller design is required or provides better solutions. Two popular application scenarios where nonlinear control design is common are model-predictive control and adaptive control. Due to the nonlinear feedback controller K, the overall closed loop in these settings has nonlinear dynamics, and it is necessary and crucial to study the closed-loop behavior in the context of nonlinear control theory. Understanding the properties of the closed loop maps of F, which are also nonlinear, is also of great interest, however, they are far less studied in control literature so far, partially due to the lack of analytical tools.

The operator-theoretic tools developed in the previous sections open up new ways of studying the closed-loop dynamics in this setting by analyzing the structure of the closed-loop maps. It turns out that the linearity of F imposes structure on the entire space of CLMs $\Phi_{\mathrm{CL}}^{\mathrm{w}\mapsto\mathrm{xu}}[F]$, not just the subspace $\Phi_{\mathrm{CL}}^{\mathrm{w}\mapsto\mathrm{xu}}[F] \cap \mathcal{LC}(\ell^{\mathcal{X}}, \ell^{\mathcal{X}\times\mathcal{U}})$ of linear ones. In this section, we show that the set of CLMs $\Phi_{\mathrm{CL}}^{\mathrm{w}\mapsto\mathrm{xu}}[F]$ is closed under a particular type of combination, which we refer to as *blending*:

Definition 2.11 (Blend of a Family of Operators). *For a fixed family of operators* $\mathcal{F} = \{\Psi_i\}_{i\in\mathcal{I}} \in \mathcal{C}(\ell^{\mathcal{X}}, \ell^{\mathcal{X}\times\mathcal{U}})$, *we call an operator* $\Psi_\Sigma \in \mathcal{C}(\ell^{\mathcal{X}}, \ell^{\mathcal{X}\times\mathcal{U}})$ *a* <u>*blend of*</u> $\mathcal{F}$ *if it can be expressed as a sum* $\Psi_\Sigma = \sum_{i\in\mathcal{I}} \Psi_i G_i$ *for some collection of operators* $\mathcal{G} = \{G_i\}_{i\in\mathcal{I}} \subset \mathcal{C}(\ell^{\mathcal{X}})$, *called* <u>*weights*</u>, *which satisfy the identity* $\sum_{i\in\mathcal{I}} G_i = I_{\mathcal{X}}$.

Thus, a blend can be thought of as a sum of operators Ψ_i of a family $\mathcal{F}$, "weighted" by the operators G_i of another family $\mathcal{G}$, which, as expressed by condition $\sum_{i\in\mathcal{I}} G_i = I_{\mathcal{X}}$, represents a sum decomposition of the identity operator. In this section, we show that, for linear system dynamics, the CLM space $\Phi_{\mathrm{CL}}^{\mathrm{w}\mapsto\mathrm{xu}}[F]$ is closed under the combination of blending described above. In particular, we can combine a family $\mathcal{F} = \{\Psi_i\}_{i\in\mathcal{I}} \subset \mathcal{LC}(\ell^{\mathcal{X}}, \ell^{\mathcal{X}\times\mathcal{U}})$ of linear CLMs with a suitable family $\mathcal{G}$ of nonlinear "weight" operators to merge the desirable properties of different linear CLMs into one nonlinear CLM. We will demonstrate the power of this approach in the next chapter, where we consider control applications with large-scale linear systems subjected to state- and input constraints and actuator saturation. The enabling theoretical result underlying these methods is stated and proven below:

Theorem 12. *Let* F *be linear,* $\{G_i : \ell^{\mathcal{X}} \mapsto \ell^{\mathcal{X}}\}$ *be* N *causal operators, and* $\{\Psi_i\}_{i=1}^{N}$ *be* N *candidate CLMs. Then,*

$$\forall i : \ \Psi_i \in \Phi_{\mathrm{CL}}^{\mathrm{w}\mapsto\mathrm{xu}}[F] \ and \ \sum_{i=1}^{N} G_i = I \quad \implies \quad \sum_{i=1}^{N} \Psi_i G_i \in \Phi_{\mathrm{CL}}^{\mathrm{w}\mapsto\mathrm{xu}}[F]. \quad (2.51)$$

Proof. Denote Ψ_Σ as the cCLM $\Psi_\Sigma = \sum_{i=1}^{N} \Psi_i G_i$ and evaluate the residual

$\Delta[\boldsymbol{F}, \boldsymbol{\Psi}_\Sigma]$:

$$\Delta[\boldsymbol{F}, \boldsymbol{\Psi}_\Sigma] = \boldsymbol{F}^+(\boldsymbol{\Psi}_\Sigma) + \boldsymbol{I} - \boldsymbol{\Psi}_\Sigma^\times = \boldsymbol{F}^+(\sum_{i=1}^N \boldsymbol{\Psi}_i \boldsymbol{G}_i) + \boldsymbol{I} - \sum_{i=1}^N \boldsymbol{\Psi}_i^\times \boldsymbol{G}_i$$

$$\overset{a)}{=} \sum_{i=1}^N \boldsymbol{F}^+(\boldsymbol{\Psi}_i \boldsymbol{G}_i) + \boldsymbol{I} - \sum_{i=1}^N \boldsymbol{\Psi}_i^\times \boldsymbol{G}_i$$

$$\overset{b)}{=} \sum_{i=1}^N \boldsymbol{F}^+(\boldsymbol{\Psi}_i \boldsymbol{G}_i) + \sum_{i=1}^N \boldsymbol{G}_i - \sum_{i=1}^N \boldsymbol{\Psi}_i^\times \boldsymbol{G}_i$$

$$\overset{c)}{=} \sum_{i=1}^N \boldsymbol{F}^+(\boldsymbol{\Psi}_i \boldsymbol{G}_i) + \boldsymbol{G}_i - \boldsymbol{\Psi}_i^\times \boldsymbol{G}_i = \sum_{i=1}^N \left(\boldsymbol{F}^+(\boldsymbol{\Psi}_i) + \boldsymbol{I} - \boldsymbol{\Psi}_i^\times\right) \boldsymbol{G}_i$$

$$\overset{d)}{=} \sum_{i=1}^N \Delta[\boldsymbol{F}, \boldsymbol{\Psi}_i] \boldsymbol{G}_i = 0.$$

The first equality $a)$ follows by linearity of $\boldsymbol{F}$. Step $b)$ follows by the definition of operators $\boldsymbol{G}_i$. Step $c)$ uses the right-distributive property of the operator product. In $d)$ we use the fact that each map $\boldsymbol{\Psi}_i$ is a CLM of $\boldsymbol{F}$. This establishes the desired result, since we showed $\Delta[\boldsymbol{F}, \boldsymbol{\Psi}_\Sigma] = 0$ and can invoke Theorem 8. $\qquad\square$

As an immediate corollary of the above result, we conclude that the space of CLMs of linear dynamics operators $\boldsymbol{F}$ is closed under blending:

Corollary 13. *If $\boldsymbol{F}$ is linear, then for any blend $\boldsymbol{\Psi}_\Sigma \in \mathcal{C}(\ell^\mathcal{X}, \ell^{\mathcal{X} \times \mathcal{U}})$ of a family $\mathcal{F} = \{\boldsymbol{\Psi}_i\}_{i \in \mathcal{I}} \subset \Phi_{\mathrm{CL}}^{\mathrm{w} \mapsto \mathrm{xu}}[\boldsymbol{F}]$, holds $\boldsymbol{\Psi}_\Sigma \in \Phi_{\mathrm{CL}}^{\mathrm{w} \mapsto \mathrm{xu}}[\boldsymbol{F}]$.*

This result provides concrete instructions on how to construct new CLMs $\boldsymbol{\Psi}_\Sigma$ from existing ones $\boldsymbol{\Psi}_i$. If we interpret the operators $\boldsymbol{G}_i$ as "weighting" operators (because they sum up to "one" in the operator space), then $\boldsymbol{\Psi}_\Sigma = \sum_{i=1}^N \boldsymbol{\Psi}_i \boldsymbol{G}_i$ resembles a weighted average of the individual CLMs $\boldsymbol{\Psi}_i$. We refer to the resulting CLM $\boldsymbol{\Psi}_\Sigma$ as a *non-linear blend* of the CLMs $\boldsymbol{\Psi}_i$ and call a finite collection of operators $\{\boldsymbol{G}_i\}$ *weight operators / weights* if $\sum_i \boldsymbol{G}_i = \boldsymbol{I}$.

It is also important to note that the above result only requires linearity of $\boldsymbol{F}$, so both $\boldsymbol{\Psi}_i$ and $\boldsymbol{G}_i$ can be nonlinear causal operators. In particular, this allows us to construct blended CLMs composed out of other sets of blended CLMs; one could call that a layered blend:

Corollary 14. *Let $\boldsymbol{F}$ be linear, $\{\boldsymbol{G}_i : \ell^{\mathcal{X}} \mapsto \ell^{\mathcal{X}}\}$, $\{\boldsymbol{G}'_j : \ell^{\mathcal{X}} \mapsto \ell^{\mathcal{X}}\}$ be N causal operators, and $\{\boldsymbol{\Psi}_{ij}\}_{i,j=1}^{N}$ be N^2 candidate CLMs. Then,*

$$\forall i, j : \ \boldsymbol{\Psi}_{ij} \in \Phi_{\mathrm{CL}}^{\mathrm{w} \mapsto \mathrm{xu}}[\boldsymbol{F}] \ \text{and} \ \sum_{i=1}^{N} \boldsymbol{G}_i = \boldsymbol{I}, \sum_{j=1}^{N} \boldsymbol{G}'_j = \boldsymbol{I}$$

$$\implies \sum_{j=1}^{N} \sum_{i=1}^{N} \boldsymbol{\Psi}_{ij} \boldsymbol{G}'_j \boldsymbol{G}_i \in \Phi_{\mathrm{CL}}^{\mathrm{w} \mapsto \mathrm{xu}}[\boldsymbol{F}]. \tag{2.52}$$

Naturally, the above corollary can be applied arbitrarily many times, and it is an open question whether there is a generating set of CLMs $\{\boldsymbol{E}_1, \boldsymbol{E}_2, \dots, \} \subset \Phi_{\mathrm{CL}}^{\mathrm{w} \mapsto \mathrm{xu}}[\boldsymbol{F}]$ from which all of $\Phi_{\mathrm{CL}}^{\mathrm{w} \mapsto \mathrm{xu}}$ can be constructed in this way.

For our discussion, we will focus on simple blends using static weighting operators $\{\boldsymbol{G}_i\}$. We refer to static operators as those that act on sequences by applying the same function $f : \mathbb{R}^n \to \mathbb{R}^m$ to each element. For notational convenience, we refer to the corresponding operator via the Kronecker product $\boldsymbol{I} \otimes f$ as defined below:

Definition. *For a fixed function $f : \mathbb{R}^n \mapsto \mathbb{R}^m$, let $\boldsymbol{I} \otimes f : \ell^{\mathcal{X}} \to \ell^{\mathcal{U}}$ be defined as the map*

$$(x_0, x_1, \dots) \mapsto (f(x_0), f(x_1), \dots).$$

For a matrix $M \in \mathbb{R}^{n \times m}$, the operator $\boldsymbol{I} \otimes f_M$, where $f_M : x \mapsto Mx$, will be referred to as $\boldsymbol{I} \otimes M$.

In the later discussion, we make use of the following types of static weighting operator:

Definition 2.12. *Any collection of operators $\{\boldsymbol{G}_i\}_{i=1}^{N}$ defined as below satisfies $\sum_{k=1}^{N} \boldsymbol{G}_k = \boldsymbol{I}$:*

1. *Linear Orthonormal Projections: $\boldsymbol{G}_i = \boldsymbol{I} \otimes U_i U_i^{\top}$, where $U_i \in \mathbb{R}^{n \times r_i}$, $U_i^{\top} U_i = I_{r_i}$, and $\sum_{i=1}^{N} U_i U_i^{\top} = I_n$.*

2. *N-Zone Saturation: Let $\mathcal{W}_1 \subset \mathcal{W}_2 \cdots \subset \mathcal{W}_{N-1}$ be a collection of nested convex sets in some normed vector space $(|\cdot|, \mathbb{R}^n)$ and denote $\Pi_k : x \mapsto \arg\min_{u \in \mathcal{W}_k} |x - u|, \forall k \in \{1, \dots, N-1\}$ as the projection maps onto set $\mathcal{W}_k$. With the abbreviations $\Pi_0 := 0$, $\Pi_N := I_n$, define $\boldsymbol{G}_k = \boldsymbol{I} \otimes (\Pi_k - \Pi_{k-1})$ for all $k \in \{1, \dots, N\}$.*

3. *Sum Inversion: For any collection of causal operators $\{\boldsymbol{G}'_i\}_{i=1}^{N}$ for which the sum $\boldsymbol{G}'_{\Sigma} = \sum_{i=1}^{N} \boldsymbol{G}'_i$ is causally invertible, define $\boldsymbol{G}_k := \boldsymbol{G}'_k (\boldsymbol{G}'_{\boldsymbol{\Psi}_{\Sigma}})^{-1}$.*

Next, we investigate how to implement the realizing controller $K = \Psi_\Sigma^{\mathrm{u}}(\Psi_\Sigma^{\mathrm{x}})^{-1}$. This leads to a controller implementation scheme which we call a "blended" system-level controller.

2.9 Blending SL Controllers

We investigate the realizing controller corresponding to candidate CLMs of the type $\Psi_\Sigma = \sum_{i=1}^N \Psi_i G_i$ formed from a collection of candidate CLMs $\{\Psi_i\}_{i=1}^N$ and weight operators $\{G\}_{i=1}^N$, $\sum_{i=1}^N G = I$.

As usual, we decompose the computation of the realizing control law $u = \Psi_\Sigma^{\mathrm{u}}(\Psi_\Sigma^{\mathrm{x}})^{-1}x$ into the computation of an internal state, called the lumped (or effective) disturbance $\hat{w} := (\Psi^{\mathrm{x}})_\Sigma^{-1}x$, and the final control action $u = \Psi_\Sigma^{\mathrm{u}}\hat{w}$. Since we no longer assume that Ψ_Σ^{x} is a CLM of F, we have to at least verify that it is a candidate CLM to ensure that $(\Psi^{\mathrm{x}})_\Sigma^{-1}$ exists and is causal. This is easy to verify from the fact that we assumed $\{\Psi_i\}_{i=1}^N$ to be all cCLMs:

Lemma 10. *Given some operators $\{G\}_{i=1}^N$ such that $\sum_{i=1}^N G = I$, if $\{\Psi_i\}_{i=1}^N$ are all cCLMs, then $\Psi_\Sigma = \sum_{i=1}^N \Psi_i G_i$ is a cCLM as well.*

Proof. Since the composition of causal operators is causal, it is clear that Ψ_Σ is causal. It is left to verify that $\Psi_\Sigma^{\mathrm{x}} - I$ is strictly causal. To this end, write the former as

$$\Psi_\Sigma^{\mathrm{x}} - I = \sum_{i=1}^N \Psi_i^{\mathrm{x}}G_i - I = \sum_{i=1}^N (\Psi_i^{\mathrm{x}}G_i - G_i) = \sum_{i=1}^N (\Psi_i^{\mathrm{x}} - I)G_i$$

and notice that $(\Psi_i^{\mathrm{x}} - I) \in \mathcal{C}_s(\ell^{\mathcal{X}}, \ell^{\mathcal{X}}) \implies (\Psi_i^{\mathrm{x}} - I)G_i \in \mathcal{C}_s(\ell^{\mathcal{X}}, \ell^{\mathcal{X}})$. $\qquad\square$

From the above, we can realize K as a system-level implementation $\mathrm{SL}[\Psi_\Sigma^{\mathrm{x}}, \Psi_\Sigma^{\mathrm{u}}]$ using the components of Ψ_Σ:

$$\hat{w}_t = x_t - \Psi_{\Sigma,t}^x(0, \hat{w}_{t-1:0}), \qquad\qquad u_t = \Psi_{\Sigma,t}^u(\hat{w}_{t:0}).$$

However, we derive an implementation in terms of the components of Ψ_i.

Rewrite the relation $\Psi_\Sigma^{\mathrm{x}}(\hat{w}) = x$ as

$$\Psi_\Sigma^{\mathrm{x}}(\hat{w}) = x$$

$$\Leftrightarrow \qquad \hat{w} = x + (I - \Psi_\Sigma^{\mathrm{x}})\hat{w} = x + (I - \sum_{i=1}^{N} \Psi_i^{\mathrm{x}} G_i)\hat{w}$$

$$\Leftrightarrow \qquad \hat{w} = x + (\sum_{i=1}^{N} G_i - \sum_{i=1}^{N} \Psi_i^{\mathrm{x}} G_i)\hat{w} = x + (\sum_{i=1}^{N}(G_i - \Psi_i^{\mathrm{x}} G_i))\hat{w}$$

$$\Leftrightarrow \qquad \hat{w} = x + (\sum_{i=1}^{N}(I - \Psi_i^{\mathrm{x}})G_i)\hat{w} = x + \sum_{i=1}^{N}(I - \Psi_i^{\mathrm{x}})G_i\hat{w} \qquad (2.53)$$

$$\Leftrightarrow \qquad \hat{w} = x + \sum_{i=1}^{N}(I - \Psi_i^{\mathrm{x}})\tilde{w}^i$$

and decompose the control action $u = \Psi_\Sigma^{\mathrm{u}}(\hat{w})$ as

$$u = \Psi_\Sigma^{\mathrm{u}}(\hat{w}) = (\sum_{i=1}^{N} \Psi_i^{\mathrm{u}} G_i)\hat{w} = \sum_{i=1}^{N}(\Psi_i^{\mathrm{u}} G_i \hat{w}) = \sum_{i=1}^{N} \Psi_i^{\mathrm{u}} \tilde{w}^i, \qquad (2.54)$$

where we define $\tilde{w}^i = G_i(\hat{w})$ as a partial disturbance term corresponding to the ith cCLM Ψ_i. This leads to the following implementation of the realizing controller K:

Definition 2.13. *Let* $K = \Psi_\Sigma^{\mathrm{u}}(\Psi_\Sigma^{\mathrm{x}})^{-1}$ *for a cCLM of the form* $\Psi_\Sigma = \sum_{i=1}^{N} \Psi_\Sigma G_i$, *where* $\{\Psi_i\}_{i=1}^{N}$ *are cCLMs, and* $\{G_i\}_{i=1}^{N}$ *are a collection of causal operators such that* $\sum_{i=1}^{N} G_i = I$. *Then, the following realization of* K *is defined as its blended SL implementation:*

$$\hat{w}_t = x_t - \sum_{i=1}^{N} \Psi_{i,t}^x(0, \tilde{w}_{t-1:0}^i) \qquad (2.55a)$$

$$\tilde{w}_t^i = G_t^i(\hat{w}_{t:0}) \qquad (2.55b)$$

$$u_t = \sum_{i=1}^{N} \Psi_{i,t}^u(\tilde{w}_{t:0}^i). \qquad (2.55c)$$

In the later sections, we will consider the special case where Ψ_Σ is a blend of linear cCLMs $\{\Psi_i\}_{i=1}^{N}$ and static nonlinearities $G_{i,t}(z_{t:0}) := g_i(z_t)$. The corresponding blended SL implementation takes the form:

$$\hat{w}_t = x_t - \sum_{i=1}^{N} \sum_{k=2}^{t+1} R_{t,k}^i \tilde{w}_{t+1-k}^i \qquad (2.56a)$$

$$\tilde{w}_t^i = g_i(\hat{w}_t) \qquad (2.56b)$$

$$u_t = \sum_{i=1}^{N} \sum_{k=1}^{t+1} M_{t,k}^i \tilde{w}_{t+1-k}^i \qquad (2.56c)$$

where the matrices $R^i_{t,k} \in \mathbb{R}^{n \times n}$ and $M^i_{t,k} \in \mathbb{R}^{n \times m}$ are associated with the linear component functions of the cCLMs Ψ_i as follows:

$$\Psi^{\mathrm{x}}_{i,t}(z_{1:t+1}) = \sum_{k=1}^{t+1} R^i_{t,k} z_k, \qquad \Psi^{\mathrm{u}}_{i,t}(z_{1:t+1}) = \sum_{k=1}^{t+1} M^i_{t,k} z_k. \tag{2.57}$$

Internal Stability of the Closed Loop

To investigate the stability of the overall closed loop, as described in the previous section, we need to add the internal states of the implementation $\hat{w}$ and $\tilde{w}^i$ to our closed loop model and analyze the stability of the corresponding closed loop maps.

To derive the dynamics of the lumped disturbance, we substitute $x = F^+(x, u) + w$ into $\hat{w} = x + (I - \Psi^{\mathrm{x}}_\Sigma)\hat{w}$. Furthermore, since $(x, u) = \Psi_\Sigma(\hat{w})$ we have $x = F^+ \Psi_\Sigma \hat{w} + w$ which yields:

$$\hat{w} = (F^+ \Psi_\Sigma)\hat{w} + w + (I - \Psi^{\mathrm{x}}_\Sigma)\hat{w} = (F^+ \Psi_\Sigma + I - \Psi^{\mathrm{x}}_\Sigma)\hat{w} + w$$

$$= w + \Delta[F, \Psi_\Sigma]\hat{w} = \sum_{i=1}^{N} \Delta[F, \Psi_i]G_i \hat{w} + w.$$

Finally, since we established that Ψ_Σ is a candidate CLM, $\Delta[F, \Psi_\Sigma] = F^+(\Psi_\Sigma) + (I - \Psi^{\mathrm{x}})$ is strictly causal and assures that $I - \Delta[F, \Psi_\Sigma]$ is causally invertible. Hence, the mapping $w \mapsto \hat{w}$ is defined by the following equation:

$$\hat{w} = (I - \Delta[F, \Psi_\Sigma])^{-1}w = (I - \sum_{i=1}^{N} \Delta[F, \Psi_i]G_i)^{-1}w \tag{2.58}$$

$$= (\sum_{i=1}^{N}(I - \Delta[F, \Psi_i])G_i)^{-1}w. \tag{2.59}$$

Since F is assumed linear, we can invoke Corollary 10 of Theorem 8 to obtain conditions for closed loop internal stability. Recall the perturbed closed loop model δCL with $\mathrm{SL}[\Psi^{\mathrm{x}}_\Sigma, \Psi^{\mathrm{u}}_\Sigma]$ as the controller. We have an internal ℓ_p-stable closed loop under the following conditions:

Theorem 15. *Let Ψ_Σ be a $\mathcal{G}$-blended cCLM corresponding to a family of ℓ_p-stable cCLMs $\mathcal{F} = \{\Psi_i\}_{i=1}^{N}$ and a family of ℓ_p-stable weights $\mathcal{G} = \{G_i\}_{i=1}^{N}$. Consider the closed loop system δCL described by the equations (2.40) for a fixed ℓ_p-stable $F \in \mathcal{LC}(\ell^{\mathcal{X} \times \mathcal{U}}, \ell^{\mathcal{X}})$ and controller $K_\Sigma = \Psi^{\mathrm{u}}_\Sigma(\Psi^{\mathrm{x}}_\Sigma)^{-1}$ with implementation $\mathrm{SL}[\Psi^{\mathrm{x}}_\Sigma, \Psi^{\mathrm{u}}_\Sigma]$.*

Then, the map $\Phi_{\delta\mathrm{CL}} : (w, v, d) \mapsto (\hat{w}, x, u)$ of the closed loop δCL is ℓ_p-stable if the operator $(\sum_{i=1}^{N}(I - \Delta[F, \Psi_i])G_i)^{-1}$ is ℓ_p-stable.

Proof. Apply Theorem 10. $\qquad\square$

In the next section, we discuss how blending inspires a nonlinear control synthesis procedure for the constrained LQR problem, which strictly outperforms any linear controller.

2.10 Conclusion

This chapter highlights a general and fundamental relationship between closed loop maps and corresponding realizing controllers in general nonlinear discrete-time systems. The key findings are as follows: 1. All closed loop maps are solutions to an operator equation, and all solutions of the equation are achievable closed loop maps. 2. Given a solution of the operator equation, we can obtain a realizing controller by parameterizing a system-level controller with the solution. This controller then imposes the given solution as the closed loop map of the system. 3. This same procedure produces robust closed loop stability even when the system-level controllers are parameterized with approximate solutions of the operator equation.

We discuss an important consequence for the special case of LTV system dynamics and nonlinear dynamics in the controller: The space of (nonlinear) CLMs is closed under "blending"-type combinations. This observation informs a procedure for constructing complex (or more expressive) CLMs from a collection of simpler CLMs. This idea can be leveraged for nonlinear SL-controller synthesis and provides, for example, systematic instructions to synthesize nonlinear controllers by "blending" multiple linear SL controllers into one.

In the next chapter, we will see that this technique has important implications for the problem setting of linear systems with actuator saturation and provides new ways to perform simple stability and performance analysis of the closed loop.

2.A Proofs

Let $(\mathcal{X}, |\cdot|_{\mathcal{X}})$ and $(\mathcal{Y}, |\cdot|_{\mathcal{Y}})$ be some finite-dimensional banach spaces, $\mathcal{X}^{\mathbb{N}}, \mathcal{Y}^{\mathbb{N}}$ be the corresponding vector space of sequences over $\mathbb{N}$. Let $\mathcal{X}_p^{\mathbb{N}} \subset \mathcal{X}^{\mathbb{N}}$ and $\mathcal{Y}_p^{\mathbb{N}} \subset \mathcal{Y}^{\mathbb{N}}$ denote the normed vectorspaces w.r.t. to the usual ℓ_p-norms defined for sequences in $\mathbb{N}$. Let $\mathcal{C}(\mathcal{X}^{\mathbb{N}}, \mathcal{Y}^{\mathbb{N}})$ be the space of all causal operators $\mathcal{X}^{\mathbb{N}} \to \mathcal{Y}^{\mathbb{N}}$ and define $\||\cdot\||_p^\star : \mathcal{C}(\mathcal{X}^{\mathbb{N}}, \mathcal{Y}^{\mathbb{N}}) \to \mathbb{R}_0^+ \cup \infty$ and $\||\cdot\||_p^\star : \mathcal{C}(\mathcal{X}^{\mathbb{N}}, \mathcal{Y}^{\mathbb{N}}) \to \mathbb{R}_0^+ \cup \infty$, with $p \in \{1, 2, \ldots, \infty\}$ for some operator $A \in \mathcal{C}(\mathcal{X}^{\mathbb{N}}, \mathcal{Y}^{\mathbb{N}})$:

$$\||A\||_p^\star := \sup_{x \in \mathcal{X}_p^{\mathbb{N}}: x \neq 0} \frac{\|A(x) - A(0)\|_p}{\|x\|_p} \qquad \||A\||_p := \||A\||_p^\star + \|A(0)\|_p.$$

Lemma 11. $\interleave \cdot \interleave_p$ *is a norm on* $\mathcal{C}(\mathcal{X}_p^{\mathrm{N}}, \mathcal{Y}_p^{\mathrm{N}})$ *and* $\interleave \cdot \interleave_p^\star$ *is a norm on* $\{A \in \mathcal{C}(\mathcal{X}_p^{\mathrm{N}}, \mathcal{Y}_p^{\mathrm{N}}) \mid A(0) = 0\}$ *or the quotientspace* $\mathcal{C}(\mathcal{X}_p^{\mathrm{N}}, \mathcal{Y}_p^{\mathrm{N}})/\sim$ *with the equivalence relation* $A \sim B \Leftrightarrow A - B \equiv const.$ *Furthermore the following holds:*

(i) $\interleave A - B \interleave_p^\star = 0$ *if and only if* A *and* B *are the same up to a constant offset.*

(ii) *Submultiplicativety holds for products* TA *and* AB *if* T *is linear and* $B(0) = 0$*:*

$$\interleave TA \interleave_p^\star \leqslant \interleave T \interleave_p^\star \interleave A \interleave_p^\star \qquad \interleave AB \interleave_p^\star \leqslant \interleave A \interleave_p^\star \interleave B \interleave_p^\star.$$

Proof. If $\interleave A \interleave_p^\star = 0$, then for all $x \in \mathcal{X}_p^{\mathrm{N}}, x \neq 0$ holds $y = A(x) = y_0 := A(0)$ because otherwise $(\|y - y_0\|_p)/\|x\|_p \neq 0$ leads to a contradiction. Let $r \neq 0$ then $rA = (rI) \circ A$ means that $\|rA(x) - rA(0)\|_p = \|r(A(x) - A(0))\|_p = |r|\|(A(x) - A(0))\|_p$ and therefore $\interleave rA \interleave_p^\star = |r| \interleave A \interleave_p^\star$. It remains to show a notion of sub-additivity. Let $A, B \in \mathcal{C}(\mathcal{X}^{\mathrm{N}}, \mathcal{Y}^{\mathrm{N}})$, then for all $x \in \mathcal{X}_p^{\mathrm{N}}$ holds $(A + B)(x) - (A + B)(0) = A(x) - A(0) + B(x) - B(0)$ and by triangle inequality of the ℓ_p-norm follows

$$\|(A + B)(x) - (A + B)(0)\|_p \leqslant \|A(x) - A(0)\|_p + \|B(x) - B(0)\|_p.$$

Let $\mathcal{X} = \mathcal{Y}$ and consider $AB \in \mathcal{C}(\mathcal{X}_p^{\mathrm{N}}, \mathcal{X}_p^{\mathrm{N}})$ with $B(0) = 0$. Then

$$
\begin{aligned}
\interleave A \interleave_p^\star &= \sup_{x \in \mathcal{X}_p^{\mathrm{N}}: \, x \neq 0} \frac{\|AB(x) - AB(0)\|_p}{\|x\|_p} \\
&= \sup_{x \in \mathcal{X}_p^{\mathrm{N}}: \, B(x) \neq 0} \frac{\|AB(x) - A(0)\|_p}{\|B(x)\|_p} \frac{\|B(x)\|_p}{\|x\|_p} \\
&\leqslant \sup_{y \in \mathcal{X}_p^{\mathrm{N}}: \, y \neq 0} \frac{\|A(y) - A(0)\|_p}{\|y\|_p} \sup_{x \in \mathcal{X}_p^{\mathrm{N}}: \, x \neq 0} \frac{\|B(x) - B(0)\|_p}{\|x\|_p} = \interleave A \interleave_p^\star \interleave B \interleave_p^\star.
\end{aligned}
$$

Now, let $T \in \mathcal{LC}(\mathcal{X}_p^{\mathrm{N}}, \mathcal{X}_p^{\mathrm{N}})$ be a linear operator and notice that $TA(x) - TA(0) = T(A(x) - A(0))$. We repeat the same process as above and obtain $\interleave TA \interleave_p^\star \leqslant \interleave T \interleave_p^\star \interleave A \interleave_p^\star$. $\qquad\square$

Chapter 3

NONLINEAR BLENDING APPROACH TO CONTROL OF LARGE-SCALE SYSTEMS

The theory developed in the previous chapter provides a framework to systematically "blend" multiple (possibly) nonlinear controllers into one stabilizing nonlinear controller. Blending happens at the controller implementation level and can be used to optimize the performance of the nonlinear closed-loop map by combining the desired properties of multiple closed-loop maps. In this chapter, we explore three applications to demonstrate the utility of this approach in control settings challenged by complex dynamics and constraints in the system or controller.

3.1 Introduction

As a warm-up, we discuss the time-varying design of the SL controller and the local stability analysis for closed loops involving nonlinear continuous-time systems, which is the first rigorous discussion of this topic. As an example, we use the problem of trajectory tracking for nonlinear continuous-time systems through discrete-time zero-order hold feedback control. Our empirical case study evaluates the SL controller on the cart-pole system and demonstrates that despite using only a rough model for synthesis, the resulting controller demonstrates very robust closed-loop performance.

In the later part, we consider nonlinear control settings, where complexity in the system dynamics is caused by sheer scale and distributedness of the system, rather than high-order nonlinearity in the dynamics. In particular, we show that the framework is particularly well-suited for large-scale systems subjected to input saturation and state constraints, but with otherwise linear dynamics. We investigate some first application scenarios where our approach naturally provides significant benefits over existing methods: distributed constrained LQR and distributed anti-windup control.

In the case of constrained LQR, we derive a synthesis procedure for blended SLS controllers that outperforms any optimal linear controller for the constrained LQR problem [36, 86, 91, 140]. As a second application, we discuss how the blended SLS technique provides a natural remedy for controller-windup in a way that is easily scalable for use in large-scale control systems. We discuss the efficacy of the methods with simulations and show that synthesis and implementation enjoy the same benefits as previous SLS synthesis methods: both are distributed, handle delays, sparse actuation, and allow for localized disturbance rejection. The presented work is based on [64] and [137].

Setup and Recap of Key Concepts

For this chapter, we consider linear dynamical systems described by the operator equation

$$x = \mathcal{S}^+ F(x, u) + w =: F^+(x, u) + w, \tag{3.1}$$

where $F^+ = \mathcal{S}^+(F)$ is a stricly causal operator formed by a right shift of causal linear operator F which we call dynamics. As linearity allows, we split F into two causal linear operators $A : \ell^n \mapsto \ell^n$, $B : \ell^m \mapsto \ell^n$ and write F as $F(x, u) := (A(x) + B(u))$. We represent A and B by a sequence of matrices $\{A_{t,k} \in \mathbb{R}^{n \times n} \,|\, t \geqslant 0,\ k \geqslant 1\}$, $\{B_{t,k} \in \mathbb{R}^{n \times n} \,|\, t \geqslant 0,\ k \geqslant 1\}$ parametrizing the component functions of A_t and B_t as:

$$A_t(x_{t:0}) = \sum_{k=0}^{t} A_{t,k+1} x_{t-k}, \qquad B_t(u_{t:0}) = \sum_{k=0}^{t} B_{t,k+1} u_{t-k}.$$

Correspondingly, F has the component functions $F_t(x_{t:0}, u_{t:0}) = A_t(x_{t:0}) + B_t(x_{t:0})$, the components of the dynamics F^+ are

$$F_t^+(x_{t:0}, u_{t:0}) = \sum_{k=1}^{t} A_{t-1,k} x_{t-k} + \sum_{k=1}^{t} B_{t-1,k} u_{t-k} \tag{3.2}$$

and the difference equation of the system dynamics takes the form:

$$x_t = \sum_{k=1}^{t} A_{t-1,k} x_{t-k} + \sum_{k=1}^{t} B_{t-1,k} u_{t-k} + w_t. \tag{3.3}$$

For the sake of defining the closed-loop maps, we view w as an input and the pair (x, u) as an output. Then, as in the previous chapter, for a causal controller $K : \ell^n \mapsto \ell^m$ and fixed plant F, the corresponding w $\mapsto$ {x, u} map of the closed-loop CL is the operator $\Psi = \left[\begin{smallmatrix}\Psi^x\\\Psi^u\end{smallmatrix}\right]$ defined by the partial maps Ψ^x and Ψ^u:

$$\Psi^x := \left(I - \mathcal{S}^+ F(I, K)\right)^{-1}, \qquad \Psi^u := K \left(I - \mathcal{S}^+ F(I, K)\right)^{-1}.$$

An operator $\Psi = (\Psi^x, \Psi^u)$ with partial maps Ψ^x and Ψ^u is a candidate closed-loop map (cCLM) of F, if it is causal and of conforming domain and co-domains, i.e., $\Psi^x : \ell^n \mapsto \ell^n$ and $\Psi^u : \ell^n \mapsto \ell^m$ and $\Psi^x - I$ is strictly causal. The residual Δ is a map $\Delta : \mathcal{C}(\ell^{\mathcal{X}\times\mathcal{U}}, \ell^{\mathcal{X}}) \times \mathcal{C}(\ell^{\mathcal{X}}, \ell^{\mathcal{X}\times\mathcal{U}}) \to \mathcal{C}(\ell^{\mathcal{X}}, \ell^{\mathcal{X}})$ where for $F \in \mathcal{C}(\ell^{\mathcal{X}\times\mathcal{U}}, \ell^{\mathcal{X}})$ and $\Psi = \left[\begin{smallmatrix}\Psi^{\mathrm{x}}\\\Psi^{\mathrm{u}}\end{smallmatrix}\right] \in \mathcal{C}(\ell^{\mathcal{X}}, \ell^{\mathcal{X}\times\mathcal{U}})$, $\Delta[F, \Psi]$ is defined as:

$$\Delta[F, \Psi] = F^{+}(\Psi) + I - \Psi^{x} =: \mathcal{S}^{+}F(\Psi) + I - \Psi^{x}.$$

The set of all closed-loop maps of F is denoted as $\Phi_{\mathrm{CL}}^{\mathrm{w}\mapsto\mathrm{xu}}[F]$ and by Theorem 8 can be equivalently defined as:

$$\Phi_{\mathrm{CL}}^{\mathrm{w}\mapsto\mathrm{xu}}[F] = \{\Psi \mid \Delta[F, \Psi] = 0\}.$$

3.2 Discrete-Time Trajectory-Tracking Control for Nonlinear Continuous-Time Systems

We derived that, in theory, we only need operators $(\Psi^{\mathrm{x}}, \Psi^{\mathrm{u}})$ to approximately satisfy the CLM condition (2.33) to obtain robustly stabilizing controllers $\mathrm{SL}[\Psi^{\mathrm{x}}, \Psi^{\mathrm{u}}]$. The generality of the robustness result argues that system level controllers could be a promising design tool in practical control applications. However, more research is needed to quantify the trade-off between the approximation grade of the condition (2.33) and the corresponding achievable control performance. As a first step towards that, we present some first empirical results that show that system-level controllers can achieve good robust control performance in challenging-to-control nonlinear systems while using only crude models of the system for synthesis. Figure 3.1 shows simulation results of using a system level controller $\mathrm{SL}[\Psi^{\mathrm{x}}, \Psi^{\mathrm{u}}]$ at 30 Hz sampling time to swing up a cart pole system under small and large closed-loop perturbations. Rather than satisfying the CLM condition (2.33) of the zero-order hold actuated cart pole system, the maps Ψ are synthesized using the following approximations:

- Ψ are taken to be affine operators, where the affine term is a sampled continuous-time desired trajectory $(x^{d}(t), u^{d}(t))$ for the system and linear part is chosen to be finite memory (2 s window in continuous-time).

- The continuous-time trajectories $(x^{d}(t), u^{d}(t))$ are low-grade approximations of swing-up motions of the cartpole.

- Ψ are chosen as CLMs of an *approximation* of the linearized system around the desired trajectory.

The above simplifications make clear that Ψ serves only as a very coarse approximation of the exact CLM condition (2.33). On the other hand, the above approximations allow to synthesize the linear part of Ψ analytically and in parallel, allowing for

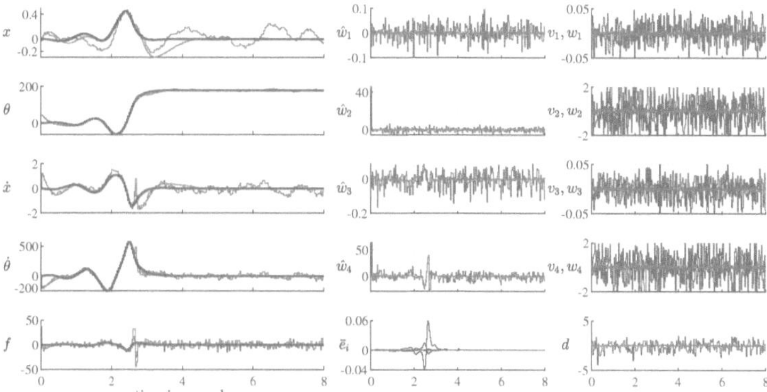

Figure 3.1: Swing-up control for cart-pole system with system level controller at 30Hz sampling rate. $(x, \dot{x}, \theta, \dot{\theta}, f)$ stand for cart position / velocity, pole angle / velocity, and cart force with units $(m, m/s, deg, deg/s, N)$. $\theta = 180°$ stands for the upward pole position. The weight of the cart and the pole is chosen as 1 kg and 0.1 kg, the length of the pole is chosen as 0.5 m. Consult [122] for detailed system description and equations. *Left*: Desired trajectory (red) vs. closed-loop performance under initial condition error $\psi(0) = 45°$ and two scenarios of small (orange) and large (blue) system perturbations $\boldsymbol{w}, \boldsymbol{v}, \boldsymbol{d}$. *Middle*: Evolution of internal state $\hat{w}$ for both scenarios and normalized disturbance due to trajectory error $\bar{e}_i$. *Right*: i.i.d Gaussian perturbations $\boldsymbol{w}, \boldsymbol{v}, \boldsymbol{d}$.

efficient computation.

Leaning on the discussion in Section 2.7, the closed-loop S_2 of the cart pole simulation can be written as

$$x_t = \phi_{t_s}(x_{t-1}, u_{t-1}) + w_t + e_t, \tag{3.4a}$$

$$\hat{w}_t = x_t + v_t - x^d(t\tau_s) - \sum_{k=2}^{t+1} R_{t,k}\hat{w}_{t+1-k}, \tag{3.4b}$$

$$u_t = u^d(t\tau_s) + \sum_{k=1}^{t+1} M_{t,k}\hat{w}_{t+1-k} + d_t, \tag{3.4c}$$

where w, d and v are state, input and internal controller state perturbations and e_t is due to errors in the trajectory synthesis. The matrices $R_{t,k} \in \mathbb{R}^{n\times n}$, $M_{t,k} \in \mathbb{R}^{m\times n}$ parameterize the linear part of $\boldsymbol{\Psi}$. Due to the approximation steps taken in the synthesis of $\boldsymbol{\Psi}$, there is a considerable gap between the real system and the model used for the synthesis. Nevertheless, Figure 3.1 shows that despite the large uncertainty of the model, the closed loop provides robust performance against a variety of perturbations: Large initial condition errors, large perturbation signals, and errors in trajectory.

Derivation and Nonlinear Stability Analysis

Using the cart pole system as an example, we demonstrate how to develop a system-level controller to track trajectories for nonlinear continuous-time systems. We use the description of the cart pole as presented in [122] and refer to the same reference for detailed derivations. The dynamic equations of the cart pole are

$$(m_c + m_p)\ddot{x}_c + m_p l \ddot{\theta}_p \cos \theta_p - m_p l \dot{\theta}_p^2 \sin \theta_p = f \tag{3.5a}$$

$$m_p l \ddot{x}_c \cos \theta + m_p l^2 \ddot{\theta}_p + m_p g l \sin \theta_p = 0 \tag{3.5b}$$

where x_c and θ_p stand for cart position and pole angle in counterclockwise direction and f represents the force exerted on the cart. Furthermore, $\theta = 0$ denotes the downward position. The parameters (m_c, m_p, l, g) are chosen as (1 kg, 0.1 kg, 0.5 m, 9.81m/s^2) and represent the mass of the cart and the pole, the length of the pole and the gravity constant, respectively. Furthermore, (3.5) can be converted into the input affine standard form

$$\dot{x} = F(x) + g(x)u \tag{3.6}$$

where $x = [x_c, \theta_p, \dot{x}_c, \dot{\theta}_p]^T$, $u = f$, (see [122] for description of $F(x)$ and $g(x)$). As in practice, controllers are usually implemented digitally, we assume zero-order hold on the input u with a sampling time of $\tau_s = 0.033\text{sec}$ $(1/\tau_s = 30\text{Hz})$. Because of this discretization, we can equivalently represent the system (3.6) at sampling times through the discrete-time system

$$x_t = \phi_{\tau_s}(x_{t-1}, u_{t-1}), \quad \phi_{\tau_s}(x, u) := \alpha(\tau_s), \text{ s.t. : } \dot{\alpha} = F(\alpha, u), \alpha(0) = x, \tag{3.7}$$

where we denote $x_t := x(t\tau_s)$ and $u_t := u(t\tau_s)$ ($t \in \mathbb{N}$) to be samples of the continuous-time signals $x(\tau)$, $u(\tau)$ at time $t\tau_s$. To put (3.7) in operator form, define $F^\phi \in \mathcal{C}_s(\ell^n \times \ell^m, \ell^n)$ with component functions $F_t^\phi(x_{t:0}, u_{t:0}) := \phi_{\tau_s}(x_{t-1}, u_{t-1})$ and equation (3.7) can be written in terms of the trajectories $(\boldsymbol{x}, \boldsymbol{u})$ as $\boldsymbol{x} = \boldsymbol{F}^\phi(\boldsymbol{x}, \boldsymbol{u})$.

Remark 16. *We use the variable τ to indicate that a variable $a(\tau)$ is a continuous-time signal and use a_t to refer to the discrete-time samples $a_t := a(t\tau_s)$.*

We use a continuous-time trajectory $x^d(\tau), u^d(\tau)$ as a reference, which is shown in red in Figure 3.1. The trajectories approximately satisfy the continuous-time dynamics, i.e., $\dot{x}^d(\tau) \approx F(x^d(\tau)) + g(x^d(\tau))u^d(\tau)$, and are designed to swing up the pole in 3 seconds and then keep the cart pole at $x_c = 0$, $\theta_p = \pi$.

Remark 17. *Notice that even if $\dot{x}^d(\tau) = F(x^d(\tau)) + g(x^d(\tau))u^d(\tau)$, it still does not hold $x_t^d \neq \phi_{T_s}(x_{t-1}^d, u_{t-1}^d)$, since the continuous-time trajectories are computed without consideration of the zero-order hold actuation.*

We take the following linear approximation of ϕ_{T_s} around the reference trajectory:

$$\phi_{T_s}(x_{t-1}, u_{t-1}) \approx x_t^d + \underbrace{\exp(\nabla F|_{x_{t-1}^d} T_s)}_{=:\hat{A}_{t-1}} *(x_{t-1} - x_{t-1}^d) \ldots \tag{3.8}$$

$$+ \underbrace{\int_0^{T_s} \exp(\nabla F|_{x_{t-1}^d} \tau)g(x_{t-1}^d)d\tau}_{=:\hat{B}_{t-1}} *(u_{t-1} - u_{t-1}^d). \tag{3.9}$$

Denote $[A_{t-1}, B_{t-1}] := \nabla_{x,u}\phi_{T_s}|_{(x_{t-1}^d, u_{t-1}^d)}$ the true linearization of ϕ_{T_s} at (x_{t-1}^d, u_{t-1}^d) and notice that the above approximation (3.8) is only an approximation of the linearization, since $[\hat{A}_{t-1}, \hat{B}_{t-1}] \neq [A_{t-1}, B_{t-1}]$. Using Taylor's theorem and assuming ϕ_{T_s} is differentiable, we can write ϕ_{T_s} as

$$\phi_{T_s}(x_{t-1}, u_{t-1}) = \phi_{T_s}(x_{t-1}^d, u_{t-1}^d) + A_{t-1}(x_{t-1} - x_{t-1}^d) + B_{t-1}(u_{t-1} - u_{t-1}^d) \ldots$$
$$+ r_{t-1}(x_{t-1} - x_{t-1}^d, u_{t-1} - u_{t-1}^d)$$

where $\lim_{|z| \to 0} |r_{t-1}(z)|/|z| = 0$. We can factor out equation (3.7) into the following components

$$x_t = x_t^d + \hat{A}_{t-1}(x_{t-1} - x_{t-1}^d) + \hat{B}_{t-1}(u_{t-1} - u_{t-1}^d) + e_t + e_t'(x_{t-1}, u_{t-1}) \tag{3.10}$$

where e_t and $e_t'(x_{t-1}, u_{t-1})$ are disturbance terms introduced due to errors in the reference trajectory and linearization

$$e_t = \phi_{T_s}(x_{t-1}^d, u_{t-1}^d) - x_t^d$$
$$e_t'(x_{t-1}, u_{t-1}) = (A_{t-1} - \hat{A}_{t-1})(x_{t-1} - x_{t-1}^d) + (B_{t-1} - \hat{B}_{t-1})(u_{t-1} - u_{t-1}^d)$$
$$\cdots + r_{t-1}(x_{t-1} - x_{t-1}^d, u_{t-1} - u_{t-1}^d)$$

and the remaining terms represent our linear approximation of the dynamics. we can express this more compactly in operator form: Define $\tilde{F}^\phi \in \mathcal{C}_s(\ell^n, \ell^n)$ with the components

$$\tilde{F}_t^\phi(x_{t:0}, u_{t:0}) := x_t^d + \hat{A}_{t-1}(x_{t-1} - x_{t-1}^d) + \hat{B}_{t-1}(u_{t-1} - u_{t-1}^d)$$

and the residual $\boldsymbol{\Delta}^\phi \in \mathcal{C}(\ell^n, \ell^n)$ with the components $\Delta_t^\phi(x_{t:0}, u_{t:0}) = e_t + e_t'(x_{t-1}, u_{t-1})$, then we can factor out the original equations of (3.7) as

$$x = \tilde{\boldsymbol{F}}^\phi(\boldsymbol{x}, \boldsymbol{u}) + \boldsymbol{\Delta}^\phi(\boldsymbol{x}, \boldsymbol{u}) \tag{3.12}$$

where per definition we have the decomposition $\tilde{\boldsymbol{F}}^\phi + \boldsymbol{\Delta}^\phi = \boldsymbol{F}^\phi$.

Our approach for synthesis is now to use $\tilde{\boldsymbol{F}}^\phi$ as a model to design a system level controller and treat $\boldsymbol{\Delta}^\phi$ as disturbance terms we want to be robust against. We do this, first solving for the CLMs $\tilde{\boldsymbol{\Psi}} = (\tilde{\boldsymbol{\Psi}}^{\mathrm{x}}, \tilde{\boldsymbol{\Psi}}^{\mathrm{u}})$ of $\tilde{\boldsymbol{F}}^\phi$ and then choosing our feedback controller as $\mathrm{SL}[\tilde{\boldsymbol{\Psi}}^{\mathrm{x}}, \tilde{\boldsymbol{\Psi}}^{\mathrm{u}}]$.

Due to Theorem 8, $\tilde{\boldsymbol{\Psi}}$ is a CLM of $\tilde{\boldsymbol{F}}^\phi$ if and only if it satisfies the CLM equation (2.33) for $\tilde{\boldsymbol{F}}^\phi$, i.e., $\tilde{\boldsymbol{\Psi}}^{\mathrm{x}} = \tilde{\boldsymbol{F}}^\phi(\tilde{\boldsymbol{\Psi}}) + \boldsymbol{I}$. We restrict $\tilde{\boldsymbol{\Psi}}$ to be of affine form $\tilde{\boldsymbol{\Psi}} : \boldsymbol{w} = \left[\begin{smallmatrix} R \\ M \end{smallmatrix}\right](\boldsymbol{w}) + \left[\begin{smallmatrix} r \\ m \end{smallmatrix}\right]$ with $\left[\begin{smallmatrix} r \\ m \end{smallmatrix}\right] \in \ell^n \times \ell^m$ and $\boldsymbol{R} \in \mathcal{LC}(\ell^n, \ell^n)$, $\boldsymbol{M} \in \mathcal{LC}(\ell^n, \ell^m)$. Thus the component functions of $\tilde{\boldsymbol{\Psi}}$ take the from

$$\tilde{\Psi}_t^x(\alpha_{1:t+1}) = \sum_{k=1}^{t+1} R_{t,k}\alpha_k + r_t \tag{3.13}$$

$$\tilde{\Psi}_t^u(\alpha_{1:t+1}) = \sum_{k=1}^{t+1} M_{t,k}\alpha_k + m_t \tag{3.14}$$

where $R_{t,j} \in \mathbb{R}^{n\times n}$, $M_{t,j} \in \mathbb{R}^{n\times m}$, r_t, m_t are some fix sequences and $\alpha_{1:t+1}$ denote the $t + 1$ arguments of the component function. Structuring $\tilde{\boldsymbol{\Psi}}$ in this form and using linearity of $\tilde{\boldsymbol{F}}^\phi$ reduces the original operator equation $\tilde{\boldsymbol{\Psi}}^{\mathrm{x}} = \tilde{\boldsymbol{F}}^\phi(\tilde{\boldsymbol{\Psi}}) + \boldsymbol{I}$ simply to the following set of linear equations for $R_{t,j}$, $M_{t,j}$ and r_t, m_t:

$$R_{t,k} = \hat{A}_{t-1}R_{t-1,k-1} + \hat{B}_{t-1}M_{t-1,k-1} \text{ for all } k \leq t, \quad R_{t,1} = I \tag{3.15}$$

$$r_t = x_t^d, \quad m_t = u_t^d. \tag{3.16}$$

Equation (3.15) is an affine subspace constraint and opens up many possible ways to synthesize for solutions $R_{t,k}$, $M_{t,k}$. In fact, (3.15) matches the linear time-varying formulation of SLS as discussed in [66], [11] and for our case-study here, we are synthesizing for $R_{t,k}$, $M_{t,k}$ by solving the following $\mathcal{H}_2/$ LQR problem for the LTV system $\tilde{\boldsymbol{F}}^\phi$ subject to an FIR constraint with horizon $T = 60$ time-steps:

$$\begin{aligned} \min_{R_{t,k}, M_{t,k}} \quad & \sum_{0\leq t\leq H} \sum_{1\leq k\leq T} \|R_{t,k}\|_F^2 + \|M_{t,k}\|_F^2 \\ s.t. \quad & R_{t,k} = \hat{A}_{t-1}R_{t-1,k-1} + \hat{B}_{t-1}M_{t-1,k-1} \\ & R_{t,1} = I, \quad R_{t,T} = 0. \end{aligned} \tag{3.17}$$

Remark 18. *See [11] for details of the $\mathcal{H}_2$/LQR problem setup and derivation of the convex optimization problem. The FIR horizon can be understood as a time window $[t, t+T]$ given to the controller to kill off the disturbance $\hat{w}_t$. Considering the sampling time of our example, $T = 60$ here translates to a 2-second window in continuous time.*

Furthermore, H denotes the length of the trajectory in sampling time steps. The above problem can be solved in closed form since it is a QP without inequality constraints. Moreover, the change of variables $R_{j+h,j+1}, M_{j+h,j+1}$ with $0 \leqslant j \leqslant T-1$, $0 \leqslant h \leqslant H$ shows that (3.17) can be decomposed over h into H separate QP's that can be solved analytically and in parallel, hence showing that the computational complexity of our synthesis approach is **independent** of the trajectory length H.

The solutions of (3.17) are taken to parameterize the operators (3.13) which give us the system-level controller $\mathrm{SL}[\tilde{\boldsymbol{\Psi}}^{\mathrm{x}}, \tilde{\boldsymbol{\Psi}}^{\mathrm{u}}]$. The resulting closed loop of the cart pole system (3.7) and controller $\mathrm{SL}[\tilde{\boldsymbol{\Psi}}^{\mathrm{x}}, \tilde{\boldsymbol{\Psi}}^{\mathrm{u}}]$ can be put into the form of our robust stability analysis in Section 2.7:

$$x_t = \phi_{t_s}(x_{t-1}, u_{t-1}) + w_t \tag{3.18a}$$

$$\hat{w}_t = x_t + v_t - x^d(t\tau_s) - \sum_{k=2}^{t+1} R_{t,k}\hat{w}_{t+1-k} \tag{3.18b}$$

$$u_t = u^d(t\tau_s) + \sum_{k=1}^{t+1} M_{t,k}\hat{w}_{t+1-k} + d_t. \tag{3.18c}$$

Furthermore, referring to Theorem 11, it can be verified that the residual operator $\boldsymbol{\Delta}^\phi$ we defined earlier matches the residual operator of Theorem 11, i.e., $\boldsymbol{\Delta}[\tilde{\boldsymbol{F}}, \tilde{\boldsymbol{\Psi}}] = \boldsymbol{\Delta}^\phi$. Thus, Theorem 11 applies directly to our problem setting. More specifically, the local result Lem. 7 can be used to obtain robust stability guarantees. If the lumped residual terms are (γ, β) ℓ_p-stable with $\gamma < 1$, then the closed-loop system is f.g. ℓ_p-stable for small enough perturbations.

3.3 The Constrained LQR Problem

We use the idea of blended CLMs to derive a novel distributed synthesis procedure that outperforms any optimal linear controller for the constrained LQR problem [36, 86, 91, 140]. A significant advantage of the approach is that despite being a nonlinear synthesis method it naturally enjoys the same benefits as the linear system level approach introduced in [11], which allows for localized controller implementation, making it scalable to large networks. For the following discussion, we assume that $\boldsymbol{F}$ is a standard linear time-invariant system, that is: $\boldsymbol{F} = \boldsymbol{Ax} + \boldsymbol{Bu}$, where $\boldsymbol{A} = \boldsymbol{I} \otimes A$ and $\boldsymbol{B} = \boldsymbol{I} \otimes B$ for some fixed matrices $A \in \mathbb{R}^{n \times n}$ and $B \in \mathbb{R}^{n \times m}$.

Consider a control problem where we wish to minimize an *average* LQR cost, but also want that the closed loop meets certain safety guarantees against a set of rare yet possible *worst-case* disturbances. Ideally, we would like to synthesize a controller that can guarantee the necessary safety constraints without too much loss in performance compared to the unconstrained LQR controller. We will phrase this design goal as the following constrained LQR problem:

$$\min_{\boldsymbol{K}} \lim_{T \to \infty} \frac{1}{T} \sum_{t=1}^{T} \mathbb{E}_{w_t^i \sim p(w)}[\mathcal{J}(x_t, u_t)] \tag{3.19a}$$

$$s.t. \qquad x_t = Ax_{t-1} + Bu_{t-1} + w_t \tag{3.19b}$$

$$u_t = K_t(x_{t:0}) \tag{3.19c}$$

$$\forall \boldsymbol{w} : \|\boldsymbol{w}\|_\infty \leqslant \eta_{max} : \tag{3.19d}$$

$$\sup_k |x_k| \leqslant x_{max}, \qquad \sup_k |u_k| \leqslant u_{max}$$

where $\mathcal{J}$ abbreviates the quadratic stage cost $\mathcal{J}(x, u) = x^T Q x + u P u$ with $Q, P > 0$. We will assume that the disturbance is stochastic but bounded such that $\|\boldsymbol{w}\|_\infty \leqslant \eta_{max}$ with known distribution which satisfies the following

Assumption 3.4. *Disturbance w_t^i are i.i.d. drawn from the scalar centered distribution $p(w)$ and uncorrelated in time t and coordinate i.*

We can equivalently phrase the optimal control problem (3.19) in terms of closed loop maps as defined in chapter 2. Recalling Def. 2.6, the optimal control problem (3.19) can be described as an optimization over the set of feasible CLMs $\boldsymbol{\Psi} \in \Phi_{CL}^{w \to xu}[\boldsymbol{F}]$, where $\boldsymbol{F} : (\boldsymbol{x}, \boldsymbol{u}) \mapsto \boldsymbol{Ax} + \boldsymbol{Bu}$ and by using the characterization Theorem 8 we obtain:

$$\min_{\boldsymbol{\Psi}^x, \boldsymbol{\Psi}^u} \lim_{T \to \infty} \frac{1}{T} \sum_{t=1}^{T} \mathbb{E}[\mathcal{J}(\Psi_t^x(w_{t:0}), \Psi_t^u(w_{t:0}))] \tag{3.20a}$$

$$s.t. \qquad \Psi_t^x(w_{t:0}) = \Psi_t^x(0, w_{t-1:0}) + w_t \tag{3.20b}$$

$$\Psi_{t+1}^x(0, w_{t:0}) = A\Psi_t^x(w_{t:0}) + B\Psi_t^u(w_{t:0})$$

$$\forall t, |w_t| \leqslant \eta_{max} : \qquad |\Psi_t^x(w_{t:0})| \leqslant x_{max} \tag{3.20c}$$

$$\forall t, |w_t| \leqslant \eta_{max} : \qquad |\Psi_t^u(w_{t:0})| \leqslant u_{max}. \tag{3.20d}$$

As in the linear SLS case [11], we do not need to have the controller K be a decision variable, since we can always realize the optimal solution $(\Psi^{(x)*}, \Psi^{(u)*})$ to (3.20) with a system level controller $\mathrm{SL}[\Psi^{(x)*}, \Psi^{(u)*}]$.

Conservativeness of Linear Solutions

We will first discuss properties of solutions to our original problem (3.19), if we restrict ourselves to only LTI controllers K. Consider the equivalent problem formulation (3.20) with the CLMs $(\Psi^{\mathrm{x}}, \Psi^{\mathrm{u}})$ restricted to being linear. This poses a convex problem and, as shown in [36], it can be approximately solved by searching for FIR CLMs $(\Psi^{\mathrm{x}}, \Psi^{\mathrm{u}})$ with enough large horizon T. However, the corresponding linear optimal CLMs $(\Psi^{\mathrm{x,lin}*}, \Psi^{\mathrm{u,lin}*})$ come with undesirable restrictions:

- $(\Psi^{\mathrm{x,lin}*}, \Psi^{\mathrm{u,lin}*})$ impose stricter safety restrictions than the required restrictions (3.20c) and (3.20d).

- $(\Psi^{\mathrm{x,lin}*}, \Psi^{\mathrm{u,lin}*})$ do not depend on the disturbance distribution $p(w)$.

To see the first point, we have the following result as a consequence of linearity:

Lemma 12. *For any linear* $(\Psi^{\mathrm{x,lin}}, \Psi^{\mathrm{u,lin}})$, *the constraint* (3.20c),(3.20d) *is equivalent to*

$$\sup_t |\Psi_t^{x,lin}(w_{t:0})| \leq \sup_t \frac{x_{max}}{\eta_{max}} |w_t| \tag{3.21a}$$

$$\sup_t |\Psi_t^{u,lin}(w_{t:0})| \leq \sup_t \frac{u_{max}}{\eta_{max}} |w_t|. \tag{3.21b}$$

Proof. Clearly, (3.21) implies (3.20c),(3.20d). The reverse implication follows from the assumed linearity of $(\Psi^{\mathrm{x,lin}}, \Psi^{\mathrm{u,lin}})$ and the homogeneity of the norms. $\square$

Lem. 12 shows that the linearity restriction in CLMs imposes stricter safety conditions (3.21) than (3.20c),(3.20d). To elaborate on the second point, notice that for linear CLMs $(\Psi^{\mathrm{x,lin}}, \Psi^{\mathrm{u,lin}})$, the objective function (3.20a) can be expressed equivalently as

$$(3.20a) = \sigma^2 \left\| \begin{matrix} Q^{1/2}\Psi^{\mathrm{x}} \\ P^{1/2}\Psi^{\mathrm{u}} \end{matrix} \right\|_{\mathcal{H}_2}^2, \quad \sigma^2 := \mathbb{E}_{w \sim p(w)}[w^2] \tag{3.22}$$

where σ^2 denotes the variance of the scalar distribution $p(w)$ and $\|.\|_{\mathcal{H}_2}$ denotes the $\mathcal{H}_2$ norm for linear operators. Since the objective function only gets scaled by a constant factor σ^2 for different distributions $p(w)$, this shows that for linear CLMs, the solutions $(\Psi^{\mathrm{x,lin}}, \Psi^{\mathrm{u,lin}})$ to (3.20) are independent of the distribution $p(w)$.

Taking the Nonlinear System Level Approach

Now we extend the search space of the problem (3.20), by allowing candidate CLMs Ψ that are blends of linear CLMs of F.

We structure the search space of blended cCLMs as follows. Let $\{\Psi_i\}_{i=1}^N$ be a collection of linear-time invariant cCLMs of finite horizon T. Correspondingly, we parametrize their component functions of in terms of the matrices $R_k^{(i)} \in \mathbb{R}^{n \times n}$ and $M_{t,k}^{(i)} \in \mathbb{R}^{n \times m}$ as follows:

$$\Psi_{i,t}^{\mathrm{x}}(z_{1:t+1}) = \sum_{k=1}^{\min\{T,t+1\}} R_k^{(i)} z_k \qquad \Psi_{i,t}^{\mathrm{u}}(z_{1:t+1}) = \sum_{k=1}^{\min\{T,t+1\}} M_k^{(i)} z_k. \tag{3.23}$$

We assume the saturation weighting of the N zone mentioned in Def. 2.12. For fixed $\eta_{N-1} \geqslant \cdots \geqslant \eta_1$, $P_{\eta_i} : \mathbb{R}^n \to \mathbb{R}^n$ denotes functions that parameterize the weights $G_i := I \otimes g_i$ where $g_0 = P_{\eta_0}$, $g_N(x) := x - P_{\eta_{N-1}}(x)$ and g_i for $i \in \{1, \ldots, N-1\}$ are defined as $g_i = P_{\eta_i} - P_{\eta_{i-1}}$. For P_η and we consider two specific classes functions, both representing different types of projection maps:

Definition 3.1 (Saturation Projection). *Let vector* $w = [w^1, \ldots, w^n]^T \in \mathbb{R}^n$. *The saturation projection is an element-wise projection:*

$$P_\eta(w) := \begin{bmatrix} \mathrm{sat}(w^1, \eta) \\ \vdots \\ \mathrm{sat}(w^n, \eta) \end{bmatrix} \tag{3.24}$$

where $\mathrm{sat}(w, \eta) = sign(w) \max\{|w|, \eta\}$.

Definition 3.2 (Radial Projection). *The radial projection is defined as:*

$$P_\eta(w) := \frac{\mathrm{sat}(|w|/\eta, 1)}{|w|/\eta} w. \tag{3.25}$$

Unless otherwise specified, the results derived in the rest of the chapter hold for both projections.

Remark 19. *For* $n = 1$, *the radial projection and the saturation projection coincide with each other. The radial and saturation projection operator act as the identity whenever* $|w| \leqslant \eta$. *Otherwise, the radial projection rescales* w *so that* $|P_\eta(w)| = \eta$ *whereas the saturation projection performs the elemental radial projection.*

This completes our setup and we arrive at the structure of our candidate CLMs as:

$$\Psi_t^x(\cdot) = \sum_{i=1}^{N} \sum_{k=1}^{\min\{T,t+1\}} R_k^{(i)}(P_{\eta_i} - P_{\eta_{i-1}})(w_{t+1-k})$$

$$\Psi_t^u(\cdot) = \sum_{i=1}^{N} \sum_{k=1}^{\min\{T,t+1\}} M_k^{(i)}(P_{\eta_i} - P_{\eta_{i-1}})(w_{t+1-k}). \tag{3.26}$$

The corresponding blended SL implementation of Ψ_Σ takes the form:

$$u_t = \sum_{i=1}^{N} \sum_{k=1}^{\min\{T,t+1\}} M_k^{(i)} \tilde{w}_{t+1-k}^i$$

$$\tilde{w}_t^{(i)} = g_i(\hat{w}_t)$$

$$\hat{w}_{t+1} = x_{t+1} - \sum_{i=1}^{N} \sum_{k=2}^{\min\{T,t+2\}} R_k^{(i)} \tilde{w}_{t+2-k}^i.$$

With regards to our optimization problem, we enforce the (linear) constraint that $\Psi^{x,i}, \Psi^{u,i}$, $i \in [N]$ are CLMs of the linear system of interests:

$$x_t = Ax_{t-1} + Bu_{t-1} + w_t, \tag{3.27}$$

with $x_t \in \mathbb{R}^n$, $w_t \in \mathbb{R}^n$, $u \in \mathbb{R}^m$.

The overall nonlinear controller $\mathrm{SL}[\Psi^x, \Psi^u]$ can be thought of as a *nonlinear blend* of the *linear* FIR controllers $\mathrm{SL}[\Psi^{x,i}, \Psi^{u,i}]$, $i \in [N]$. Although the nonlinear operator Ψ^x, Ψ^u differs from its linear components $\Psi^{x,i}$, $\Psi^{u,i}$ only by the static nonlinear function $P_{\eta_i}(w)$, the upcoming sections will demonstrate that this simple additional nonlinearity proves surprisingly useful. In particular, η_i's separate any disturbance w_t into N zones such that for each ith linear controller $\mathrm{SL}[\Psi^{x,i}, \Psi^{u,i}]$, only the portion of w_t that "falls" between η_i and η_{i-1} is acted upon. Intuitively, one could choose different behaviors for various portions of the disturbance signal, specifying either performance or safety properties.

For ease of exposition, we focus on the two-zone case of the proposed controller $\mathrm{SL}[\Psi^x, \Psi^u]$ although all the analysis naturally extends to the N-zone case.

Consider the general problem (3.20), where we now search for CLMs (Ψ^x, Ψ^u) of the form presented in (3.26) with $N = 3$ and the choice of $\eta_2 = \eta_{max}$, with some $\eta_1 < \eta_2$. However, since we have the assumption $\|w\| \leq \eta_2$, $R_j^{(3)}$ and $M_j^{(3)}$ drop out of the objective and safety constraint and can be chosen as arbitrary FIR CLMs for now. In the later section we will discuss that this extra degree of freedom can be used for anti-windup control.

Restricting ourselves to this form of CLM leads to the convex problem (3.20), which is a relaxation of the general problem (3.20).

$$\min_{R^{(i)},M^{(i)}} \left\| \begin{bmatrix} Q & 0 \\ 0 & P \end{bmatrix}^{1/2} \begin{bmatrix} R^{(1)} & R^{(2)} \\ M^{(1)} & M^{(2)} \end{bmatrix} \Sigma_w^{1/2} \right\|_F^2 \tag{3.28a}$$

$$s.t. \quad \eta_1 |R^{(1)}| + (\eta_2 - \eta_1)|R^{(2)}| \leq x_{max} \tag{3.28b}$$

$$\eta_1 |M^{(1)}| + (\eta_2 - \eta_1)|M^{(2)}| \leq u_{max} \tag{3.28c}$$

$$R_{k+1}^{(i)} = AR_k^{(i)} + BM_k^{(i)} \tag{3.28d}$$

$$R_1^{(i)} = I, \quad R_T^{(i)} = 0$$

where

$$\Sigma_w = \begin{bmatrix} \alpha_1 I & \alpha_2 I \\ \alpha_2 I & \alpha_3 I \end{bmatrix}$$

with $\alpha_1 = \mathbb{E}[P_{\eta_1}(w)^2]$, $\alpha_2 = \mathbb{E}[P_{\eta_1}(w)(P_{\eta_2}(w) - P_{\eta_1}(w))]$, and $\alpha_3 = \mathbb{E}[(P_{\eta_2}(w) - P_{\eta_1}(w))^2]$, where $w \sim p(w)$ and $\|w\|_\infty \leq \eta_{max}$. Moreover $R^{(i)}$ and $M^{(i)}$ are abbreviations for the row-wise concatenation of the matrices associated with the linear CLMs $\Psi^{x,i}$, $\Psi^{u,i}$, $i.e$, $R^{(i)} = [R_T^{(i)}, R_{T-1}^{(i)}, \ldots, R_1^{(i)}]$, $M^{(i)} = [M_T^{(i)}, M_{T-1}^{(i)}, \ldots, M_1^{(i)}]$. Therefore, only constraints (3.28b), (3.28c) are sufficient conditions of the constraint (3.20c), (3.20d) through the multiplicativity of the norm. All other equations in the above optimization are equivalent to the original problem (3.20). Finally, solving the convex problem (3.28) gives the suboptimal nonlinear CLMs (Ψ^{*x}, Ψ^{*u}) for the system dynamics (3.19b), realized by an internally stabilizing controller $SL[\Psi^{*x}, \Psi^{*u}]$. The next theorem states a main result of this chapter:

Theorem 20. *For all $\eta_1 \in [0, \eta_2]$, the nonlinear system level controller* $SL[\Psi^{*x}, \Psi^{*u}]$ *synthesized from (3.28) achieves lower optimal LQR cost for (3.19) than any linear solutions.*

Proof. First, recall that restricting K to be linear in the problem (3.19) is equivalent to restricting Ψ^x and Ψ^u to be linear in the equivalent formulation (3.20). Furthermore, notice that under the restriction of linear (Ψ^x, Ψ^u), the problem (3.20) is equivalent to (3.28) with the added constraint $R^{(1)} = R^{(2)}$, $M^{(1)} = M^{(2)}$, which shows that any solution (Ψ^{*x}, Ψ^{*u}) of the problem (3.28) achieves a lower cost than a linear solution ($\Psi^{x,lin*}$, $\Psi^{u,lin*}$) of (3.20). $\square$

Remark 21. *This argument extends directly to the N-blend case.*

Localized Controller for Constrained LQR

Thanks to the particular form of (3.26), when the projection is chosen to be the saturation projection Def. 3.1, structural constraints of the controller, such as the sparsity and delay constraints, can be added in a convex way to the synthesis procedure described in Section 3.3. This is because imposing structural constraints on the nonlinear controller (3.26) is equivalent to imposing them on the linear CLM components of (3.26). Detailed in [11], localization of disturbance, communication, and actuation delay, as well as sparsity pattern, are all convex constraints in terms of linear CLMs in the linear System Level Synthesis framework. Specifically, all the constraints mentioned could be cast as a convex subspace $\mathcal{S}_x$ and $\mathcal{S}_u$ for linear CLMs $\Psi^{\mathrm{x,i}}, \Psi^{\mathrm{u,i}}, i \in [N]$. The corresponding system-level controller $\mathrm{SL}[\boldsymbol{T}^{(u)}, \boldsymbol{T}^{(x)}]$ can then be implemented in a localized fashion conforming to the subspace constraints on $\Psi^{\mathrm{x,i}}, \Psi^{\mathrm{u,i}}$. Therefore, the nonlinear controller synthesis in Section 3.3 naturally inherits all capabilities of the linear system level controllers in terms of distributed controller synthesis and implementation.

Simulation

To corroborate the results presented in the previous sections, we demonstrate the performance of a four-zone nonlinear blending controller with radial projection compared against the optimal linear controller for the constrained LQR problem of an open-loop unstable system:

$$
x_t = \begin{bmatrix} 1 & 1 & 0 \\ 1 & 2 & 1 \\ 0 & 1 & 1 \end{bmatrix} x_{t-1} + \begin{bmatrix} 0 \\ 0 \\ 1 \end{bmatrix} u_{t-1} + w_t \tag{3.29}
$$

with $u_{max} = 40$, $x_{max} = 15$, $\eta_{max} = 1$, $Q = I_3$, $P = 10$. The disturbances w_k are chosen to be a truncated i.i.d. Gaussian random variable with variance σ^2. Figure 3.2 shows the optimal cost improvement of the presented nonlinear approach over the optimal linear controller for different choices of variance σ^2. Figure 3.2 showcases that the proposed controller can exploit the knowledge of the disturbance distribution to achieve performance improvement over the linear optimal linear controller: For small σ the proposed controller gains more than 30%.

3.5 Distributed Anti-Windup Controller for Saturated Systems

In the constrained LQR problem, we considered a linear system $x_t = A x_{t-1} + B u_{t-1} + w_t$ and designed a controller that ensures that the closed-loop system state x_t and input u_t remained within the specified bounded sets $\mathcal{X} = \{x \mid |x| \leqslant x_{max}\}$ and

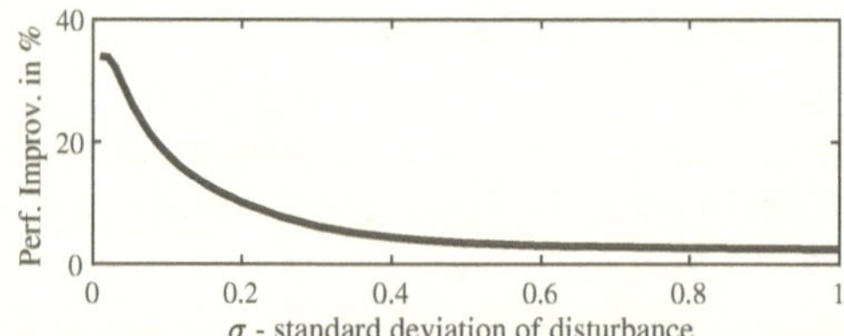

Figure 3.2: Performance improvement of nonlinear controller $\mathrm{SL}[\boldsymbol{\Psi}^{*\mathrm{x}}, \boldsymbol{\Psi}^{*\mathrm{u}}]$ over optimal linear controller $\mathrm{SL}[\boldsymbol{\Psi}^{\mathrm{x,lin}*}, \boldsymbol{\Psi}^{\mathrm{u,lin}*}]$ for different variances σ^2 of the non-truncated disturbance. The nonlinear blending controller synthesizes over 4 linear controllers w.r.t. to the projection parameters $\eta_1 = 0.05, \eta_2 = 0.1, \eta_3 = 0.2, \eta_4 = \eta_{max} = 1$

$\mathcal{U} = \{u \mid |u| \leq u_{max}\}$, for any bounded sequence $w \in \ell^\infty$, $w_t \in \mathcal{W}$, $\mathcal{W} = \{w \mid |w| \leq \eta_{max}\}$. This partial problem is one of robust set invariance [96]. In this section, we discuss this problem in a broader context, where input constraints are not part of the problem specification, but rather enforced by a saturation nonlinearity $\mathrm{sat}_{\mathcal{U}}$ in our input. Hence, we consider our system to be nonlinear and of the form:

$$H': \quad x_t = Ax_{t-1} + B\mathrm{sat}_{\mathcal{U}}(u_{t-1}) + w_t. \tag{3.30}$$

If we can be sure that the disturbances of the system and the initial condition remain within the specified bounds $\mathcal{W}$, then solving the robust invariance problem, for example, with one of the techniques from the previous chapter and the general SLS framework [36], is sufficient. However, in practice, it is more realistic to consider that our assumptions are only *mostly* true and it is possible that our assumptions are temporarily violated on rare occasions. In such a scenario, we have to accept some degradation in our guarantees, such as state constraints and performance bounds. However, we would like this degradation to happen gracefully and at least ensure that some basic properties of the closed-loop such as stability are still preserved.

In this section, we discuss how control design through blended CLMs provides a new perspective on this topic. In particular, blending with N-zone saturations naturally ensures graceful degradation in the presence of saturation and offers an elegant solution to the issue of controller wind-up [70, 78, 81], an important problem that commonly arises when input saturation is ignored during linear control design.

The main results of this section show that with the appropriate choice of blending and weight operators, closed-loop stability and convergence to the target set $\mathcal{X}$ are guaranteed even in the saturated regime, provided that disturbance w violates our assumption "occasionally"; mathematically, we phrase this as $w = w_1 + w_2$, where $\|w_1\|_\infty \leq \eta_{max}$ and $w_2 \in \ell_p$.

Setup of Candidate CLMs for Unsaturated Regime

First, we consider the requirements of the unsaturated regime and assume that it is a feasible control problem. That is, for the given tuple of sets $(\mathcal{X}, \mathcal{U}, \mathcal{W})$, we can design a linear controller such that for any disturbance sequence w in $\mathcal{W}$, (that is, $w_t \in \mathcal{W}$ for all t), the state x_t is always guaranteed to stay within the set $\mathcal{X}$ if $x_0 \in \mathcal{X}$. As shown in the previous section, we can pose this as a feasibility problem of convex constraints. Another general approach to address this problem in the context of SLS has been presented in [36], which allows polytopic sets $\mathcal{X}, \mathcal{U}$, and $\mathcal{W}$. We assume that the control design for the unsaturated regime is feasible and we formulate this in terms of cCLMs Ψ next.

We define the operators $F \in \mathcal{LC}_s(\ell^n \times \ell^m, \ell^n)$ and $F' \in \mathcal{C}_s(\ell^n \times \ell^m, \ell^n)$ to distinguish the dynamics of the linear system with and without actuator saturation. Let $F(x, u) = A(x) + B(u)$, where $A = I \otimes A$ and $B = I \otimes B$. Let $F'(x, u) = A(x) + B_\mathcal{U}(u)$, where $B_\mathcal{U} := B \circ (I \otimes \mathrm{sat}_\mathcal{U})$. In general, a saturation nonlinearity sat is characterized by the properties described in (3.3), that is, as long as the input is in the set $\mathcal{U}$, the nonlinearity has no effect, and the system behaves linearly. However, we will focus our discussion on projection-based saturations $\Pi_\mathcal{U}$ as shown in the example below.

Definition 3.3. *Given some closed bounded convex set $\mathcal{U}$ with $0 \in \mathcal{U}$, a saturation function is a Lipshitz continuous map $\mathrm{sat}_\mathcal{U} : \mathbb{R}^m \to \mathcal{U}$ onto $\mathcal{U}$, which satisfies*

$$\forall u \in \mathbb{R}^m : \ \mathrm{sat}_\mathcal{U}(\mathrm{sat}_\mathcal{U}(u)) = \mathrm{sat}_\mathcal{U}(u) \ and \ \forall u \in \mathcal{U} : \ \mathrm{sat}_\mathcal{U}(u) = u.$$

Example. *Let $\mathcal{U}$ be some convex body in a normed vector space $(| \ |, \mathbb{R}^m)$ containing the origin in its interior. Then, the map $\Pi_\mathcal{U} : u \mapsto \arg\min_{u' \in \mathcal{U}} |u - u'|$ is a saturation.*

For our blended cCLM $\Psi_\Sigma = \sum_{i=1}^N \Psi_\Sigma G_i$ we consider a collection of linear cCLM $\{\Psi_i\}_{i=1}^N$ and weight operators $\{G_i\}_{i=1}^N$ constructed from saturation functions as described in Def. 2.12.

Let $\mathcal{W}_1 \subset \mathcal{W}_2 \cdots \subset \mathcal{W}_{N-1}$ be a collection of nested convex sets and pick $\mathcal{W}_{N-1} := \mathcal{W}$. Denote $\Pi_{\mathcal{W}_k} : x \mapsto \arg\min_{u \in \mathcal{W}_k} |x - u|, \forall k \in \{1, \ldots, N-1\}$ as the projection maps onto the sets $\mathcal{W}_k$. Define the weights as $G_k = I \otimes g_k$, where $g_k = \Pi_{\mathcal{W}_k} - \Pi_{\mathcal{W}_{k-1}}$ for all $k \in \{2, \ldots, N-1\}$ and $g_0 = \Pi_{\mathcal{W}_1}, g_N : x \mapsto x - \Pi_{\mathcal{W}_{N-1}}(x)$.

We purposely picked the last weight operator such that $G_N = I \otimes (I - \Pi_W)$, hence in the unsaturated regime, the last cCLM Ψ_N does not contribute to the closed-loop state and input behavior. This is due to the fact that for any sequence $w \subset W$, it holds that

$$w \subset W: \implies w = \sum_{i=1}^{N} G_i(w) = \sum_{i=1}^{N-1} G_i(w)$$

and therefore

$$w \subset W: \implies \Psi_\Sigma w = \sum_{i=1}^{N-1} \Psi_i G_i(w).$$

We assume that $\{\Psi_i\}_{i=1}^{N-1}$ have been designed to satisfy the desired notion of robust set invariance in the unsaturated regime:

Assumption 3.6. $\{\Psi_i\}_{i=1}^{N-1}$ *are CLMs of the linear system F, and for all $w \in W$, the following holds:*

$$\sum_{i=1}^{N-1} \Psi_i^{\mathsf{x}} G_i(w) \in \mathcal{X} \qquad \sum_{i=1}^{N-1} \Psi_i^{\mathsf{u}} G_i(w) \in \mathcal{U}. \tag{3.31}$$

For notational convenience, we split the blended CLM into two parts. Let $\left[\begin{smallmatrix} R \\ M \end{smallmatrix}\right]$ stand for

$$R := \sum_{i=1}^{N-1} \Psi_i^{\mathsf{x}} G_i \qquad\qquad M := \sum_{i=1}^{N-1} \Psi_i^{\mathsf{u}} G_i$$

so we can write:

$$\Psi_\Sigma^{x} = R + \Psi_N^{\mathsf{x}} G_N \qquad\qquad \Psi_\Sigma^{u} = M + \Psi_N^{\mathsf{u}} G_N.$$

It is clear from this setup that for sequences $w \in W$, the map $\left[\begin{smallmatrix} R \\ M \end{smallmatrix}\right]$ is a CLM for the linear system F, and *if* we restricted the domain of the CLMs to sequences in W, it would also be a CLM of the nonlinear system F'.

Design Approach for Stability in the Saturated Regime

We now consider the case where the disturbance leaves the set W occasionally, leading to actuator saturation. It is commonly known that graceful performance degradation cannot be taken for granted, as instability phenomena like the "wind-up" effect can occur if the controller synthesis improperly deals with actuator saturation.

For stable system matrices A, a modification based on the IMC principle of the controller is shown in [36], which guarantees closed-loop stability when $w \notin \mathcal{W}$.

In the case of our blended SL controller $\Psi_\Sigma^{\mathrm{u}}(\Psi_\Sigma^x)^{-1}$, it turns out that stability in the saturated regime is ensured if we make two simple modifications. First, we require $\Psi_N^{\mathrm{u}} = 0$, and Ψ_N^x to be an LTI FIR cCLM with the linear components

$$\Psi_{N,t}^x(w_{t:0}) = \sum_{k=0}^{\min\{T-1,t\}} A^k w_{t-k} \qquad\qquad \Psi_{N,t}^{\mathrm{u}} = 0, \qquad (3.32)$$

for some finite horizon $T \geq 1$. Second, we prepend the projection to $\Pi_\mathcal{W}$ to R and M.

Next, we show that with the above rule, ℓ_p stability of the nonlinear closed loop is guaranteed. Moreover, global stability results (for stable A) and local stability results (for unstable A), along with the corresponding transient bounds, can be derived for the closed loop. The convergence to $\mathcal{X}$ in finite time is shown for ℓ_p perturbations with $p < \infty$.

Dynamics of Lumped Disturbances in the Saturated Regime

The basis for our stability analysis is the provided by the following lemma:

Lemma 13. *Given assumption (3.6) and assuming Ψ_N satisfies (3.32), then for $\Psi_\Sigma' = \left[\begin{smallmatrix} R \\ M \end{smallmatrix}\right]\Pi_\mathcal{W} + \Psi_N G_N$ holds:*

$$\Delta[F', \Psi_\Sigma'] = \Delta[F, \Psi_N]G_N.$$

Proof.

$$\begin{aligned}
\Delta[F', \Psi_\Sigma] &= F'^+\Psi_\Sigma' + I - \Psi_\Sigma'^x \\
&= A^+ R\Pi_\mathcal{W} + A^+\Psi_N^x G_N + B_\mathcal{U}^+(M\Pi_\mathcal{W} + \Psi_N^{\mathrm{u}} G_N) \\
&\quad \cdots + I - R\Pi_\mathcal{W} - \Psi_N^x G_N.
\end{aligned}$$

Now since, $Mw \in \mathcal{U}$ for all $w \in \mathcal{W}$ and $\Pi_\mathcal{W} w \in \mathcal{W}, \forall w$, we can conclude $M\Pi_\mathcal{W} w \in \mathcal{U}$ for *all* w. Furthermore, since $\Psi_N^{\mathrm{u}} = 0$, we have $\Psi_N^{\mathrm{u}} G_N = 0$. It is therefore trivial to rewrite the term $B_\mathcal{U}^+(M\Pi_\mathcal{W} + \Psi_N^{\mathrm{u}} G_N)$ as:

$$B_\mathcal{U}^+(M\Pi_\mathcal{W} + \Psi_N^{\mathrm{u}} G_N) = B_\mathcal{U}^+(M\Pi_\mathcal{W}) = B^+(M\Pi_\mathcal{W}) + B^+\Psi_N^{\mathrm{u}} G_N.$$

Moreover, using the fact that $I = G_N + \Pi_\mathcal{W}$ we substitute in the above equation and yield the decomposition:

$$\Delta[F', \Psi_\Sigma'] = (A^+ R + B^+ M - R + I)\Pi_\mathcal{W} + \Delta[F, \Psi_N]G_N.$$

Since, the first term equals $\Delta[F, [{}^R_M]]\Pi_\mathcal{W}$ and the $[{}^R_M]$ are CLMs of the linear system F we have $\Delta[F, [{}^R_M]]\Pi_\mathcal{W} = 0$. This leaves us with the desired result: $\Delta[F', \Psi'_\Sigma] = \Delta[F, \Psi_N]G_N$. $\qquad\qquad\square$

Applying Theorem 11 from the previous chapter, we obtain the dynamics of the lumped disturbance as:

$$\hat{w} = \Delta[F, \Psi_N]G_N\hat{w} + w.$$

We analyze the stability of these dynamics by rewriting them as the difference equation:

$$\hat{w}_t = \begin{cases} A^T(\hat{w}_{t-T} - \Pi_\mathcal{W}(\hat{w}_{t-T})) + w_t & \text{for } t \geq T \\ w_t & \text{else} . \end{cases} \tag{3.33}$$

Here is a quick derivation for clarity: $\Psi_N^{\mathrm{u}} = 0$, means $\Delta[F, \Psi_N] = A^+\Psi_N^{\mathrm{x}} + I - \Psi_N^{\mathrm{x}}$. Define the auxiliary variables $\tilde{w}^N = G_N\hat{w}$, $\alpha = \Psi_N^{\mathrm{x}}\tilde{w}^N$ $\beta = A^+\Psi_N^{\mathrm{x}}\tilde{w}^N$, $\gamma = (I - \Psi_N^{\mathrm{x}})\tilde{w}^N$ and following calculations to compute $\beta + \gamma$

$$
\begin{aligned}
\hat{w}_t &= \beta_t + \gamma_t + w_t \\
\alpha_t &= \sum_{k=0}^{\min\{T-1,t\}} A^k \tilde{w}^N_{t-k} \\
\beta_t &= A\alpha_{t-1} \\
\gamma_t &= \sum_{k=1}^{\min\{T-1,t\}} -A^k \tilde{w}^N_{t-k}
\end{aligned}
\quad \Leftrightarrow \quad
\begin{aligned}
\alpha_t &= \sum_{k=1}^{\min\{T,t+1\}} A^{k-1} \tilde{w}^N_{t+1-k} \\
\beta_t &= \sum_{k=1}^{\min\{T,t\}} A^k \tilde{w}^N_{t-k} \\
\beta_t + \gamma_t &= A^T \tilde{w}^N_{t-T}, \text{ if } t \geq T, \text{ else } 0 \\
\hat{w}_t &= A^T \tilde{w}^N_{t-T} + w_t, \text{ if } t \geq T, \text{ else } w_t.
\end{aligned}
$$

Finally, substituting $\tilde{w}^N_t = \hat{w}_t - \Pi_\mathcal{W}(\hat{w}_t)$ into the last expression on the right yields the difference equation (3.33).

Stability and Convergence in Saturated Regime

Next, we analyze the stability of the dynamic system (3.33) and its implications for the overall closed-loop. We show that we can always choose T to achieve closed-loop stability. The choice of T depends on the matrix A and the norm used in projection $\Pi_\mathcal{W}$.

From equation (3.33), it is easy to see that we can decompose the sequence $\hat{w}$ and w into T subsequences $\hat{\omega}^{[i]}_k := \hat{w}_{i+kT}$, $i \in \{0, \ldots, T-1\}$, $\omega^{[i]}_k := w_{i+kT}$, whose dynamics can be analyzed entirely separately. In fact, for each i and k holds

$$\hat{\omega}^{[i]}_k = A^T(\hat{\omega}^{[i]}_{k-1} - \Pi_\mathcal{W}(\hat{\omega}^{[i]}_{k-1})) + \omega^{[i]}_k, \text{ for all } t \geq 1 \text{ and } \hat{\omega}^{[i]}_0 = \omega^{[i]}_0. \tag{3.34}$$

Notice that f.g. ℓ_p/ℓ_∞-stability of the above system is sufficient for fg. ℓ_p/ℓ_∞-stability of the original system (3.33), since:

$$\|\hat{\boldsymbol{\omega}}^{[i]}\|_p \leqslant \gamma\|\boldsymbol{\omega}^{[i]}\|_p \quad \text{for all } i \in \{0,\dots,T-1\}$$

$$\implies \|\hat{\boldsymbol{w}}\|_p = \sqrt[p]{\sum_{i=0}^{T-1}\|\hat{\omega}^{[i]}\|_p^p} \leqslant \sqrt[p]{\sum_{i=0}^{T-1}\gamma^p\|\hat{\omega}^{[i]}\|_p^p} = \gamma\sqrt[p]{\sum_{i=0}^{T-1}\|\hat{\omega}^{[i]}\|_p^p} = \gamma\|\boldsymbol{w}\|_p.$$

Hence, we drop the index i and analyze the auxiliary system

$$\hat{\omega}_{k+1} = A^T(\hat{\omega}_k - \Pi_{\mathcal{W}}(\hat{\omega}_k)) + \omega_k.$$

For the next lemma, define $\mathcal{B}_\eta := \{w|\ |w| < \eta\}$ as the ball of radius η corresponding to the norm $|\cdot|$ used in the definition of the projection $\Pi_{\mathcal{W}}$. The next result proves the conditions for closed-loop stability dependent on the design parameter T.

Lemma 14. *Define $\bar{\eta} := \sup\{\eta|\ \mathcal{B}_\eta \subset \mathcal{W}\}$ and define $|A| := \sup_{|x|=1}|Ax|$. If $\bar{\eta} > 0$ and assuming ℓ_p, ℓ_∞ are formulated also w.r.t. the norm $|\cdot|$, then:*

1. *If $|A^T| < 1$, then the following bound holds for all $\boldsymbol{w}$:*

$$\|\hat{\boldsymbol{w}}\|_p \leqslant \frac{1}{1-|A^T|}\|\boldsymbol{w}\|_p.$$

2. *For any $0 \leqslant \gamma < \min\{1,|A^T|\}$ holds:*

$$\|\boldsymbol{w}\|_p \leqslant (1-\gamma)\frac{|A^T|\bar{\eta}}{|A^T|-\gamma} \quad \implies \quad \|\hat{\boldsymbol{w}}\|_p \leqslant \frac{1}{1-\gamma}\|\boldsymbol{w}\|_p.$$

Proof. First observe the following property of the projection map $\Pi_{\mathcal{W}}$:

$$
\begin{aligned}
|\Pi_{\mathcal{W}}(w) - w| = \min \quad & |w'| & \leqslant \min \quad & |w'| \\
\text{s.t.} \quad & w + w' \in \mathcal{W} & \text{s.t.} \quad & |w+w'| < \bar{\eta} \\
\overset{a)}{\leqslant} \min_{t \geqslant 0} \quad & t|w| & = \max&\{0, |w|-\bar{\eta}\} \\
\text{s.t.} \quad & (1-t)|w| < \bar{\eta} & &
\end{aligned}
$$

where inequality $a)$ follows by restricting the search to the line $w' = tw, t \geqslant 0$. This shows that $|\Pi_{\mathcal{W}}(w) - w| < |w|$ for all $w \in \mathbb{R}^n$. Hence, it follows that for any ω holds:

$$|A^T(\hat{\omega} - \Pi_{\mathcal{W}}(\hat{\omega}))| \leqslant |A^T|\max\{0, |\omega|-\bar{\eta}\} \leqslant \max\{0, |A^T||\omega|-|A^T|\bar{\eta}\}.$$

This proves the first statement. On the other hand, for any $\gamma < \min\{1, |A^T|\}$, the following local small-gain property holds:

$$\|\boldsymbol{a}\|_p < \frac{|A^T|\bar{\eta}}{|A^T| - \gamma} \quad \Longrightarrow \quad \|(\boldsymbol{I} \otimes A^T(\mathrm{id} - \Pi_\mathcal{W}))\boldsymbol{a}\|_p \leqslant \gamma\|\boldsymbol{a}\|_p \tag{3.35}$$

where $\mathrm{id} : x \mapsto x$. Lastly, the desired local and global results follow by direct application of the small gain result lemma 7. $\qquad\qquad\square$

Lemma 14 formulates stability conditions in terms of T and a requirement of $|A^T| < 1$ to ensure global stability. Due to Gelfand's theorem, a standard result from Linear Algebra, we are guaranteed that as long as A is Schur, it is always possible to pick a T large enough such that $|A^T| < 1$, however T might depend on the particular norm $|\cdot|$ we chose in the projection $\Pi_\mathcal{W}$.

Lemma 15 (Gelfand's Theorem). *Denote by $\rho(A)$ the spectral radius (max absolute value of eigenvalue) of $A \in \mathbb{R}^{n\times n}$, then for any matrix norm $|.|$ holds* $\lim_{k\to\infty}(|A^k|)^{1/k} = \rho(A)$.

The lemma 14 tells us that if A is Schur, we can always choose T such that $|A^T| < 1$, then we are guaranteed f.g. ℓ_p stability for the lumped disturbances $\hat{w}$ and by Theorem 11, this shows overall closed-loop stability for our system since $\boldsymbol{F}$ is trivially Lipschitz and therefore i.f.g. ℓ_p stable everywhere. Furthermore, even in the case where $|A^T| > 1$, the second result offers at least local ℓ_p-stability for the closed-loop. Hence, we have stability for small disturbances in the saturated regime, even if the system is open-loop unstable.

As a corollary of the above result, we can prove that as long as our disturbance only exits the set $\mathcal{W}$ a finite number of times, then we are also guaranteed to violate the state constraint $\mathcal{X}$ a finite number of times. Our strategy is to show that there exists a time t' for which x_t is guaranteed to stay in $\mathcal{X}$ for all time $t > t'$ if w is composed of $w_1 + w_2$ where $\|w_1\|_\infty < \bar{\eta}$ and $w_2 \in \ell_p$.

Corollary 16. *Assume $\bar{\eta} > 0$ and $|A^T| \leqslant 1$ as used in Lemma 14. If $w = w_1 + w_2$ where $\|w_1\|_\infty \leqslant \bar{\eta} - \varepsilon$ for some $\varepsilon > 0$ and $w_2 \in \ell_p$, then there exists a time t' such that for all $t > t'$: $x_t \in \mathcal{X}$.*

Proof. Recall the relationship $\hat{w} = (\mathbf{\Psi}^\times)^{-1}x \Leftrightarrow \mathbf{\Psi}^\times \hat{w} = x$ and $\mathbf{\Psi}^x_\Sigma = R(I - G_N) + \mathbf{\Psi}^\times_N(G_N)$ to decompose x into the terms s and s':

$$x = \underbrace{R\Pi_{\mathcal{W}}(\hat{w})}_{=:s} + \underbrace{\mathbf{\Psi}^\times_N(\hat{w} - \Pi_{\mathcal{W}}(\hat{w}))}_{=:s'}. \tag{3.36}$$

Since $Rw \in \mathcal{X}$ for all $w \in \mathcal{W}$ and $\Pi_{\mathcal{W}}$ maps onto $\mathcal{W}$, it is clear that $s = R\Pi_{\mathcal{W}}(\hat{w}) \in \mathcal{X}$ for all $\hat{w}$. Denote $(s)_+ := \max\{0, s\}$ and retrace the proof of Lem. 14, to arrive at the inequality:

$$|\hat{w}_t| \leq (|\hat{w}_{t-1}| - \bar{\eta})_+ + |w_t|.$$

From the above, we obtain the following inequalities

$$|\hat{w}_t| \leq (|\hat{w}_{t-1}| - \bar{\eta})_+ + |w_t|$$

$$\Leftrightarrow \quad |\hat{w}_t| - \bar{\eta} \leq (|\hat{w}_{t-1}| - \bar{\eta})_+ + |w_t| - \bar{\eta}$$

$$\Leftrightarrow \quad \leq (|\hat{w}_{t-1}| - \bar{\eta})_+ + |w_{2,t}| - \varepsilon$$

$$\Rightarrow \quad \leq (|\hat{w}_{t-1}| - \bar{\eta})_+ + (|w_t| - \bar{\eta})_+$$

$$\Rightarrow \quad (|\hat{w}_t| - \bar{\eta})_+ \leq (|\hat{w}_{t-1}| - \bar{\eta})_+ + (|w_t| - \bar{\eta})_+$$

$$\Rightarrow \quad \forall \tau_f : \sum_{t=0}^{\tau_f}(|\hat{w}_t| - \bar{\eta})_+ - (|\hat{w}_{t-1}| - \bar{\eta})_+ \leq \sum_{k=0}^{\tau_f}(|w_t| - \bar{\eta})_+$$

$$\Leftrightarrow \quad \forall \tau_f : (\hat{w}_{\tau_f} - \bar{\eta})_+ \leq \sum_{k=0}^{\tau_f}(|w_t| - \bar{\eta})_+ \leq \sum_{k=0}^{\tau_f}(|w_{2,t}| - \varepsilon)_+.$$

Recall $w = w_2 + w_1$, where $w_2 \in \ell_p$ and $|w_1|_\infty < \bar{\eta}$. The last line shows $\|\hat{w}\|_\infty < \infty$, since there exists some timestep T_f such that $\forall t > T_f, |w_t| - \bar{\eta} \leq -\frac{\varepsilon}{2}$. Furthermore, for all $k \geq T_f$ holds:

$$|\hat{w}_k| - \bar{\eta} \leq (|\hat{w}_{k-1}| - \bar{\eta})_+ - \tfrac{\varepsilon}{2}.$$

The shows that after some finite time T_2, $|\hat{w}_t| < \bar{\eta}$ and therefore $\hat{w}_t - \Pi_{\mathcal{W}}(\hat{w}_t) = 0$ for all $t \geq T_2$. Finally, since cCLM $\mathbf{\Psi}^\times_N$ is FIR, it also implies $s'(t) = 0$ after at most T time steps afterward. This proves that eventually, that is, after some $t' > 0$, $x_t \in \mathcal{X}$ for all $t \geq t'$.

$\square$

3.7 Example

As an example, we revisit the structure of the optimal blended SL controller of the constrained LQR problem and augment it with the anti-windup technique discussed in the previous section. In the control design procedure, we discussed that the last CLM $\boldsymbol{\Psi}_N$ (in the small example, it was $\boldsymbol{\Psi}_3$) was a degree of freedom that does not affect the LQR cost. However, in light of our previous discussion, we can use that extra degree of freedom to ensure graceful degradation in the event that the disturbance violates our assumptions.

As discussed in our previous results, we pick $\boldsymbol{\Psi}_N^{\mathrm{u}} = 0$ and pick $\boldsymbol{\Psi}_N^{\mathrm{x}}$ according to the rule (3.32). The blended cCLM $\boldsymbol{\Psi}_\Sigma$ changes only the component $\boldsymbol{\Psi}_\Sigma^{\mathrm{x}}$:

$$\boldsymbol{\Psi}_t^{\mathrm{x,a}}(w_{t:0}) = \sum_{i=1}^{N} \left(\sum_{k=1}^{\min\{T,t+1\}} R_k^{(i)}(P_{\eta_i} - P_{\eta_{i-1}})(w_{t+1-k}) \right) \\ + \sum_{k=1}^{\tau+1} A^{k-1}\left(w_{t+1-k} - P_{\eta_N}(w_{t-k+1})\right). \tag{3.37}$$

In the above, τ is a design parameter, and recall that by design, we have chosen $\eta_N = \eta_{max}$, the expected norm bound on disturbances. This extra term added to the CLM accounts for residual disturbances that are not attenuated by the original controller $\mathrm{SL}[\boldsymbol{\Psi}^{\mathrm{x}}, \boldsymbol{\Psi}^{\mathrm{u}}]$ because the disturbances are larger than expected by projection mapping, that is, $|w_t| > \eta_{max}$. Therefore, $\mathrm{SL}[\boldsymbol{\Psi}^{\mathrm{x,a}}, \boldsymbol{\Psi}^{\mathrm{u}}]$ considers the τ-step propagation of the unaccounted for disturbances from $\mathrm{SL}[\boldsymbol{\Psi}^{\mathrm{x}}, \boldsymbol{\Psi}^{\mathrm{u}}]$.

The resulting lumped dynamics under the augmented controller $\mathrm{SL}[\boldsymbol{\Psi}^{\mathrm{x,a}}, \boldsymbol{\Psi}^{\mathrm{u}}]$ take the form:

$$\hat{w}_t = A^{\tau+1}(\hat{w}_{t-\tau} - P_{\eta_{max}}(\hat{w}_{t-\tau})) + w_t \tag{3.38}$$

where $P_{\eta_{max}}()$ is the entry-wise saturation used in the constrained LQR problem. $P_{\eta_{max}}()$ can be equivalently viewed as an instance of a projection $\Pi_{\mathcal{W}}$ with respect to the ∞-norm $|\cdot|_\infty$, and where $\mathcal{W}$ denotes the scaled ∞-norm ball $\eta_{max}\{x \mid \|x\|_\infty \leqslant 1\}$. Leveraging the stability theorem Lem. 14 and using Gelfand's Lemma, closed-loop stability for the saturated regime is ensured under the following condition stated in the lemma below:

Lemma 17. *Assume $\rho(A) < 1$ and pick τ such that $|A^{\tau+1}|_\infty < 1$. Then the internal dynamics (3.38) are globally finite-gain ℓ_∞-stable, where for all $\boldsymbol{w} \in \ell_\infty^n$,*

$$\|\hat{\boldsymbol{w}}\|_\infty \leqslant \frac{1}{1 - |A^{\tau+1}|_\infty}\|\boldsymbol{w}\|_\infty.$$

Proof. Pick $|\cdot| := |\cdot|_\infty$ and choose τ such that $|A^{\tau+1}|_\infty < 1$. Then apply the lemma (14). $\qquad\square$

Therefore, if the system is open-loop stable, $\mathrm{SL}[\mathbf{\Psi}^{\mathrm{x,a}}, \mathbf{\Psi}^{\mathrm{u}}]$ guarantees graceful degradation when the closed-loop is saturated. Similarly, we obtain local stability results in the case where A is open-loop unstable. Similarly to the large-scale constrained LQR case in Section 3.3, the augmentation for anti-windup allows for distributed and localized implementation.

Simulation

We end this chapter with some simulation experiments demonstrating the effectiveness of anti-windup augmentation. We consider a bi-directional chain system with the ith node's dynamics being

$$x_{t+1}^i = (1 - 0.4|\mathcal{N}(i)|)x_t^i + 0.4 \sum_{j \in \mathcal{N}_i} x_t^j + \mathrm{sat}(u_t^i, u_{max}) + w_t^i$$

where $\mathcal{N}(i)$ denotes the set of vertices that has an edge connected to ith vertex and w_t^i is the ith coordinate of disturbance vector at time t. In particular, $\|w\|_\infty \leqslant 1$ and $x_0 = 0$. One can check that the overall chain system is open-loop marginally (un)stable. In this chain example, we allow 1 time step communication delay between nodes and actuation delay with 50%.

We illustrate the anti-windup property of the nonlinear controller (3.26) in the decentralized setting with additional sparsity, locality, and delay constraints in Figure 3.3c. First, a nominal integral controller for this system is designed and called the *Integral Controller*. Due to its integral structure, the *Integral Controller* for the unconstrained closed loop guarantees convergence of the state to the origin under persistent disturbance, i.e., step rejection. In comparison, a second linear controller synthesized from the standard constrained LQR problem is generated that guarantees stability for all admissible w under saturation. We refer to this linear controller as the *non-integral controller* since the states only stay bounded under persistent admissible disturbance.

The nonlinear controller with the saturation projection here is chosen to be a two-zone blending controller. The simulation shows the anti-windup property as well as the preservation of step rejection in both large- and small-disturbance schemes of the proposed method. Figure 3.3c shows that the blended SL-controller stabilizes the system while the integral controller becomes unstable under worst-case bounded disturbance. On the other hand, in Figure 3.4c, the proposed blending controller preserves the performance of step rejection while the linear Non-integral Controllers

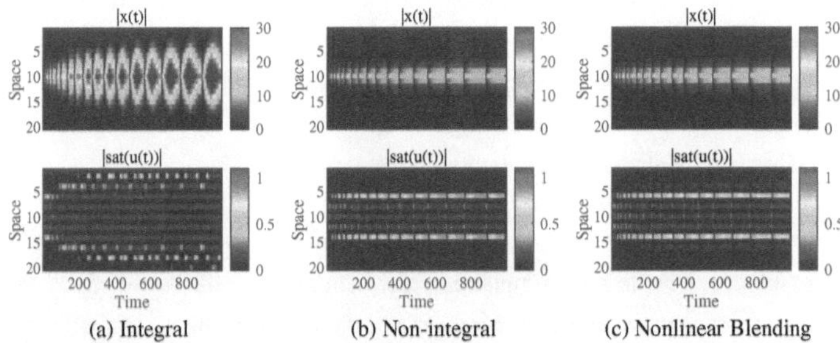

(a) Integral   (b) Non-integral   (c) Nonlinear Blending

Figure 3.3: *Worst-case Response*: The heatmaps show how a worst-case disturbance is propagated through space-time for the saturated chain system. The integral controller becomes unstable due to saturation and the naive blending controller possesses has the anti-windup property of the non-integral controller.

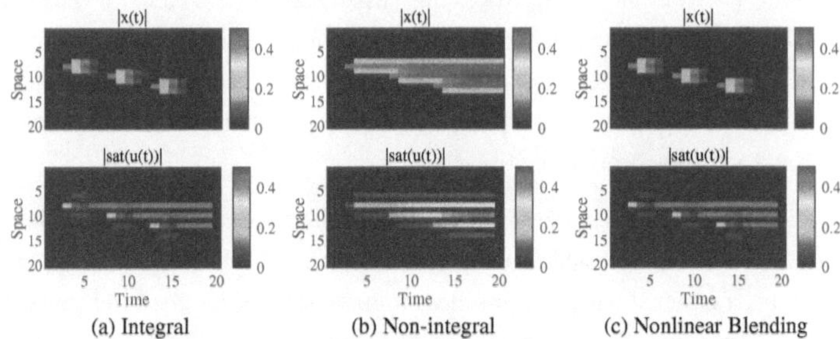

(a) Integral   (b) Non-integral   (c) Nonlinear Blending

Figure 3.4: Step disturbance rejection: Response to small step disturbances at node 8,10,12 entering at time 2,6,10, respectively. As in the scalar case, the proposed blending controller not only stabilizes under saturation but also recovers the performance objective of rejecting small step disturbances. This contrasts against the non-integral controller, which sacrifices small-signal performance for stability.

forfeit the performance objective in order to preserve stability in the saturated closed loop. For more details on the simulation, see [137].

3.8 Conclusion

We showcase the nonlinear system-level approach developed in [64] and illustrate the use cases for a class of nonlinear system-level controllers. We propose a tractable nonlinear control synthesis method that outperforms any optimal linear controller for constrained LQR problems. It is further shown that such a controller naturally possesses an anti-windup property for linear systems with input saturation. A key highlight is that the presented approach is compatible with locality/delay constraints and distributed implementation, similar to the linear system-level approach [11]. Overall, this chapter is a first step in exploring the full potential of the new nonlinear control synthesis framework developed in [64] and highlights that even just the presented special case of the framework, called "nonlinear blending" of linear controllers, offers many benefits.

Chapter 4

ROBUSTNESS AND SENSITIVITY OF CERTAINTY-EQUIVALENT ADAPTATION

This chapter investigates closed-loop in learning-to-control problem settings, where the decision-making of the algorithm is structured in a certainty equivalent way. Certainty Equivalence (CE) is a common principle underlying many learning and control algorithms: At each time step, we hypothesize a model for the system dynamics and then act according to the *nominal* control law designed for that model. In other words, we pretend to have found a model that will remain accurate going forward. We show that there is a natural way to perform certainty-equivalent adaptation for SL controllers, which simplifies the analysis and is instructive for the design of the overall learning-to-control algorithm. The particular structure of the nonlinear closed-loop maps reveals three general design principles sufficient for designing stable learning-to-control algorithms:

1. **Smooth parametrizations of nominal CLMs and Dynamics**: The nominal CLMs and dynamics chosen by CE-adaptation should change smoothly with small changes in the nominal model.
2. **Consistency of Selected Models**: The sequence of models that we hypothesize should be consistent with the online data (up to some ℓ_p-bounded error).
3. **Efficient Model Selection**: Our model hypothesis should eventually converge; however, it is <u>not</u> required to converge to the true model.

In the second part of the chapter, we explore applications of this result for problem settings related to online learning of optimal controllers. To this end, we focus on the setting of linear time-invariant systems and linear-quadratic costs, a problem setting that has received immense recent attention in the learning and control literature. Guided by the theoretical findings in the first part, we follow the principle of certainty equivalence to design a learning-to-control scheme with nominal LQ-optimal system-level controllers. We analyze the closed-loop stability of the learning-to-control scheme and provide conditions for model selections that are sufficient for closed-loop stability, which are closely related to consistent model chasing, a core topic in Part 2 of the book. The main technical result underlying this analysis is on perturbation analysis of LQ-optimal CLMs. The result in itself is new and characterizes the sensitivity (i.e., analytic bounds on the Lipschitz constant) of LQ-optimal closed-loop maps for LTI systems in terms of system-theoretic properties such as controllability and observability.

4.1 Closed-Loop Dynamics in the Presence of CE-Adaptation

We assume a nonlinear system in the standard form we introduced in Chapter 2 and that the disturbance w belongs to $\mathcal{W}^{\mathbb{N}} := \{w \in \ell^{\mathcal{X}} \mid \forall k \in \mathbb{N} : w_k \in \mathcal{W}\}$ for some closed-bounded set $\mathcal{W} \subset \mathbb{R}^n$.

$$x = F^+(x, u) + w, \quad w \in \mathcal{W}^{\mathbb{N}}.$$

We restrict ourselves to diagonal operators F, i.e., F can be decomposed as:

$$F = \sum_{t=0}^{\infty} P^{[t]} F P^{[t]} \tag{4.1}$$

and let $\{f_k : \mathbb{R}^{n+m} \mapsto \mathbb{R}^n\}_{k=0}^{\infty}$ be the unique functions formed by dropping the obsolete arguments of the component functions; alternatively, each f_τ is well-defined by the mapping $f_\tau : (x_\tau, u_\tau) \mapsto F_\tau(x_{\tau:0}, u_{\tau:0})$. Moreover, we assume that the dynamics F are unknown, however, belong to a known set of dynamics $\{\hat{F}(\omega) \mid \omega \in \Omega\} =: \hat{F}[\Omega]$ parametrized over some compact metric space (Ω, d_Ω) with some fixed map $\hat{F} : \Omega \to \mathcal{C}(\ell^{\mathcal{X} \times \mathcal{U}}, \ell^{\mathcal{X}})$. Correspondingly with our assumption of F, we assume that $\mathbb{D}[\Omega]$ consists of diagonal operators and there exists some $\omega^* \in \Omega$ such that $\mathbb{D}(\omega^*) = F$. We summarize these assumptions and definitions below: Moreover, we assume that the dynamics F are unknown; however, they belong to a known set of dynamics $\{\hat{F}(\omega) \mid \omega \in \Omega\} =: \hat{F}[\Omega]$ parameterized over some compact metric space (Ω, d_Ω) with some fixed map $\hat{F} : \Omega \to \mathcal{C}(\ell^{\mathcal{X} \times \mathcal{U}}, \ell^{\mathcal{X}})$. Similarly to our assumption of F, we assume that $\mathbb{D}[\Omega]$ consists of diagonal operators and that there exists some $\omega^* \in \Omega$ such that $\mathbb{D}(\omega^*) = F$. We summarize these assumptions and definitions below:

Assumption 4.2. *We are given a compact metric space (Ω, d_Ω) and a map $\hat{F} : \Omega \to \mathcal{C}(\ell^{\mathcal{X} \times \mathcal{U}}, \ell^{\mathcal{X}})$, where each $\omega \in \Omega$ represents a dynamic system with diagonal dynamics operator $\hat{F}[\omega]$.*

As we discussed, taking the CE-approach is the most common (often also efficient [92]) way to tackle learning of feedback control in the closed-loop. The defining feature of CE-adaptation is the simple structure of the decision-making process: At each time-step k, we hypothesize a model parameter θ_k and evaluate the *nominal* controller $u_k = K_k[\theta_k](x_{k:0})$ designed for the assumed dynamics $\hat{F}[\theta_k] := \mathbb{D}(\theta_k)$; one can view the nominal controllers as a parametrized collection $\{\mathbb{K}(\omega) \mid \omega \in \Omega\}$, $\mathbb{K} : \Omega \to \mathcal{C}(\ell^{\mathcal{X}}, \ell^{\mathcal{U}})$ of specified (or synthesized online) control laws, where each control law $K[\omega] := \mathbb{K}(\omega)$ achieves some desired nominal closed-loop behavior, assuming the system dynamics are $\hat{F}[\omega]$. Selecting θ_k is commonly performed via online system identification [16], or gradient-based adaptation rules [71], and we shall denote this process as the causal operator $S \in \mathcal{C}(\ell^{\mathcal{X} \times \mathcal{U}}, \Omega^{\mathbb{N}})$, $\Omega^{\mathbb{N}} := \{\omega \mid \omega_k \in \Omega, \forall k \in N\}$ which we refer to as the *model selection/selector* S.

To summarize, under CE-adaptive control, the closed-loop dynamics are governed by the following equations at each time-step k:

$$x_k = f_{k-1}(x_{k-1}, u_{k-1}) + w_k \tag{4.2a}$$

$$\theta_k = S_k(x_{k:0}, u_{k-1:0}) \tag{4.2b}$$

$$u_k = K_k[\theta_k](x_{k:0}). \tag{4.2c}$$

It is hard to gauge the behavior of the closed-loop dynamics from the above equations; however, that changes with a slight tweak to how we perform adaptation. Next, we formulate CE-adaptation in terms of nominal closed-loop maps and system-level controllers.

CE-Adaptation in System Level Implementations

Instead of encoding the desired nominal behavior as a parameterization $\mathbb{K} : \omega \mapsto K[\omega]$ of control laws $\{K[\omega] \mid \omega \in \Omega\}$, we can equivalently consider a representation in the form of nominal CLMs $\{\Psi[\omega] \mid \omega \in \Omega\}$, $\Psi[\omega] \in \mathcal{C}(\ell^{\mathcal{X}}, \ell^{\mathcal{X} \times \mathcal{U}})$. Hence, we assume each nominal CLM $\Psi[\omega]$ is a CLM of the dynamics $\hat{F}[\omega]$:

$$\forall \omega \in \Omega : \qquad \Psi[\omega] \in \mathcal{C}(\ell^{\mathcal{X}}, \ell^{\mathcal{X} \times \mathcal{U}}) \text{ s.t.: } \Psi[\omega] \in \Phi_{\mathrm{CL}}^{\mathrm{w} \mapsto \mathrm{xu}}[\hat{F}[\omega]] \tag{4.3}$$

and corresponds to the nominal control law $K[\omega] = \Psi^u[\omega](\Psi^x[\omega])^{-1}$. However, the System Level Implementation $\mathrm{SL}[\Psi^u[\omega], \Psi^x[\omega]]$ of the nominal controller offers alternative (to (4.2c)) ways to perform adaptation, simply by swapping out the nominal CLMs in the SL control structure. The resulting closed-loop dynamic equations over time k take the form:

$$x_k = f_{k-1}(x_{k-1}, u_{k-1}) + w_k \tag{4.4a}$$

$$\hat{w}_k = x_k - \Psi_k^x[\theta_k](0, \hat{w}_{k-1:0}) \tag{4.4b}$$

$$u_k = \Psi_k^u[\theta_k](\hat{w}_{k:0}) \tag{4.4c}$$

$$\theta_k = S_k(x_{k:0}, u_{k-1:0}). \tag{4.4d}$$

Despite the correspondence $K[\omega] = \Psi^u[\omega](\Psi^x[\omega])^{-1}$, the closed-loops (4.4) and (4.2) are *not* the same because CE-adaptation is performed differently.

Remark. *In fact, the adaptive control law* (4.2c) *parameterized by CLMs takes the form* $u = \sum_{t=0}^{\infty} P^{[t]}\Psi^u[\theta_t]P^t\Psi^{-x}[\theta_t]P^t x$.

However, the above closed-loop dynamics can be manipulated into a form that turns out to be insightful, particularly for the design of S. In the next section, we discuss sufficient conditions for closed-loop stability provided by appropriate design of the model selection.

Model Selection Conditions for Closed-Loop Stability

For this section, we consider S a design variable, and therefore we treat the parameter sequence θ as an input of the closed-loop model (the set of all $(x, u, \theta, \hat{w}, w)$ conforming with equations (4.4)). Regardless of θ, the usual identity still remains for all k:

$$x_k = \Psi_k^x[\theta_k](\hat{w}_{k:0}) \qquad\qquad u_k = \Psi_k^u[\theta_k](\hat{w}_{k:0}). \tag{4.5}$$

Moreover, treating the sequence θ as an index set, we denote $\Psi_{|\theta}$ as the operator with component functions $(\Psi_{|\theta})_k : (\hat{w}_{k:0}) \mapsto \Psi_t[\theta_t](\hat{w}_{k:0})$. We can rewrite (4.5) in sequence space as:

$$x = \Psi_{|\theta}^x \hat{w} \qquad\qquad u = \Psi_{|\theta}^u \hat{w}. \tag{4.6}$$

An intuitive design criterion for model selection and choice of θ is the *1-step prediction error* e. At time step k, the prediction error is defined as $e_k = x_k - \hat{f}_{k-1}[\theta_k](x_{k-1}, u_{k-1})$, i.e., it measures how well the current selected model θ_k matches with the most recent system transition $(x_{k-1}, u_{k-1}) \mapsto x_k$. However, the selection of θ_k is allowed to depend on e_k, since we get to observe x_k before choosing θ_k. In particular, it is easy to see that we can always control the size of this error to be at most the size of w_k. Since there exists a true $\omega^* \in \Omega$ such that $F = \hat{F}[\omega^*]$, given x_k and x_{k-1}, u_{k-1} at time k, ω^* accurately models the latest state transition $x_{k-1}, u_{k-1} \mapsto x_k$. That is, $e_k = x_k - \hat{f}_k[\theta_k](x_{k-1}, u_{k-1}) = w_k$, therefore it is always possible to find a parameter θ_k such that $|e_k| \leqslant |w_k|$. The same argument applies to the entire history of observed transitions $(x, u)_{k-1:0} \mapsto x_{k:1}$, and therefore it is always possible to select a parameter θ_k such that the prediction error in hindsight

$$e_{[k:0]|k} := x_{k:1} - \hat{F}_{k-1:0}[\theta_k](x_{k-1:0}, u_{k-1:0})$$

is always smaller in norm than $w_{k-1:0}$. Hence, at each time k, the history of past observations $x_{k:0}, u_{k-1:0}$ provides sufficient knowledge to find some θ_k^{*p}, for example, by explicit minimization, such that $\|e_{[k:0]|k}\|_p \leqslant \|w_{k-1:0}\|_p$, *regardless* of what value w takes. We will call a sequence θ for which the hindsight prediction errors are uniformly bounded over time, i.e., $\sup_k\{\|e_{[k:0]|k}\|_p\} < \infty$, to be an ℓ_p-*consistent parameter selection*, and $\theta^{*p} = (\theta_0^{*p}, \theta_1^{*p}, \dots)$ to be ℓ_p-*optimal*. Similarly, we call a sequence θ to be a finite horizon ℓ_p-consistent selection for some T, if the norm of the truncated hindsight prediction errors

$$e_{\mathcal{I}_k|k} := x_{\mathcal{I}_k} - \hat{F}_{\mathcal{I}_k-1}[\theta_k](x_{k-1:0}, u_{k-1:0}), \quad \mathcal{I}_k = [k - T, k]$$

forms a scalar ℓ_p-bounded sequence $(|e_{\mathcal{I}_0|0}|, |e_{\mathcal{I}_1|1}|, \dots)$. We summarize these definitions below and define causal operators S to be ℓ_p-*consistent model selectors* if the output parameter sequence $\theta = S(x, u)$ is always a ℓ_p-consistent selection, whenever $x = F^+(x, u) + w, w \in \ell_p^{\mathcal{X}}$.

Definition 4.1. *Let (Ω, d) be a parameter space, $\hat{F} : \Omega \ni \omega \mapsto \hat{F}[\omega] \in \mathcal{C}((\mathcal{X} \times \mathcal{U})^{\mathbb{N}}, \mathcal{X}^{\mathbb{N}})$ be parameterization of dynamic functions, and $S \in \mathcal{C}((\mathcal{X} \times \mathcal{U})^{\mathbb{N}}, \Omega^{\mathbb{N}})$ be a parameter selector. Denote $E_T : (\Omega \times \mathcal{X} \times \mathcal{U})^{\mathbb{N}} \to \mathbb{R}_0^+$ as the causal operator defined by its component functions as*

$$P^{[k]} E_T : (\theta, x, u) \mapsto \|P^{\mathcal{I}_k}(x - \hat{F}^+[\theta_k](x, u))\|_p, \quad \mathcal{I}_k := [k - T, k].$$

(i) *S is called an ℓ_p-consistent parameter selector if for all $x \in \ell^{\mathcal{X}}, u \in \ell^{\mathcal{U}}, w \in \ell_p^{\mathcal{X}}$ such that $x = \hat{F}^+[\omega](x, u) + w$ for some $\omega \in \Omega$, $E_\infty(S(x, u), x, u) \in \ell_\infty$.*

(ii) *S is called an ℓ_p-consistent parameter selector with a finite T-horizon if for all $x \in \ell^{\mathcal{X}}, u \in \ell^{\mathcal{U}}, w \in \ell_p^{\mathcal{X}}$ such that $x = \hat{F}^+[\omega](x, u) + w$ for some $\omega \in \Omega$, $E_T(S(x, u), x, u) \in \ell_p$.*

Remark. *We discuss consistent model selections in depth in Chapter 6.*

As alluded to previously, the following decision rule always defines a corresponding ℓ_p-consistent selector:

$$S_t : P^t(x, u) \mapsto \theta_t \in \arg\min_\omega \ \|P^{\mathcal{I}_k}(P^t x - \hat{F}^+[\omega] P^t(x, u))\|_p$$

$\theta^{*p} = (\theta_0^{*p}, \theta_1^{*p}, \dots)$, and therefore the model selectors defined above always exist. Next, we consider S, which are just ℓ_∞-consistent selectors of horizon $T = 1$, and consider F, which are diagonal operators, i.e., for any t, $P^{[t]} F = P^{[t]} F P^{[t]}$.

In other words, the selector S is assumed to guarantee the uniform boundedness of the 1-step prediction error $e_k = f_{k-1}(x_{k-1}, u_{k-1}) + w_k - \hat{f}_{k-1}[\theta_k](x_{k-1}, u_{k-1})$. Provided that assumption, e can replace the role of w as input, and as summarized in Theorem 22, we can analyze the closed-loop stability in terms of the stability of the map $(e, \theta) \mapsto (\hat{w}, x, u)$.

To derive these equations, first, notice that the following statement is vacuous:

$$x_k = \hat{f}_{k-1}[\theta_k](x_{k-1}, u_{k-1}) + e_k.$$

By substituting the above into (4.4b) and using the identity (4.5), we arrive at the following set of equations:

$$\hat{w}_k = \hat{f}_{k-1}[\theta_k](\Psi_{k-1}[\theta_{k-1}](\hat{w}_{k-1:0})) - \Psi_k^x[\theta_k](0, \hat{w}_{k-1:0}) + e_k. \tag{4.7}$$

For a fixed k, the right-hand side of the equation corresponds with the k-th component of the operator $\hat{F}^+[\theta_k]\Psi[\theta_{k-1}] + \overline{\Psi^x}[\theta_k]$, where $\overline{\Psi^x} := I - \Psi^x$ hence we can write the above set of equations in sequence space as:

$$\hat{w} = \sum_{k=0}^{\infty} P^{[k]}(\hat{F}^+[\theta_k]\Psi[\theta_{k-1}] + \overline{\Psi^x}[\theta_k])(\hat{w}) + e$$

$$= \sum_{k=0}^{\infty} P^{[k]}(\hat{F}^+[\theta_k]\Psi[\theta_{k-1}] + \overline{\Psi^x}[\theta_k])(P^{k-1}\hat{w}) + e$$

where the second equation follows from the strict causality of $\hat{F}^+$ and $\overline{\Psi^x}$. The operator on the right-hand side is what we refer to as $\Delta_{|\theta}$, and it determines the stability of the disturbance dynamics. The terms of the sum are parameterized by the function $\Lambda^+[\omega, \nu] := \hat{F}^+[\omega]\Psi[\nu] + \overline{\Psi^x}[\omega]$.

Theorem 22 below summarizes our findings so far:

Theorem 22. *Consider a fixed sequence $\theta \in \Omega^{\mathbb{N}}$ and let $(\hat{w}, x, u)$ be governed by the closed-loop equations (4.4a)-(4.4c) where F and $\hat{F}[\omega], \omega \in \Omega$ are all diagonal operators. Then $(\hat{w}, x, u)$ satisfy:*

$$x = \Psi^x_{|\theta}\hat{w} \qquad u = \Psi^u_{|\theta}\hat{w} \qquad \hat{w} = (I - \Delta^+_{|\theta})^{-1}e \tag{4.8}$$

where $\Psi_{|\theta} \in C(\mathcal{X}^{\mathbb{N}}, (\mathcal{X} \times \mathcal{U})^{\mathbb{N}})$ and $\Delta^+_{|\theta} \in C_s((\mathcal{X} \times \mathcal{U})^{\mathbb{N}}, \mathcal{X}^{\mathbb{N}})$ denote the operators:

$$\Psi_{|\theta} = \sum_{t=0}^{\infty} P^{[t]}\Psi[\theta_t]P^t \qquad \Delta^+_{|\theta} = \sum_{k=1}^{\infty} P^{[t]}\Lambda^+[\theta_t, \theta_{t-1}]P^{t-1} \tag{4.9}$$

and $\Lambda^+ : \Omega^2 \to \mathcal{C}_s((\mathcal{X} \times \mathcal{U})^{\mathbb{N}}, \mathcal{X}^{\mathbb{N}})$, $\Omega^2 := \Omega \times \Omega$, *for fixed* $\omega, \nu \in \Omega$, *is defined as*

$$\Lambda^+[\omega, \nu] := \hat{F}^+[\omega]\Psi[\nu] + \overline{\Psi^x}[\omega]$$

.

The stability of the closed-loop systems is determined by the stability of the maps $\Psi|_\theta$ and $(I - \Delta_{|\theta}^+)^{-1}$ in equation (4.8). If $\Psi[\omega], \omega \in \Omega$ are all CLMs of the corresponding dynamics $\hat{F}[\omega], \omega \in \Omega$ and $\hat{F}$ are linear we can split $\Delta_{|\theta}^+$ into two components, as stated below:

Lemma 18. *If* $\Psi[\omega] \in \Phi_{\mathrm{CL}}^{\mathrm{w}\mapsto\mathrm{xu}}(\hat{F}[\omega])$, $\forall\omega \in \Omega$, *the operator* $\Delta|_\theta$ *can be written as the sum* $\Delta_{|\theta}^{+\varphi} + \Delta_{|\theta}^{+f}$, *where where* $\Delta_{|\theta}^{+f}$, $\Delta_{|\theta}^{+\varphi}$ *are defined by their component functions for* $k \in \mathbb{N}$ *as:*

$$U^{*[k]}\Delta_{|\theta}^{+f}U^k : (\hat{w}_{k:0}) \mapsto (\hat{f}_{k-1}[\theta_k] - \hat{f}_{k-1}[\theta_{k-1}]) \circ (\Psi_{k-1}[\theta_{k-1}])(\hat{w}_{k-1:0})$$
$$U^{*[k]}\Delta_{|\theta}^{+\varphi}U^k : (\hat{w}_{k:0}) \mapsto (\overline{\Psi}_k[\theta_k] - \overline{\Psi}_k[\theta_{k-1}])(0, \hat{w}_{k-1:0}).$$

Moreover, if $\{\hat{f}_k[\omega] \mid \omega \in \Omega, \, k \in \mathbb{N}\}$ *are linear functions, then the components of* $\Delta_{|\theta}^+$ *become*

$$U^{*[k]}\Delta_{|\theta}^+U^k = \hat{f}_{k-1}[\theta_k] \circ (\Psi_{k-1}[\theta_{k-1}] - \Psi_{k-1}[\theta_k]). \tag{4.10}$$

Proof. Per definition, $\Psi[\theta_k] \in \Phi_{\mathrm{CL}}^{\mathrm{w}\mapsto\mathrm{xu}}[F[\theta_k]]$ and $\Psi[\theta_{k-1}] \in \Phi_{\mathrm{CL}}^{\mathrm{w}\mapsto\mathrm{xu}}[F[\theta_{k-1}]]$ holds, and therefore the operators have to satisfy the respective CLM equation:

$$F^+[\theta_{k-1}](\Psi[\theta_{k-1}]) = \Psi^x[\theta_{k-1}] - I \qquad F^+[\theta_k](\Psi[\theta_k]) = \Psi^x[\theta_k] - I.$$

With the above, we can rewrite (4.7) as

$$\hat{w}_k = \Delta_{k-1}^f(\hat{w}_{k-1:0}) + \Delta_{k-1}^\varphi(\hat{w}_{k-1:0}) + e_k$$

with the operators defined below:

$$\Delta_{k-1}^f := (\hat{f}_{k-1}[\theta_k] - \hat{f}_{k-1}[\theta_{k-1}]) \circ (\Psi_{k-1}[\theta_{k-1}])$$
$$\Delta_{k-1}^\varphi(\hat{w}_{k-1:0}) := (\Psi_k[\theta_{k-1}] - \Psi_k[\theta_k])(0, \hat{w}_{k-1:0})$$
$$e_k := x_k - \hat{f}_{k-1}[\theta_k](x_{k-1}, u_{k-1}).$$

Lastly, if $\{\hat{f}_k[\omega] \mid \omega \in \Omega, \, k \in \mathbb{N}\}$ are linear functions, the components of $\Delta_{|\theta}^f + \Delta_{|\theta}^{\varphi+}$ take on a simpler a form:

$$(\Delta_{|\theta}^f + \Delta_{|\theta}^{\varphi+})_{k-1} = \hat{f}_{k-1}[\theta_k] \circ (\Psi_{k-1}[\theta_{k-1}] - \Psi_{k-1}[\theta_k]).$$

$\square$

Hence, CE-adaptation using the SL implementation of the nominal controllers allows us to represent the closed-loop dynamics in the form:

$$x = \mathbf{\Psi}_{|\theta}^{x}\hat{w} \qquad u = \mathbf{\Psi}_{|\theta}^{u}\hat{w} \qquad \hat{w} = \left(I - \mathcal{S}^{+}\mathbf{\Delta}_{|\theta}^{f} - \mathcal{S}^{+}\mathbf{\Delta}_{|\theta}^{\varphi+}\right)^{-1}e$$

From this representation, we can derive conditions for closed-loop stability in terms of θ. As expected, a necessary condition for closed-loop stability is sensible design of the nominal CLMs that assures at least some type of stability of the maps $\mathbf{\Psi}[\omega]$. Moreover, boundedness of the closed-loop trajectory (θ, w, x, u) is implied, if θ is chosen such that the derived operator $(I - \mathcal{S}^{+}\mathbf{\Delta}_{|\theta}^{f} - \mathcal{S}^{+}\mathbf{\Delta}_{|\theta}^{\varphi+})^{-1}$ is $\ell_{\infty}^{\mathcal{X}}$-stable and the prediction error e is bounded. This observation provides an objective for the design of the model selection: Finding a causal selection rule S which guarantees the former conditions for any selected sequence $\theta = S(x, u)$. A natural design approach is to view S as a means to prove stability via the small-gain theorem. For example, if S guarantees sufficiently small gains of $\mathbf{\Delta}_{|\theta}^{\varphi+}$ and $\mathbf{\Delta}_{|\theta}^{f+}$, then we can prove stability of $(I - \mathcal{S}^{+}\mathbf{\Delta}_{|\theta}^{f} - \mathcal{S}^{+}\mathbf{\Delta}_{|\theta}^{\varphi+})^{-1}$ via small-gain theorems such as Theorem 2. From this representation, we can derive conditions for closed-loop stability in terms of θ. As expected, a necessary condition for closed-loop stability is a sensible design of the nominal CLMs that guarantees at least some type of stability of the maps $\mathbf{\Psi}[\omega]$. Moreover, the boundedness of the closed-loop trajectory (θ, w, x, u) is implied if θ is chosen such that the derived operator $(I - \mathcal{S}^{+}\mathbf{\Delta}_{|\theta}^{f} - \mathcal{S}^{+}\mathbf{\Delta}_{|\theta}^{\varphi+})^{-1}$ is $\ell_{\infty}^{\mathcal{X}}$ stable and the prediction error e is bounded. This observation provides an objective for the design of the model selection: Finding a causal selection rule S that guarantees the former conditions for any selected sequence $\theta = S(x, u)$. A natural design approach is to view S as means to prove stability via the small-gain theorem. For example, if S guarantees sufficiently small gains of $\mathbf{\Delta}_{|\theta}^{\varphi+}$ and $\mathbf{\Delta}_{|\theta}^{f+}$, then we prove the stability of $(I - \mathcal{S}^{+}\mathbf{\Delta}_{|\theta}^{f} - \mathcal{S}^{+}\mathbf{\Delta}_{|\theta}^{\varphi+})^{-1}$ using small-gain theorems such as Theorem 2. However, we obtain less restrictive design criteria if we conduct analysis with the adapted small-gain conditions of Lem. 8:

Theorem 23. *Let w, x, u, θ be a trajectory of the closed-loop dynamical system described by the equations (4.4) for some disturbance $w \in \ell_{\infty}^{\mathcal{X}}$ and sequence of selected parameters θ. The state and input sequence x, u are bounded if the following conditions are met:*

    i) Consistent model selection: S is an ℓ_{∞}-consistent model selector with a horizon of at least $T = 1$.

    ii) Efficient selection: the selection θ converges in Ω.

iii) *CE-nominal design:* $\boldsymbol{\Psi}$ *is a parametrization of CLMs: for each* $\omega \in \boldsymbol{\Omega}$ *holds* $\boldsymbol{\Delta}[\boldsymbol{\Psi}[\omega], \hat{\boldsymbol{F}}[\omega]] = 0.$

iv) *Smooth/robust parametrization of CLMs: The "oracle" map* $\boldsymbol{\Psi} : \boldsymbol{\Omega} \to C(\ell^{\mathcal{X}}, \ell^{\mathcal{X} \times \mathcal{U}})$ *is continuous over the metric space* $(\boldsymbol{\Omega}, d)$ *w.r.t. the operator norm* $\||\cdot\||_\infty.$

v) *Smooth/robust parametrization of Dynamics : The map* $\boldsymbol{D}$ *defined by the correspondence*

$$\boldsymbol{D} : (\omega_1, \omega_2) \mapsto \hat{\boldsymbol{F}}[\omega_1] \circ \boldsymbol{\Psi}[\omega_2]$$

between the domain $(\boldsymbol{\Omega}^2, d_{\Omega^2})$, $d_{\Omega^2}(x, y) := d(x_1, y_1) \vee d(x_2, y_2)$ *and codomain* $(C(\ell^n, \ell^n), \||(\cdot) - (\cdot)\||_\infty)$ *is continuous at every point p of the subset* $\{(\omega, \omega) \,|\, \omega \in \boldsymbol{\Omega}\} \subset \boldsymbol{\Omega}^2.$

Proof. We assumed $(\boldsymbol{\Omega}, d)$ is a compact metric space and defined for the family of dynamics $\{\hat{\boldsymbol{F}}[\omega] \,|\, \omega \in \boldsymbol{\Omega}\}$ and CLMs $\{\boldsymbol{\Psi}[\omega] \,|\, \omega \in \boldsymbol{\Omega}\}$ the map $\boldsymbol{\Lambda}^+ : \boldsymbol{\Omega}^2 \to C_s(\ell^{\mathcal{X}}, \ell^{\mathcal{X}})$, where

$$\boldsymbol{\Lambda}^+[\omega, \nu] := \boldsymbol{F}^+[\omega]\boldsymbol{\Psi}[\nu] + \overline{\boldsymbol{\Psi}^x}[\omega] = \boldsymbol{D}^+[\omega, \nu] + \boldsymbol{I}_n - \overline{\boldsymbol{\Psi}^x}[\omega]$$

for each $(\omega, \nu) \in \boldsymbol{\Omega}^2$ in the metric space $(\boldsymbol{\Omega}^2, d_{\Omega^2})$. Recall the definitions of the operator norm $\||\cdot\||_p$ and pseudo-norm $\||\cdot\||_p^\star$ from Lem. 11:

$$\||\boldsymbol{A}\||_p^\star := \sup_{x \in \mathcal{X}_p^{\mathbb{N}}:\, x \neq 0} \frac{\|\boldsymbol{A}(\boldsymbol{x}) - \boldsymbol{A}(0)\|_p}{\|\boldsymbol{x}\|_p} \qquad \||\boldsymbol{A}\||_p := \||\boldsymbol{A}\||_p^\star + \|\boldsymbol{A}(0)\|_p$$

and let $\mathbb{E}_0 : C(\ell^{\mathcal{X}}, \ell^{\mathcal{X}}) \to \ell^{\mathcal{X}}$ denote the linear evaluation map such that for each $\forall \boldsymbol{A} \in C(\ell^{\mathcal{X}}, \ell^{\mathcal{X}}) : \mathbb{E}_0(\boldsymbol{A}) = \boldsymbol{A}(0)$. Since per assumption, $\boldsymbol{D} : (\boldsymbol{\Omega}^2, d_{\Omega^2}) \to (C(\ell^n, \ell^n), \||\cdot\||_\infty)$ is continuous for all $p = (p_1, p_2) \in \boldsymbol{\Omega}^2$, s.t.: $p_1 = p_2$ and $\boldsymbol{\Psi}[\cdot]$ is a continuous map $(\boldsymbol{\Omega}, d) \to \||\cdot\||_\infty$, it follows that $\boldsymbol{\Lambda}^+ : (\boldsymbol{\Omega}^2, d_{\Omega^2}) \to (C(\ell^n, \ell^n), \||\cdot\||_\infty)$ has to be continuous for all $p = (p_1, p_2) \in \boldsymbol{\Omega}^2$, s.t.: $p_1 = p_2$ as well.

Per assumption, we are assured that our parameter sequence converges to some parameter $\theta_\infty = \lim_{t \to \infty} \theta_t$, $\theta_\infty \in \boldsymbol{\Omega}$. Pick $\varepsilon = \mu < 1$ and invoke continuity of $\boldsymbol{\Lambda}^+$ at θ_∞. Thus, there exists some $\delta > 0$ such that

$$\forall \omega, \nu \in \mathsf{B}_\delta[\theta_\infty] : \quad \||\boldsymbol{\Lambda}^+[\omega, \nu] - \boldsymbol{\Lambda}^+[\theta_\infty, \theta_\infty]\||_\infty \leqslant \varepsilon. \qquad (4.11)$$

Since $\lim_{t \to \infty} \theta_t = \theta_\infty$, there exists some $N \in \mathbb{N}$ such that $\forall k > N$ holds $d(\theta_{k-1}, \theta_\infty) < \delta$. Now decompose $\boldsymbol{\Delta}|_\theta$ into the sum

$$\boldsymbol{\Delta}|_\theta = \sum_{k=1}^\infty \boldsymbol{P}^k \boldsymbol{\Lambda}^+[\theta_k, \theta_{k-1}] = \sum_{k=1}^\infty \boldsymbol{P}^{[k]} \boldsymbol{\Lambda}^+[\theta_k, \theta_{k-1}] \boldsymbol{P}^{k-1}$$

and assume some arbitrary $z \in \ell_\infty^\mathcal{X}$. Next, rewrite $\|\overline{P^N}\Delta|_\theta z\|_\infty$ as

$$\|\overline{P^N}\Delta|_\theta z\|_\infty = \sup_{h>N} \|P^{[h]}\Delta|_\theta z\|_\infty = \sup_{h>N} \|P^{[h]}\sum_{k=1}^\infty P^{[k]}\Lambda^+[\theta_k, \theta_{k-1}]P^{k-1}z\|_\infty \tag{4.12}$$

$$= \sup_{h>N} \|P^{[h]}\Lambda^+[\theta_h, \theta_{h-1}]P^{h-1}z\|_\infty. \tag{4.13}$$

Next we bound each individual term $\|P^{[h]}\Lambda^+[\theta_h, \theta_{h-1}]P^{h-1}z\|_\infty$:

$$\|P^{[h]}\Lambda^+[\theta_h, \theta_{h-1}]P^{h-1}z\|_\infty$$
$$\leqslant \|\Lambda^+[\theta_h, \theta_{h-1}]P^{h-1}z - \Lambda^+[\theta_h, \theta_{h-1}](0)\|_\infty + \|\mathbb{E}_0\Lambda^+[\theta_h, \theta_{h-1}]\|_\infty. \tag{4.14}$$

By construction $h > N$, $\theta_h, \theta_{h-1} \in B_\delta[\theta_\infty]$ and by continuity of Λ^+ we have

$$\|\mathbb{E}_0\Lambda^+[\theta_h, \theta_{h-1}]\|_\infty \leqslant \|\mathbb{E}_0\Lambda^+[\theta_h, \theta_{h-1}] - \mathbb{E}_0\Lambda^+[\theta_\infty, \theta_\infty]\|_\infty \leqslant \varepsilon,$$

where $\Lambda^+[\theta_\infty, \theta_\infty] = \Delta[\hat{F}, \Psi][\theta_\infty] = 0$, since $\Psi[\theta_\infty]$ is designed as a CLM of $\hat{F}[\theta_\infty]$. Similarly continuity of Λ^+ implies that

$$\|\Lambda^+[\theta_h, \theta_{h-1}]P^{h-1}z - \Lambda^+[\theta_h, \theta_{h-1}](0)\|_\infty \leqslant \mu\|P^{h-1}z\|_\infty$$
$$\leqslant \mu\|P^N z\|_\infty + \mu\|P^{(N,h-1]}z\|_\infty$$

and by substituting this bound into (4.14) we obtain a bound for each term on the right-hand side of (4.13):

$$\|P^{[h]}\Lambda^+[\theta_h, \theta_{h-1}]P^{h-1}z\|_\infty \leqslant \mu\|P^{(N,h-1]}z\|_\infty + \mu\|P^N z\|_\infty + \varepsilon. \tag{4.15}$$

Finally, we apply the above to (4.13) and bound $\|\overline{P^N}\Delta|_\theta z\|_\infty$ from above as:

$$\|\overline{P^N}\Delta|_\theta z\|_\infty$$
$$\leqslant \sup_{h>N} \mu\|P^{(N,h-1]}z\|_\infty + \mu\|P^N z\|_\infty + \varepsilon$$
$$\leqslant \mu\sup_{h>N}\sup_{j\in(N,h-1]}|z_j| + \mu\|P^N z\|_\infty + \varepsilon$$
$$\leqslant \mu\sup_{h>N}|z_h| + \mu\|P^N z\|_\infty + \varepsilon = \mu\|\overline{P^N}z\|_\infty + \mu\|P^N z\|_\infty + \varepsilon. \tag{4.16}$$

We are ready to apply our findings for stability analysis of the lumped disturbance dynamics. Recall from our derivations, that $\hat{w}$ is governed by the dynamic equations $\hat{w} = \Delta^+|_\theta\hat{w} + e$ and therefore:

$$\overline{P^N}\hat{w} = \overline{P^N}\Delta^+|_\theta\hat{w} + \overline{P^N}e.$$

Denote the sequence of the intervals $\mathcal{I}_k = [N+1, k], k \geqslant N+1$ and apply the truncation $\boldsymbol{P}^{\mathcal{I}_t}$ to the above equation for some $t > N$ and use causality of $\boldsymbol{\Delta}|_\theta$ to arrive at the equality:

$$\boldsymbol{P}^{\mathcal{I}_t}\hat{\boldsymbol{w}} = \boldsymbol{P}^{\mathcal{I}_t}\overline{\boldsymbol{P}^N}\boldsymbol{\Delta}^+|_\theta\boldsymbol{P}^t\hat{\boldsymbol{w}} + \boldsymbol{P}^{\mathcal{I}_t}\boldsymbol{e}.$$

Now, we apply (4.16) by setting $\boldsymbol{z} = \boldsymbol{P}^t\hat{\boldsymbol{w}}$ and obtain the inequality

$$\|\boldsymbol{P}^{\mathcal{I}_t}\hat{\boldsymbol{w}}\|_\infty \leqslant \mu\|\overline{\boldsymbol{P}^N}\boldsymbol{P}^t\hat{\boldsymbol{w}}\|_\infty + \mu\|\boldsymbol{P}^N\boldsymbol{P}^t\hat{\boldsymbol{w}}\|_\infty + \|\boldsymbol{P}^{\mathcal{I}_t}\boldsymbol{e}\|_\infty + \varepsilon$$

$$\leqslant \mu\|\boldsymbol{P}^{\mathcal{I}_t}\hat{\boldsymbol{w}}\|_\infty + \mu\|\boldsymbol{P}^N\hat{\boldsymbol{w}}\|_\infty + \|\boldsymbol{P}^{\mathcal{I}_t}\boldsymbol{e}\|_\infty + \varepsilon$$

$$\Leftrightarrow \quad (1-\mu)\|\boldsymbol{P}^{\mathcal{I}_t}\hat{\boldsymbol{w}}\|_\infty \leqslant \mu\|\boldsymbol{P}^N\hat{\boldsymbol{w}}\|_\infty + \|\boldsymbol{P}^{\mathcal{I}_t}\boldsymbol{e}\|_\infty + \varepsilon.$$

The constant $\varepsilon = \mu$ was chosen such that $\mu < 1$. Hence from the above we get the inequality

$$\|\boldsymbol{P}^{\mathcal{I}_t}\hat{\boldsymbol{w}}\|_\infty \leqslant \frac{\mu\|\boldsymbol{P}^N\hat{\boldsymbol{w}}\|_\infty + \|\boldsymbol{e}\|_\infty + \mu}{1-\mu}.$$

The right-hand side of the above inequality is bounded because: 1. we assumed a bounded prediction error $\boldsymbol{e} \in \ell_\infty^{\mathcal{X}}$ and 2. N is finite and thus $\|\boldsymbol{P}^N\boldsymbol{P}^t\hat{\boldsymbol{w}}\|_\infty$ [1]. Since the scalar sequence $(s_N, s_{N+1}, \dots)$, where $s_k = \|\boldsymbol{P}^{\mathcal{I}_t}\hat{\boldsymbol{w}}\|_\infty$ is non-decreasing and bounded above, it converges and proves that $\overline{\boldsymbol{P}^N}\hat{\boldsymbol{w}} \in \ell_\infty^{\mathcal{X}}$ and therefore $\hat{\boldsymbol{w}} \in \ell^{\mathcal{X}}$. Moreover, $\|\hat{\boldsymbol{w}}\|_\infty$ can be bounded above in terms of the norm of its finite T-truncation as:

$$\|\hat{\boldsymbol{w}}\|_\infty \leqslant \left\{\|\boldsymbol{P}^N\hat{\boldsymbol{w}}\|_\infty, \frac{\mu\|\boldsymbol{P}^N\hat{\boldsymbol{w}}\|_\infty + \|\boldsymbol{e}\|_\infty + \varepsilon}{1-\mu}\right\}.$$

Finally, we utilize the continuity of the CLM parametrization map $\boldsymbol{\Psi} : (\boldsymbol{\Omega}, d) \to (\ell^{\mathcal{X} \times \mathcal{U}}, \|\cdot\|_\infty)$. Since $\boldsymbol{\Psi}$ is continuous, it also follows that $\mathbb{E}_0 \circ \boldsymbol{\Psi}$ is continuous. Due to the compactness of $\boldsymbol{\Omega}$, there exists some $C > 0$ and $c_0 > 0$ such that $\|\boldsymbol{\Psi}[\omega]\|_\infty \leqslant C$ and $\|\boldsymbol{\Psi}[\omega](0)\|_\infty \leqslant c_0 \ \forall \omega \in \boldsymbol{\Omega}$. This also implies that the component functions have to obey the inequality:

$$|\Psi_t[\omega](z_{t:0})| \leqslant C\|z_{t:0}\|_\infty + c_0.$$

This proves the boundedness of $\boldsymbol{x}$ and $\boldsymbol{u}$, since $(x_t, u_t) = \Psi_t[\theta_t](\hat{w}_{t:0}) \leqslant C\max_{k \leqslant t}|\hat{w}_k| + c_0$, and therefore $\boldsymbol{\Psi} \in \mathcal{C}(\ell_\infty^{\mathcal{X}}, \ell_\infty^{\mathcal{X} \times \mathcal{U}})$. $\qquad\square$

The stability conditions above admit a natural interpretation and provide us with a concrete guideline for nominal control design and the design of model selectors. To

---

[1] Since we are in the discrete-time setting, we do not have to worry about finite-time escapes.

assure closed-loop stability, the design and parametrization of the nominal control behavior, represented by the maps $\hat{F}$ and Ψ, and the online adaptation strategy, denoted by the model selector S, have to fulfill the following requirements:

- Smooth parameterization maps for dynamics and CLMs: The set of system dynamics $\mathcal{F} = \{\hat{F}[\omega] \mid \omega \in \Omega\}$, which describe the model uncertainty, is usually dictated by the specific problem setting and is therefore fixed. However, we have design freedom in how we choose to cover the space $\mathcal{F}$, that is, the choice of (Ω, d) and the parametrization mapping $\hat{F} : \Omega \to \mathcal{C}((\mathcal{X} \times \mathcal{U})^{\mathrm{N}}, \mathcal{X}^{\mathrm{N}})$ such that $\bigcup_{\omega \in \Omega} \hat{F}[\omega] = \mathcal{F}$ is not unique. Moreover, for a given Ω and $\hat{F}$, there is even more freedom in how we assign nominal-CLMs $\Psi[\omega]$ for each $\hat{F}[\omega]$. Conditions iii), iv), and v) of Theorem 23 state that the parameter space Ω and the maps $\hat{F}$, Ψ should be designed so that the parametrization of nominal CLMs and dynamics is as smooth as possible over the space Ω. An intuitive notion of smoothness is presented in the form of continuity of Ψ and the map D, which parametrizes the composition of $\hat{F}$ and Ψ. Notice that this is a weaker requirement than enforcing continuity of $\hat{F}$.

- ℓ_∞-consistent selector S with convergent selection θ: Conditions i) and ii) of Theorem 23 impose intuitive design requirements for the model selector S. The operator S, i.e., the algorithm in charge of the adaptation, should always select parameters θ_t that are consistent with our observations up until time t (allowing for some small ℓ_∞-bounded error), and our selection strategy should be efficient in that it eventually settles on a fixed parameter θ_∞, i.e., S should guarantee the convergence of the sequence θ. As discussed in Part 2 of the book, designing such selectors, and even ones with stronger properties, is possible and requires us to consider the general problem of *Consistent Model Chasing*, which we discuss in Chapter 6.

CE-based adaptive control schemes are commonly used in problem settings concerned with learning optimal controllers [1, 5, 92, 94]. In such scenarios, it is common that each nominal CLM $\Psi[\omega]$ is chosen to be an optimal closed-loop map with respect to the specific optimal control problem (OCP). A general OCP can be succinctly defined by a dynamics operator $F \in \mathcal{C}(\ell^{\mathcal{X} \times \mathcal{U}}, \ell^{\mathcal{X}})$ and a functional $\mathcal{J} : \mathcal{C}(\ell^{\mathcal{W}}, \ell^{\mathcal{X} \times \mathcal{U}}) \mapsto \mathbb{R}^+,$[2]

---

[2] In our formulation, we always set $\mathcal{W} = \mathcal{X}$.

as the following infinite-dimensional optimization problem

$$\mathrm{OCP}[\mathcal{J}, \boldsymbol{F}]: \quad \min_{\boldsymbol{\Phi}} \quad \mathcal{J}(\boldsymbol{\Phi}) \tag{4.17}$$

$$\mathrm{s.t.:} \quad \boldsymbol{\Phi}^{\times} = \boldsymbol{F}^{+}(\boldsymbol{\Phi}) + \boldsymbol{I}$$

$$\boldsymbol{\Phi} \in \mathcal{C}(\ell_p^{\mathcal{W}}, \ell_q^{\mathcal{X} \times \mathcal{U}})$$

over all stable closed-loop maps $\boldsymbol{\Phi} \in \mathcal{C}(\ell_p^{\mathcal{W}}, \ell_q^{\mathcal{X} \times \mathcal{U}})$ of $\boldsymbol{F}^{+}$, where p and q represent some specified notion of closed-loop stability. Correspondingly, the nominal CLM parametrization $\boldsymbol{\Psi}$ is a map $\Omega \ni \omega \mapsto \boldsymbol{\Phi}^{*} \in \arg\min \ \mathrm{OCP}[\mathcal{J}, \hat{\boldsymbol{F}}[\omega]]$ which corresponds dynamics $\hat{\boldsymbol{F}}[\omega]$ with the respective optimal CLM solutions $\boldsymbol{\Phi}^{*} \in \arg\min \ \mathrm{OCP}[\mathcal{J}, \hat{\boldsymbol{F}}[\omega]]$. Moreover, if such solutions are unique, i.e., the set $\arg\min \ \mathrm{OCP}[\mathcal{J}, \hat{\boldsymbol{F}}[\omega]]$ is a singleton, then the nominal CLM parametrization map $\boldsymbol{\Psi}$ is entirely determined by the OCP functional $\mathcal{J}$ and the parametrization of dynamics $\hat{\boldsymbol{F}}$. In light of Theorem 23, establishing the smoothness conditions iv) and v) falls entirely on sensitivity analysis of the solutions to the optimal control problem $\mathrm{OCP}[\mathcal{J}, \boldsymbol{F}[\omega'_{+\delta}]]$ to perturbations $\omega'_{+\delta} \in \mathrm{B}_{\delta}[\omega]$ around fixed parameters $\omega \in \Omega$. Thus, having a well-conditioned OCP formulation over the parametrization $\boldsymbol{F}$ is a prerequisite for using the optimal controllers for online adaptation. This is, again, states an intuitive requirement: Assume, for example, that for some $\omega_{\maltese} \in \Omega$ the dynamics $\hat{\boldsymbol{F}}[\omega_{\maltese}]$ lose u-controllability (in an essential way important to the problem), then we can expect $\mathrm{OCP}[\mathcal{J}, \hat{\boldsymbol{F}}[\omega_{\maltese}]]$ to exhibit discontinuities in any neighborhood of $\omega_{\maltese}$. We explore this connection between the conditioning of optimal control problems and well-posedness for learning-to-control in the well-known and important class of OCPs defined by LQ-functionals $\mathcal{J}$ and LTI dynamics $\hat{\boldsymbol{F}} : \Omega \to \mathcal{LC}(\ell^{\mathcal{W}}, \ell^{\mathcal{X} \times \mathcal{U}})$, where we assume Ω to be a compact subset of some finite-dimensional Euclidean space. The main result assumes $\hat{\boldsymbol{F}}$ to be affine and states that as long as the set of linear dynamics $\{\hat{\boldsymbol{F}}\omega\}_{\omega \in \Omega}$ are all of sufficient *and* of equal degree of controllability, then the natural parametrization of LQ-optimal linear CLMs $\boldsymbol{\Psi} : \Omega \ni \omega \mapsto \boldsymbol{\Phi}^{*}[\omega] \in \mathcal{LC}(\ell^{\mathcal{W}}, \ell^{\mathcal{X} \times \mathcal{U}})$ fulfills the smoothness conditions iv) and v) of Theorem 23. Therefore, CE-adaptation with any model selector S that meets conditions i) and ii) of Theorem 23 yields stable learning-to-control in closed-loop (1).

4.3 Certainty Equivalent Adaptation with LQ-Optimal SL Controllers

LQR is a canonical and well-studied problem of optimal control [77, 85, 132]. In recent years, it has received revived attention in the context of learning and control problems [2, 37, 38, 92]. The general problem setup assumes an LTI system of the form $x_k = Ax_k + Bu_{k-1} + w_k$, with state $x \in \mathbb{R}^n$, input $u \in \mathbb{R}^m$, i.i.d standard Gaussian disturbance $w_t \sim \mathcal{N}(0, I_n)$, and system matrices $A \in \mathbb{R}^{n \times n}$ and $B \in \mathbb{R}^{n \times m}$. Our objective is to find a linear causal controller $\boldsymbol{K} \in \mathcal{LC}(\ell^{\mathcal{U}}, \ell^{\mathcal{X}})$ that solves the optimization problem

$$\min_{\boldsymbol{K} \in \mathcal{LC}(\ell^{\mathcal{U}}, \ell^{\mathcal{X}})} \quad \limsup_{T \to \infty} \frac{1}{T} \sum_{k=0}^{T} \mathbb{E}_{w_k^i \sim \mathcal{N}(0,I)}[x_t^\top C^\top C x_t + u_t^\top D^\top D u_t] \quad (4.18\text{a})$$

$$\text{s.t.:} \qquad x_k = Ax_{k-1} + Bu_{k-1} + w_k, \text{ for } k \geqslant 1 \quad (4.18\text{b})$$

$$u_k = K_k(x_{k:0}), \text{ for } k \in \mathbb{N}, \qquad x_0 = w_0, \quad (4.18\text{c})$$

where C and D are fixed matrices of full rank, i.e., $C^\top C > 0$, $D^\top D > 0$. As stated, it is well-known that the solution to the above problem is the static linear feedback control law $u_{lqr} : x \mapsto -K^* x$, where the gain matrix K^* is obtained by solving the Discrete Algebraic Ricatti-Equation (DARE). However, in applications with systems of large scale (such as, for example, the power grid), it is not feasible to implement the former in practice due to technical limitations in sensing, actuation, and communication. To account for that, one has to translate the former into constraints on the controller structure, which we then incorporate into the problem formulation. A common constraint is communication delay between sensors and actuators of different subsystems, which can be translated into spatial-temporal subspace constraints on the operator $\boldsymbol{K}$, as shown in the example below:

Example. *Assume we have p separate actuators and q different sensors that partition the input- and state-space into orthogonal subspaces as $\mathcal{V}_1 \oplus \cdots \oplus \mathcal{V}_p = \mathbb{R}^m$ and $\mathcal{U}_1 \oplus \cdots \oplus \mathcal{U}_q = \mathbb{R}^n$, respectively. Furthermore, let $V_i V_i^\top$ and $U_j U_j^\top$ denote orthogonal projection maps onto the subspaces $\mathcal{V}_i$ and $\mathcal{U}_j$, and denote $\boldsymbol{\Pi}_i^V = \boldsymbol{I}_m \otimes V_i V_i^\top$ and $\boldsymbol{\Pi}_j^U = \boldsymbol{I}_n \otimes U_j U_j^\top$ their diagonal extensions to $\mathcal{LC}(\ell^{\mathcal{U}}, \ell^{\mathcal{U}})$ and $\mathcal{LC}(\ell^{\mathcal{X}}, \ell^{\mathcal{X}})$, respectively. Now, if each actuator i receives the measurement from sensor j with $d(i, j)$ time-steps delay, we can formulate these $i \cdot j$ number of constraints as:*

$$\forall i, j, \tau : \qquad \boldsymbol{P}^\tau \boldsymbol{\Pi}_i^V \boldsymbol{K} \boldsymbol{\Pi}_j^U = \boldsymbol{P}^\tau \boldsymbol{\Pi}_i^V \boldsymbol{K} \boldsymbol{\Pi}_j^U \boldsymbol{P}^{\tau - d(i,j)}. \quad (4.19)$$

Adding constraints, however, such as (4.19) to the LQR problem (4.18), makes the problem hard to solve with traditional approaches. With the introduction of SLS [11, 128], this changed: By changing the search space from $\boldsymbol{K}$ to the CLMs $\boldsymbol{\Phi}$, we can formulate tractable subspace constraints on $\boldsymbol{\Phi}$, which are sufficient for $\boldsymbol{K} = \boldsymbol{\Phi}^u (\boldsymbol{\Phi}^x)^{-1}$ to fulfill (4.18). A common approximation [11] is to restrict the CLM search to FIR maps of a fixed finite horizon t. Then, we obtain the following finite-dimensional QP problem for Φ: The resulting optimization problem takes the form:

$$\text{LQ}(A, B, t): \quad \min \left\| \begin{bmatrix} C & 0 \\ 0 & D \end{bmatrix} \begin{bmatrix} R_1 & R_2 & \dots & R_t \\ M_1 & M_2 & \dots & M_t \end{bmatrix} \right\|_F^2 \tag{4.20}$$

$$\text{s.t.:} \ R_1 = I$$
$$R_{k+1} = AR_k + BM_k, \quad \forall k : 1 \leqslant k \leqslant t$$
$$R_{t+1} = 0$$

where $R_k \in \mathbb{R}^{n \times n}$ and $M_k \in \mathbb{R}^{m \times n}$ represent matrices associated with the component functions of cCLMs

$$\Phi_k^x(w_{k:0}) := \sum_{j=1}^{\min\{t,k+1\}} R_j w_{k+1-j} \quad \Phi_k^u(w_{k:0}) = \sum_{j=1}^{\min\{t,k+1\}} M_j w_{k+1-j}. \tag{4.21}$$

For fixed A, B, and horizon t, the optimal CLMs $\Phi^*(A, B, t) \in \mathcal{LC}(\ell^{\mathcal{X}}, \ell^{\mathcal{X} \times \mathcal{U}})$, exist if $[A, B]$ is t-controllable (i.e., it is possible to drive the system state from any initial condition to the origin within t time-steps) and are unique if we assume that C, D, and B have a trivial null-space.

LQ-Optimal CLMs for Nominal SL Control

Motivated by the scalability of SLS in complex system settings, such as large-scale systems, we investigate the LQ-optimal SL controllers of the problem (4.20) for CE-based learning and control.

Setup. We assume that we are given an uncertainty set of LTI dynamics, described by a compact metric space (Ω, d) of LTI systems, and an affine parameterization map $\hat{F}$: $\Omega \to \mathcal{LC}(\ell^{\mathcal{X} \times \mathcal{U}}, \ell^{\mathcal{X}})$, where each operator $\hat{F}[\omega]$ is linear, diagonal and has component functions $\{\hat{F}_k[\omega] : (x_{k:0}, u_{k:0}) \mapsto f[\omega](x_k, u_k)\}_{k \in \mathbb{N}}, f[\omega] : (x, u) \mapsto A[\omega]x + B[\omega]u$ for some fixed continuous matrix-valued functions $A : (\Omega, d) \to (\mathbb{R}^{n \times n}, |\cdot|)$ and $B : (\Omega, d) \to (\mathbb{R}^{m \times n}, |\cdot|)$. We consider a *fixed horizon t* and let the nominal CLM parameterization $\Psi : \Omega \to \mathcal{LC}(\ell^{\mathcal{X}}, \ell^{\mathcal{X} \times \mathcal{U}})$ assign to each parameter $\omega \in \Omega$ the LQ-optimal CLM $\Psi[\omega] := \Phi^*(A[\omega], B[\omega], t)$ w.r.t. to the optimization problem $\text{LQ}(A[\omega], B[\omega], t)$ described by (4.20) and (4.21). We leave the model selector $S \in \mathcal{C}(\ell^{\mathcal{X} \times \mathcal{U}}, \ell^{\Omega})$ unspecified and assume that it is ℓ_∞ consistent and guarantees a converging parameter selection $\theta \in \ell^{\Omega}$. As we discuss in detail in Chapter 6 in Part 2 of the book, this assumption is justified, as we can tackle the design of S in *isolation* from the question of nominal control design, as a problem instance of *Consistent Model Chasing*.

Algorithm 1 Learning-to-Control with LQ-optimal SL Controllers

---

Setup: model selector $S \in \mathcal{C}(\ell^{\mathcal{X} \times \mathcal{U}}, \ell^{\Omega})$, compact parameter space (Ω, d), continuous maps $A : \Omega \to \mathbb{R}^{n \times n}$ and $B : \Omega \to \mathbb{R}^{m \times n}$, specified $\mathrm{LQ}(A', B', t)$ with fixed invertible C, D, $t \in \mathbb{N}$ and online solver $\Phi^*(A', B', t)$, CLM parameterization $\Psi : \omega \mapsto \Phi^*(A[\omega], B[\omega], t)$

Initialization: choose some $\theta_0 \in \Omega$

1: **for** $k = 0, 1, \ldots$ **to** ∞ **do**
2:      observe x_k
3:      $\theta_k \leftarrow S_k(x_{k:0}, u_{k-1:0})$ $\triangleright$ update model to $\theta_k \in \Omega$
4:      solve $\mathrm{LQ}(A[\theta_k], B[\theta_k], t)$ from (4.20) $\triangleright$ Solve LQ-OCP
5:      Set $\Psi[\theta_k] \leftarrow \Phi^*(A[\theta_k], B[\theta_k], t)$ from (4.21) $\triangleright$ Synthesize optimal CLM
6:      $\hat{w}_k \leftarrow x_k - \Psi_k^x[\theta_k](0, \hat{w}_{k-1:0})$ $\triangleright$ adapt SL-controller
7:      $u_k \leftarrow \Psi_k^u[\theta_k](\hat{w}_{k:0})$
8:      apply action u_k
9: **end for**

Remark. $\hat{F}[\omega]$ *can be expressed as* $\sum_{k=0}^{\infty} U_n^{*[k]} A[\omega] U_n^{[k]} + U_n^{*[k]} B[\omega] U_m^{[k]}$.

The CE-Adaptive Control Algorithm. The learning-to-control algorithm is described in Algorithm 1 and naturally inherits many of the benefits and features of basic System Level Synthesis. [66] first recognized the importance of SL controllers for online learning and control and showed that by extending the system-level robustness analysis of [127], [11] to linear time-varying systems, one can obtain scalable and easy-to-analyze adaptive control methods via system-level controllers. A key difference from [66] is that in [66], the parameterization Ψ consisted of candidate CLMs synthesized to be robust to a limited amount of parametric uncertainty and, therefore, closed-loop stability was only guaranteed if Ω was not too large. As illustrated in Algorithm 1, in each time step k we select the system matrices $A_k = A[\theta_k]$, $B_k = B[\theta_k]$ and synthesize an LQ-optimal LQ controller by solving the problem $\mathrm{LQ}(A_k, B_k, t)$ presented in (4.20). Note that whether or not the parameterization Ψ is available offline or computed on demand online does not make any difference for closed-loop stability analysis. As described in detail in [11], we can add – at no cost – linear constraints to ensure that the controller implementation is aligned with real-world restrictions such as delay and sparsity of sensing, actuation and communication. Almost all attractive features of the SLS approach [11] carry over into our online learning and control setting.

Closed-Loop Dynamics. Appealing to our previous discussion and derivation of Theorem 22 and Lem. 22, the closed-loop dynamics are described by the following set of equations:

$$\forall k \in \mathbb{N}: \quad x_k = \Psi_k^x[\theta_k](\hat{w}_{k:0}) \tag{4.22a}$$

$$\theta_k = S_k(x_{k:0}, u_{k-1:0}) \tag{4.22b}$$

$$e_k = x_k - A[\theta_k]x_{k-1} - B[\theta_k]u_{k-1} + w_k \tag{4.22c}$$

$$\hat{w}_k = \underbrace{f[\theta_{k-1}](\Psi_{k-1}[\theta_{k-1}] - \Psi_{k-1}[\theta_k])}_{U^{*[k]}\Delta_{|\theta}^+ U^{[k-1]}}(\hat{w}_{k-1:0}) + e_k \tag{4.22d}$$

$$u_k = \Psi_k^u[\theta_k](\hat{w}_{k:0}) \tag{4.22e}$$

where $\Psi : \Omega \ni \omega \mapsto \Phi^*(A[\omega], B[\omega], t)$ and f represents the linear, time-invariant transition function $f[\omega] : (x, u) \mapsto A[\omega]x + B[\omega]u$. Let $\mathsf{M}_\Omega^{*\infty} \subset \ell^{\mathcal{X} \times \mathcal{U} \times \mathcal{W} \times \mathcal{X} \times \mathcal{W}}$ denote the model set of the above closed-loop and $\mathsf{M}_\Omega^\infty \subset \ell^{\mathcal{X} \times \mathcal{U} \times \mathcal{W}}$ the model of the uncertain open-loop dynamics for bounded disturbance w:

$$\mathsf{M}_\Omega^{*\infty} := \{\boldsymbol{\tau}^* = (\boldsymbol{x}, \boldsymbol{u}, \hat{\boldsymbol{w}}, \boldsymbol{e}, \boldsymbol{w}) \text{ s.t.: } (4.22) \text{ and } \boldsymbol{w} \in \ell_\infty^{\mathcal{W}}\} \tag{4.23}$$

$$\mathsf{M}_\Omega^\infty := \{\boldsymbol{\tau} = (\boldsymbol{x}, \boldsymbol{u}, \boldsymbol{w}) \text{ s.t.: } \boldsymbol{x} = \hat{\boldsymbol{F}}[\omega](\boldsymbol{x}, \boldsymbol{u}) + \boldsymbol{w} \text{ for some } \omega \in \Omega, \boldsymbol{w} \in \ell_\infty^{\mathcal{W}}\}. \tag{4.24}$$

Investigating closed-loop stability in the ℓ_∞-BIBO sense now means proving that all trajectories $\boldsymbol{\tau}^* \in \mathsf{M}_\Omega^{*\infty}$ are bounded, i.e., that $\mathsf{M}_\Omega^{*\infty}$ is a subset of $\ell_\infty^{\mathcal{X} \times \mathcal{U} \times \mathcal{W} \times \mathcal{X} \times \mathcal{W}}$.

ℓ_∞-BIBO-Stability of Closed-Loop Dynamics. We appeal to the conditions of Theorem 23 for stability analysis of the closed-loop dynamical system. Since we assumed that S guarantees $\boldsymbol{e} \in \ell^\infty$, $\exists \theta_\infty$ s.t.: $\lim_{t \to \infty} \theta_t = \theta_\infty$ and we have $\Delta[\hat{\boldsymbol{F}}[\omega], \Psi[\omega]] = 0, \forall \omega \in \Omega$ due to $\Psi[\omega] := \Phi^*(A[\omega], B[\omega], t)$, verifying stability reduces down to checking iv) and v). Furthermore, it can be shown that in our setup iv) actually implies v), and therefore, checking the continuity of the CLM parameterization map Ψ over Ω is sufficient for verifying both conditions. This is easily seen by first rewriting the difference $\hat{\boldsymbol{F}}[p_1]\Psi[p_2] - \hat{\boldsymbol{F}}[\omega]\Psi[\omega]$ as

$$\hat{\boldsymbol{F}}[p_1]\Psi[p_2] - \hat{\boldsymbol{F}}[\omega]\Psi[\omega]$$
$$= \hat{\boldsymbol{F}}[p_1]\Psi[p_2] - \hat{\boldsymbol{F}}[p_1]\Psi[p_1] + \hat{\boldsymbol{F}}[p_1]\Psi[p_1] - \hat{\boldsymbol{F}}[\omega]\Psi[\omega]$$
$$= \hat{\boldsymbol{F}}[p_1](\Psi[p_2] - \Psi[p_1]) + \Psi^\times[p_1] - \Psi^\times[\omega], \tag{4.25}$$

where the last equation follows from $\Delta[\hat{\boldsymbol{F}}[\omega], \Psi[\omega]] = 0$ and the linearity of $\hat{\boldsymbol{F}}$, and second by noticing that the $\|\cdot\|_\infty$-norm of $\hat{\boldsymbol{F}}[p_1](\Psi[p_2] - \Psi[p_1])$ is bounded above as

$$\|\hat{\boldsymbol{F}}[p_1](\Psi[p_2] - \Psi[p_1])\|_\infty \leqslant c_\Omega \cdot \|\Psi[p_2] - \Psi[p_1]\|_\infty \tag{4.26}$$

where $c_\Omega = \max\limits_{p \in \Omega}\{|A[p]| + |B[p]|\}$; the maximum c_Ω exists because (Ω, d) is a compact metric space and the parameterizations A and B are continuous maps on (Ω, d). Hence, from (4.25) and (4.26) and since A and B were already assumed to be continuous, the stability conditions iv) and v) would be satisfied if $\Phi^*(A, B, t)$ is continuous in A and B. Therefore, we need to investigate the sensitivity of LQ optimal CLM solutions $\Phi^*(A + \partial A, B + \partial B, t)$ with respect to small matrix perturbations $\partial A \in \mathbb{R}^{n \times n}$, $\partial B \in \mathbb{R}^{n \times m}$. This perturbation analysis constitutes the second part of this chapter and culminates in the theorem stated below:

Theorem 24 (Lipschitzness of LQ-optimal CLM oracle). *Let $t \in \mathbb{N}$ be fixed, $C \in \mathbb{R}^{n \times n}$, $D \in \mathbb{R}^{m \times m}$ be fixed invertible matrices, and $\Phi^*(A, B, t)$ represent the unique optimum of the optimal control problem $\mathrm{LQ}(A, B, t)$ described by (4.20), (4.21). Assume that $\mathcal{S}_{ab}$ is a compact subset of $\mathbb{R}^{n \times (n+m)}$ such that each pair $[A, B] \in \mathcal{S}_{ab}$ is t-controllable, i.e., the matrix $P_t(A, B) = [A^{t-1}B, A^{t-2}B, \cdots, B] \in \mathbb{R}^{n \times mt}$ is of rank n. Then there exists a fixed positive constant $L \in \mathbb{R}^+$ such that for all $[A_1, B_1], [A_2, B_2] \in \mathcal{S}_{ab}$ holds*

$$\|\Phi^*(A_1, B_1, t) - \|\Phi^*(A_2, B_2, t)\|_\infty \leqslant L(|\partial_{12}A| + |\partial_{12}B|)$$

where $\partial_{12}A := A_1 - A_2$ and $\partial_{12}B := B_1 - B_2$.

The above result, which will be stated in much greater detail later, states that the map $\Phi^*(A', B', t)$ is Lipshitz-continuous over compact sets of t-controllable pairs of matrices $[A', B']$, and therefore our CLM parametrization map Ψ is indeed continuous over (Ω, d). Hence, in summary, as long as all system matrix pairs $\{[A[\omega], B[\omega]]\}_{\omega \in \Omega}$ are t-controllable, all stability conditions imposed on the nominal control design, i.e., iii), iv), v) of Theorem 23, are met, and the closed-loop system (4.22) is ℓ_∞-BIBO stable (with a fitting choice of S). We summarize our discussion in the theorem stated below:

Theorem 25 (ℓ_∞-stability of Closed-Loop Alg.1). *Recall the Closed-Loop setup described in Algorithm 1 and the corresponding closed-loop equations (4.22) and sets $\mathrm{M}_\Omega^{*\infty}$ and M_Ω^∞ of closed-loop (4.23) and open-loop (4.24) trajectories, respectively. All closed-loop trajectories $\tau \in \mathrm{M}_\Omega^{*\infty}$ are bounded if the following conditions are met:*

1. *If $\theta = S(x, u)$, where $(x, u, w) \in \mathrm{M}_\Omega^\infty$, then θ converges in (Ω, d) and*
   $$\sup_{k \in \mathbb{N}} |x_k - A[\theta_k]x_{k-1} - B[\theta_k]u_{k-1}| < \infty.$$

2. Each pair of matrices $[A[\omega], B[\omega]]$, $\omega \in \Omega$ is t-controllable.

Proof. If the model selector S satisfies the consistency Assumption (1), then the conditions (i)) and (ii)) of Theorem 23 hold for any sequence of selections θ and consistency errors e of a closed-loop trajectory $(x, u, \hat{w}, e, w) \in M_\Omega^{*\infty}$. Condition Theorem 23(iii)) is automatically satisfied, since $\Psi[\omega] = \Phi^*(A[\omega], B[\omega], t)$ are always CLMs of the corresponding dynamics $\hat{F}[\omega]$. Since (Ω, d) is compact and A and B are continuous maps, it follows that the image $S_\Omega = \{[A[\omega], B[\omega]] \mid \omega \in \Omega\}$ is a compact subset of $\mathbb{R}^{n \times (n+m)}$. By Assumption (2), we also conclude that all pairs $[A', B'] \in S_\Omega$ are t-controllable, and therefore, by Lipshitzness of the LQ-optimal CLM map Φ^* (according to Theorem 24) and continuity of A and B, we conclude that Ψ is continuous over (Ω, d), and therefore condition Theorem 23(iv)) is met. Moreover, as derived earlier via (4.25) and (4.26), the former implies also that condition Theorem 23(v)) is true. Finally, we invoke Theorem 23 and conclude that all closed-loop trajectories $(x, u, \hat{w}, e, w) \in M_\Omega^{*\infty}$ have to be bounded. $\qquad\square$

The conditions stated above show that closed-loop stability is largely characterized by the properties of the model selection S. Moreover, it is important to notice that it is nowhere required that the selections θ converge to the true system parameter θ^*, merely that it converges to *some* $\theta_\infty \in \Omega$. This is a major distinction from most modern learning and control algorithms [3, 35, 50] which explicitly require accurate system identification in order to guarantee stability. In Part 2 of the book, we will discuss how to define such procedures S through *Consistent Model Chasing*. Moreover, aside from convergence, stronger properties (competitiveness) can be obtained for S which allow for sharper analysis of the transient and cost performance of the closed-loop. A key component towards establishing these types of guarantees is proving that the CE-LQ optimal CLMs Φ change indeed gradually with perturbations in θ. Sensitivity analysis of the (*oracle*) map $\Phi : \Omega \to \mathcal{LC}(\ell^\mathcal{X}, \ell^{\mathcal{X} \times \mathcal{U}})$ is an important part of what we later refer to as *robust oracle design*. In the remaining part of this chapter, we will prove the previously claimed Lipschitz property of the optimal CLM parametrization map $\Phi^*(A', B', t)$ and give analytic bounds of the Lipschitz constant L in terms of system-theoretic properties such as controllability and observability. We start the second part of this chapter with a discussion on basic control theory results regarding controllability, observability, and grammians, which we refine for the use-case of controlling large-scale systems.

4.4 Digression on Linear Algebra and Control Theory Basics

Here we discuss some basic linear algebra results that are frequently used in our arguments but are not always mentioned explicitly in our derivations. For the following definitions, let M denote an arbitrary matrix $M \in \mathbb{R}^{n \times m}$, where w.l.o.g. $n \geq m$.

Reduced SVD Decomposition

The factorization $M = U_r \Sigma_r V_r^\top$, $U_r \in \mathbb{R}^{n \times k}$, $\Sigma_r \in \mathbb{R}^{k \times k}$, $V^\top \in \mathbb{R}^{k \times m}$ is called a *reduced* SVD decomposition of M, if $U_r^\top U_r = I_k$, $\Sigma_r > 0$ is positive-definite and diagonal, and $V_r^\top V_r = I_k$. The diagonal entries of Σ_r are the non-zero singular values of M. We denote $\underline{\sigma}(M) \in \mathbb{R}^n$ as the vector $[\sigma_1, \sigma_2, \ldots, \sigma_k, \sigma_{k+1}, \ldots, \sigma_n]^\top$ of singular values of M, ordered in descending order; the diagonal entries (up to some permutation) of Σ_r are $\sigma_1, \ldots, \sigma_k$, while $\sigma_{k+1} = \cdots = \sigma_n = 0$. M is invertible if and only if the reduced SVD is such that $k = n$. Correspondingly, $\sigma_1 = \sigma_{\max}(M) = \|M\|_2$, $\sigma_n = \sigma_{\min}(M)$ and we define $\sigma_{-1} := \sigma_k(M)$ as the smallest *non-zero* singular value of M. If M is invertible, then $\sigma_{-1}(M) = \sigma_{\min}(M)$ and vice versa.

Moore-Penrose Inverse

The Moore-Penrose inverse of M can be uniquely defined as $M^\dagger = V_r \Sigma_r^{-1} U_r^\top$ for any reduced SVD decomposition of M. It is also the unique matrix $M^\dagger$ satisfying all of the following four conditions:

$$MM^\dagger M = M \qquad\qquad M^\dagger M M^\dagger = M^\dagger$$
$$(MM^\dagger)^\top = MM^\dagger \qquad\qquad (M^\dagger M)^\top = M^\dagger M.$$

Here some of the key-properties we use of $M^\dagger$:

Lemma 19. *For any matrix M and its pseudo-inverse $M^\dagger$ holds:*

i) $MM^\dagger$ *and* $M^\dagger M$ *are orthogonal projections with rank k.*

ii) $M^\dagger$ *is the (left/right)-inverse of M if and only if M is (left/right)-invertible.*

iii) $M^\dagger = \lim_{t \searrow 0}(M^\top M + tI)^{-1}M^\top$ *and* $M^\dagger x = x^\dagger := \lim_{t \searrow 0} x_t^\dagger$ *for any x where* $x_t^\dagger := (M^\top M + tI)^{-1}M^\top x$ *denotes the unique solution to the optimization problem*

$$\min_v \|Mv - x\|_2^2 + t\|v\|_2^2.$$

iv) *The optimal value of the least-squares problem* $\min_v \|Mv - x\|_2^2$ *is attained at* $v^* = M^\dagger x$ *and if there is more than one minimizer, then v^* is the unique one* *of smallest $\|\ \|_2$-norm.*

v) $\|M^\dagger\|_2^{-1} = \sigma_{-1}(M)$ *and* $\|M\|_2^{-1} = \sigma_{-1}(M^\dagger)$.

vi) $M^\dagger = (M^\top M)^\dagger M^\top$ *and* $M^\dagger = M^\top (MM^\top)^\dagger$.

4.5 Controllability and Observability at a Fixed Time

Large-scale systems often consist of many sparsely interconnected small systems, i.e., the dimension of the large system is $n = Nn_S$, where n_S denotes the typical dimensions of the small systems, and N denotes the number of subsystems. Denote $x_t^{(i)} \in \mathbb{R}^{n_S}$ as the state of the ith subsystem and $u^{(i_1)\top}, u^{(i_2)\top}, \ldots, u^{(i_M)\top}$ denote the inputs of the M ($< N$) systems $i_1, i_2, \ldots, i_M$ which have actuation available. Let $x = \left[x^{(1)\top}, x^{(2)\top}, \ldots, x^{(N)\top}\right]$ denote the state of the overall interconnected system and $u = \left[u^{(i_1)\top}, u^{(i_2)\top}, \ldots, u^{(i_M)\top}\right] \in \mathbb{R}^m$ the total input of dimension $m = Mm_S$, where m_S is the dimension of the inputs of the subsystems. Typically, the large dimension of the overall system is due to N and M being large numbers, i.e., $n \gg n_S$, $m \gg m_S$.

If the overall system is controllable, we know that any initial condition x_0 can be controlled to the origin within at most n time-steps. However, in the above case $n = Nn_S$, this is often a conservative statement: If communication between subsystems is fast enough, interconnection is sparse, and actuation is available in sufficiently many subsystems (M is on the order of N), it is possible to control arbitrary initial conditions to the origin within much fewer steps $t \ll n$.

Motivated by applications in large-scale control settings, we refine the notion of controllability for this purpose. We shall call $[A, B]$ to be t-controllable if within t time-steps we can control any initial condition x_0 to the origin - or equivalently we can reach any target state x_f at time t - regardless of the initial condition. We adopt basic controllability results to the finite time setting below:

Lemma 20. *The following statements are equivalent:*

- $[A, B]$ *is t-controllable.*
- *For any pair $\zeta_0, \zeta_f \in \mathbb{R}^n$, there exists $u_0, u_1, \ldots, u_{t-1}$ such that $x_0 = \zeta_0$, $x_t = \zeta_f$, where*

$$x_{k+1} = Ax_k + Bu_k, \forall \quad k \leqslant t.$$

- *The matrix $P_t = \left[A^{t-1}B, A^{t-2}B, \ldots, B\right] \in \mathbb{R}^{n \times tm}$ is full row rank.*

The fourth condition gives us a necessary condition for t-controllability:

$$t > \frac{n}{m} = \frac{N}{M}\frac{n_S}{m_S}.$$

Similarly, we define that a pair (C, A) is t-observable if the sequence of observations $y_0, \ldots, y_{t-1}$, $y_k = Cx_k$ of an impulse response $x_k = A^k\xi_0$ are sufficient to compute the initial condition ξ_0. As expected, t-observability has multiple equivalent definitions, which are dual to those of controllability.

Lemma 21. *The following statements are equivalent:*

- *(C, A) is t-observable.*
- *Let $y^1_{[0:t-1]}$, $y^2_{[0:t-1]}$ be observations $y^i_k = CA^k\zeta_i$, of the state-trajectories $x^1_k = CA^k\zeta_1$ and $x^2_k = CA^k\zeta_2$, respectively. Then it holds:*

$$y^1_{[0:t-1]} = y^2_{[0:t-1]} \Leftrightarrow \zeta_1 = \zeta_2.$$

- *The matrix $Q_t = [C^\top, (CA)^\top, \ldots, (CA^{t-1})^\top]^\top$ is full column rank.*

Controllability and Observability Grammians

The relationship between controllability and observability and their corresponding grammians, defined below, will play an important role in our derivations. In contrast to standard literature, we discuss this interplay also in the setting of partial observability and partial controllability. To this end, we make use of the Moore-Penrose inverse.

Definition 4.2 (controllability/observability grammians). *Let $C \in \mathbb{R}^{h \times n}$, $A \in \mathbb{R}^{n \times n}$, $B \in \mathbb{R}^{n \times m}$ be fixed system matrices. The sequence of positive-semidefinite matrices $\{W^c_t\}$ and $\{W^o_t\}$ are called controllability and observability grammians of the system $x_{k+1} = Ax_k + Bu_k$, $y_k = Cx_k$, respectively, if they satisfy the following set of dynamic equations:*

$$W^c_0 = BB^\top, \quad W^c_t = AW^c_{t-1}A^\top + BB^\top, \tag{4.27}$$

$$W^o_0 = C^\top C, \quad W^o_t = A^\top W^o_{t-1}A + C^\top C. \tag{4.28}$$

By substitution, we obtain the explicit form of W^c_t and W^o_t:

$$W^c_t \text{ s.t. } (4.27) \iff W^c_t = P_tP_t^\top = \sum_{k=0}^{t} A^k BB^\top A^{k\top}$$

$$W^o_t \text{ s.t. } (4.28) \iff W^o_t = Q_t^\top Q_t = \sum_{k=0}^{t} A^{k\top} C^\top C A^k.$$

It is easy to see, using standard arguments, that the controllability grammians obey the following properties:

Lemma 22. *Let $\{W_t^c\}$ be the controllability grammians of $[A, B]$. Then, for any $t \geq 1$,*

1. *$[A, B]$ is t-controllable if and only if $W_t^c > 0$.*

2. *If $[A, B]$ is t-controllable, then, given some target state ζ_f and initial condition $x_0 = 0$, we can compute the optimal (unique) input sequence that drives the system to $x_t = \zeta_f$ with minimal cost $\|\boldsymbol{u}\|_2^2 = \sum_{k=0}^{t-1} \|u_k\|_2^2$ as*

$$\boldsymbol{u}^* = P_t^\top W_t^{-1} \zeta_f, \text{ where } \boldsymbol{u}^* = [u_0^{*\top}, u_1^{*\top}, \ldots, u_{t-1}^{*\top}]^\top. \tag{4.29}$$

   The optimal cost is $\|\boldsymbol{u}^*\|_2^2 = \zeta_f^\top W_t^c \zeta_f$.

3. *More generally, for a given $\zeta_f \in \mathbb{R}^n$, $x_t^* = P_t P_t^\dagger \zeta_f$ is the state closest in 2-norm among all states reachable within t time-steps, starting from the origin. Among all input sequences that reach x_t^* at time t, $\boldsymbol{u}^* = P_t^\top (W_t^c)^\dagger \zeta_f$ is the one with the smallest ℓ_2-norm. Furthermore, it holds $\|\boldsymbol{u}^*\|_2^2 = \zeta_f W_t^c \zeta_f$.*

Proof. The first statement is obvious after recalling (4.27). For the second, notice that the desired optimal control problem reduces to the least squares problem:

$$\min_{\boldsymbol{u}} \quad \|\boldsymbol{u}\|_2^2$$
$$\text{s.t.} : \quad \zeta_f = P_t \boldsymbol{u}.$$

Since $[A, B]$ is t-controllable, it holds that P_t is full-column rank and $W_t = P_t P_t^\top > 0$, and the above problem is feasible for any $\zeta_f \in \mathbb{R}^n$. The unique solution is $\boldsymbol{u}^* = P_t^\top (P_t P_t^\top)^{-1} \zeta_f = P_t^\top W_t^{-1} \zeta_f$. The last statement follows by recalling the properties of the pseudo-inverse. $\qquad\square$

Lemma 23. *Let $\{W_t^o\}$ be the observability grammians of (C, A). Then, for any $t \geq 1$,*

1. *(C, A) is t-observable if and only if $W_t^o > 0$.*

2. *If (C, A) is t-observable, then for any sequence of observations $y_{[0:t-1]}$, $y_t = C x_t$ of an impulse response $x_t = A^t \xi_0$, it holds that $\xi_0 = (W_t^o)^{-1} Q_t^\top y_{[0:t-1]}$. Moreover, for an arbitrary $\hat{y}_{[0:t-1]}$, $\hat{\xi}_0 = (W_t^o)^\dagger Q_t^\top \hat{y}_{[0:t-1]}$ is the initial condition [3] which produces the closest feasible impulse response $y_{[0:t-1]} = Q_t \hat{\xi}_0$, i.e.,*
$$\|y - \hat{y}\|_2^2 = \min_\xi \sum_{k=0}^{t-1} \|\hat{y}_t - C A^k \xi\|_2^2 = \hat{\xi}_0^\top W_T^o \xi_0.$$

3. *$\mathcal{R}(W_t^o)$ is the subspace of observable states in $\mathbb{R}^n$ at time t.*

---

[3] And is of smallest 2-norm if there are multiple such initial conditions.

4.6 Perturbation Analysis of Open-Loop Maps over Finite Time-Horizon

In this section, we discuss how linear system behavior changes with different system matrices A, B under *feedforward/open-loop* control. We consider the following LTI system with state $x \in \mathbb{R}^n$, control input $u \in \mathbb{R}^m$, disturbance $w \in \mathbb{R}^n$, and the two outputs $y \in \mathbb{R}^n$ and $z \in \mathbb{R}^{n+m}$:

$$x_k = A_i x_{k-1} + B_i u_{k-1} + w_{k-1}, \qquad z_k = \begin{bmatrix} C x_k \\ D u_k \end{bmatrix} \qquad y_k = C x_k. \tag{4.30}$$

We consider the output matrices C and D to be fixed and invertible, and B_i to always be of full column rank. For a given pair A_i, B_i and a fixed horizon t, we define the open-loop map $\boldsymbol{G}_t$ as the causal map $\boldsymbol{G}_t(A_i, B_i)(\boldsymbol{w}, \boldsymbol{u})_{[0:t]} \mapsto (\boldsymbol{x}, \boldsymbol{y})_{[0:t]}$ between the input signals $\boldsymbol{w}, \boldsymbol{u}$ and the state and output signals $\boldsymbol{x}, \boldsymbol{y}$. Our goal is to analyze the sensitivity of the map $\boldsymbol{G}_t(A_i, B_i)$ with respect to changes in (A_i, B_i).

Setup and Notation

We represent the system (4.30) in batch form: Let $\underline{x}_t$ denote the stacked vector $[x_0^\top, x_1^\top, \ldots, x_t^\top]^\top$, and define $\underline{u}_t, \underline{w}_t, \underline{y}_t, \underline{z}_t$ accordingly. Let $\boldsymbol{E}_k^\top \in \mathbb{R}^{n \times (t+1)n}$, for $k \in \{0, 1, \ldots, t\}$, denote the map $\boldsymbol{E}_k^\top : \underline{x}_t \mapsto x_k$, i.e., $\boldsymbol{E}_k$ is a block-column matrix containing all zero matrices $0_{n \times n}$, except for the k-th block row, which contains the identity matrix I_n:

$$\boldsymbol{E}_k^\top = \begin{bmatrix} 0_{n\times n} & \cdots & \underbrace{I_n}_{k+1} & \cdots & 0_{n\times n} \end{bmatrix}.$$

We represent the open-loop map $\boldsymbol{G}_t : (\underline{w}_t, \underline{u}_t) \mapsto (\underline{x}_t, \underline{y}_t)$ as a Toeplitz matrix, and, as shown below, we decompose it into block matrices, each representing a partial open-loop map:

$$\begin{matrix} & \underline{w}_t & \underline{u}_t \\ \begin{matrix} \underline{x}_t \\ \underline{y}_t \end{matrix} & \multicolumn{2}{c}{\begin{bmatrix} \boldsymbol{G}_t^{xw} & \boldsymbol{Z}_t^+ \boldsymbol{G}_t^{xu} \\ \boldsymbol{G}_t^{yw} & \boldsymbol{Z}_t^+ \boldsymbol{G}_t^{yu} \end{bmatrix}} \end{matrix} = \boldsymbol{G}_t \qquad \begin{matrix} & 0 & 1 & \cdots & t-1 & t \\ \begin{matrix}0\\1\\2\\ \vdots \\ t\end{matrix} & \multicolumn{5}{c}{\begin{bmatrix} 0_{n\times n} & \cdots & \cdots & 0_{n\times n} & 0_{n\times n} \\ I_n & 0_{n\times n} & \cdots & 0_{n\times n} & 0_{n\times n} \\ 0_{n\times n} & \ddots & \ddots & \vdots & \vdots \\ \vdots & \ddots & \ddots & 0_{n\times n} & \vdots \\ 0_{n\times n} & \cdots & 0_{n\times n} & I_n & 0_{n\times n} \end{bmatrix}} \end{matrix} = \boldsymbol{Z}_t^+. \tag{4.31}$$

Now we can express the relationship between the system trajectories over the time-horizon $[0, t]$ as:

$$
\begin{aligned}
\underline{x}_t &= Z_t^+ G_t^{xu}(A, B)\underline{u}_t + G_t^{xw}(A)\underline{w}_t \\
\underline{y}_t &= Z_t^+ G_t^{yu}(C, A, B)\underline{u}_t + G_t^{yw}(C, A)\underline{w}_t
\end{aligned}
\qquad
\underline{z}_t = \begin{bmatrix} CG_t^{yw}(\underline{w}_t) + CG_t^{yu}(\underline{u}_t) \\ D\underline{u}_t \end{bmatrix}
$$

$$\tag{4.32}$$

where Z_t^+ denotes the delay operator, and:

$$
G_t^{xw}(A) = \begin{bmatrix}
I_n & 0 & 0 & \cdots & 0 \\
A & I_n & 0 & \cdots & 0 \\
A^2 & A & I_n & \cdots & 0 \\
& \cdots & \cdots & & \\
A^t & A^{t-1} & A^{t-2} & \cdots & I_n
\end{bmatrix}
\qquad
\begin{aligned}
G_t^{zu} &= \begin{bmatrix} G_t^{yu} \\ D \end{bmatrix} \\
G_t^{xu}(A, B) &= G_t^{xw}(A)(I_{t+1} \otimes B)
\end{aligned}
$$

$$\tag{4.33}$$

$$
G_t^{yw}(C, A) = (I_{t+1} \otimes C)G_t^{xw}(A) \qquad G_t^{yu}(C, A, B) = (I_{t+1} \otimes C)G_t^{xu}(A, B).
$$

We denote the k-th controllability matrix and grammian with respect to the pair (A, B) as $P_k(A, B)$ and $W_k^c(A, B)$, respectively. Similarly, we define the k-th observability matrix and grammian of a pair (C, A) as $Q_k(C, A)$ and $W_t^o(C, A)$:

$$
\begin{aligned}
P_k(A, B) &= [A^k B, A^{k-1}B, \cdots, B] \\
W_k^c(A, B) &= P_k(A, B)P_k^\top(A, B) \\
W_k^o(C, A) &= Q_k(C, A)^\top Q_k(C, A)
\end{aligned}
\qquad
Q_k(C, A) = \begin{bmatrix} C \\ CA \\ \vdots \\ CA^k \end{bmatrix}.
\qquad (4.34)
$$

We refer to the rank of $P_k(A, B)$ and $Q_k(C, A)$ as the degree of k-controllability and k-observability of the system (C, A, B). We quantify the level of controllability and observability in terms of the singular values [4] of the matrices $P_k(A, B)$ and $Q_k(C, A)$: $\bar{\sigma}_t^c(A, B)$ and $\bar{\sigma}_t^o(C, A)$ denote the largest singular eigenvalues of $P_k(A, B)$ and $Q_k(C, A)$, while $\underline{\sigma}_t^c(A, B)$ and $\underline{\sigma}_t^o(C, A)$ denote the smallest *non-zero* ones. We summarize these definitions below:

Definition 4.3. *For a fixed set of parameters* (A, B, C) *and time-horizon* k, *we define* $\bar{\sigma}_k^c(A, B)$, $\underline{\sigma}_k^c(A, B)$, $\bar{\sigma}_k^o(C, A)$, *and* $\underline{\sigma}_k^o(C, A)$ *as:*

$$
\bar{\sigma}_k^c(A, B) := \|P_k(A, B)\|_2 \qquad \underline{\sigma}_k^c(A, B) := \sigma_{-1}(P_k(A, B)) = \|P_k^\dagger(A, B)\|_2^{-1}
$$

$$
\bar{\sigma}_k^o(C, A) := \|Q_k(C, A)\|_2 \qquad \underline{\sigma}_k^o(C, A) := \sigma_{-1}(Q_k(C, A)) = \|Q_k^\dagger(C, A)\|_2^{-1}.
$$

---

[4]Which coincide with the eigenvalues of the grammians.

Moreover, we define $\bar{\rho}_k^{c/o}(A)$, $\underline{\rho}_k^{c/o}(A)$ as:

$$\bar{\rho}_k^c(A) := \|P_k(A, I)\|_2 \qquad\qquad \underline{\rho}_k^c(A) := \sigma_{-1}(P_k(A, I))$$

$$\bar{\rho}_k^o(A) := \|Q_k(I, A)\|_2 \qquad\qquad \underline{\rho}_k^o(A) := \sigma_{-1}(Q_k(I, A)).$$

The quantities $\bar{\rho}_k^{c/o}(A)$ and $\underline{\rho}_k^{c/o}(A)$ measure the controllability and observability of a fictitious system, where B and C are replaced with the identity matrix I, while A is being kept the same:

$$x*_{k+1} = Ax_k^* + u_k^* \qquad\qquad y_k^* = x_k^*.$$

$\bar{\rho}_k^{c/o}(A)$ and $\underline{\rho}_k^{c/o}(A)$ represent the largest and smallest non-zero singular values of the controllability and observability matrix $P_k(A, I)$ and $Q_k(I, A)$ and can be interpreted as describing the attainable level of controllability and observability for the system matrix A, if we had the ability to change the input and output matrix.

Remark 26. *If we drop mentioning the dependence on t and/or the parameters (A, B, C) in some statement, it is assumed to hold for all t and/or parameters (A, B, C). For example, if we state, "$\boldsymbol{G}^{xw}$ satisfies ... ", we implicitly mean "For all A, B and t, $\boldsymbol{G}_t^{xw}(A, B)$ satisfies ... ".*

Since $W_t^c = P_t P_t^\top$ and $W_t^o = Q_t^\top Q_t$, the following relation between controllability/observability matrices P_t and Q_t and corresponding grammians W_t^c and W_t^o is always true.

Lemma 24.

(i) $\|P_t\|_2 = (\lambda_{\max}(W_t^c))^{\frac{1}{2}}$ *and* $\|P_t^\dagger\|_2 = \left(\lambda_{-1}^{\downarrow}(W_t^c)\right)^{-\frac{1}{2}}$.

(ii) $\|Q_t\|_2 = (\lambda_{\max}(W_t^o))^{\frac{1}{2}}$ *and* $\|Q_t^\dagger\|_2 = \left(\lambda_{-1}^{\downarrow}(W_t^o)\right)^{-\frac{1}{2}}$.

The next two sections are concerned with deriving key lemmas needed for the perturbation analysis.

Open-Loop Map Norm-Bounds

Here we derive approximations for the worst-case ℓ_2-gain of the open-loop map $\boldsymbol{G}$ over a fixed time-horizon t. We will express the bounds in terms of the relation between the controllability/observability matrices P_t and Q_t and the corresponding Grammians.

With the help of the previous lemma and the following easily verified fact

Lemma 25. $(G_t^{xw}(A))^{-1} = I_t - Z_t^+ \otimes A.$

we can derive upper bounds and lower bounds of the toeplitz operators $G^{(\cdot)}$:

Lemma 26. *Any singular value σ_i of $\sigma(G_t^{xw})$ and G_t^{xu} lies in the range described by the inequalities below:*

$$\frac{1}{1 + \|A\|_2} \leqslant \sigma_i(G_t^{xw}) \leqslant \sqrt{t+1}\,\bar{\rho}_t^c(A) \tag{4.35a}$$

$$\frac{\sigma_{\min}(B)}{1 + \|A\|_2} \leqslant \sigma_i(G_t^{xu}) \leqslant \sqrt{t+1}\,\bar{\sigma}_t^c(A, B) \tag{4.35b}$$

$$\frac{\sigma_{\min}(C)}{1 + \|A\|_2} \leqslant \sigma_i(G_t^{yw}) \leqslant \sqrt{t+1}\,\bar{\sigma}_t^o(C, A). \tag{4.35c}$$

Proof. $\|G_t^{xu}\|_2^2 := \max\limits_{\|u\|_2=1} \|G_t^{xu} u\|_2^2.$ By decomposing $u = [u_0^\top, \ldots, u_t^\top]^\top$ we can rewrite this as

$$\|G_t^{xu}\|_2^2 = \max_{\|u\|_2=1} \left\| \begin{bmatrix} Bu_0 \\ ABu_0 + Bu_1 \\ \cdots \\ A^t Bu_0 + \cdots + Bu_t \end{bmatrix} \right\|_2^2 = \max_{\|u\|_2=1} \sum_{k=0}^{t} \|P_k u\|_2^2$$

$$\leqslant \sum_{k=0}^{t} \max_{\|u\|_2=1} \|P_k u\|_2^2 = \sum_{k=0}^{t} \|P_k\|_2^2 \leqslant \sum_{k=0}^{t} (\bar{\sigma}_k^c(A, B))^2$$

$$\leqslant (t+1)\|P_t\|_2^2 = (t+1)(\bar{\sigma}_t^c(A, B))^2$$

where we used the fact that $\|P_k\|_2^2$ increases in k since it is equal to the largest eigenvalue of the corresponding controllability Gramian $W_k^c = \sum_{i=0}^{k-1} A^i BB^\top A^{i\top}$. Thus, we obtain the bound

$$\|G_t^{xu}(A, B)\|_2 \leqslant \sqrt{(t+1)}\,\bar{\sigma}_t^c(A, B),$$

and the bound on $\|G^{xw}(A)\|_2$ follows by setting $B = I$. We apply a similar idea to bound $\|G_t^{yw}\|_2$. Splitting the identity operator $I_t = \sum_{k=0}^{t} E_k E_k^\top$ into a sum of the $t+1$ orthogonal projections $E_k E_k^\top$, we use the triangle inequality to obtain:

$$\|G_t^{yw}\|_2 = \|\sum_{k=0}^{t} G_t^{yw} E_k E_k^\top\|_2 \leqslant \sum_{k=0}^{t} \|G_t^{yw} E_k E_k^\top\|_2$$

$$\leqslant \sum_{k=0}^{t} \|Q_k\|_2 \leqslant \sqrt{t+1}\|Q_t\|_2 = \sqrt{t+1}\,\bar{\sigma}_t^o.$$

For the lower bound, notice that the following chain of inequalities

$$(\sigma_{-1}(\boldsymbol{G}^{xw}(A)))^{-1} = \sigma_1((\boldsymbol{G}_w(A))^{\dagger}) = \|\boldsymbol{I}_t - \boldsymbol{Z}^{+} \otimes A\|_2 \leqslant 1 + \|A\|_2$$

implies $\sigma_{-1}(\boldsymbol{G}^{xw}(A)) \geqslant (1 + \|A\|_2)$. Now for the operator $\boldsymbol{G}^{xu}(A, B)$, notice that for any vector $\underline{u}_t$ we have:

$$\|\boldsymbol{G}_t^{xu}(A, B)\underline{u}_t\|_2 = \|\boldsymbol{G}_t^{xw}(A)(I_t \otimes B)\underline{u}_t\|_2 \geqslant \sigma_{\min}(\boldsymbol{G}_t^{xw}(A))\|(I_t \otimes B)\underline{u}_t\|_2$$
$$\geqslant \sigma_{\min}(\boldsymbol{G}_t^{xw}(A))\sigma_{\min}(B)\|\underline{u}_t\|_2.$$

Since $\boldsymbol{G}_t^{xw}(A)$ is invertible and we assume $\sigma_{\min}(B)$ to be full column rank, we conclude that for all $\underline{u}_t$ it holds

$$\|\boldsymbol{G}_t^{xu}(A, B)\underline{u}_t\|_2 \geqslant \frac{\sigma_{\min}(B)}{1 + \|A\|_2}\|\underline{u}_t\|_2,$$

which yields the stated result. $\qquad\square$

Recall that $\boldsymbol{G}_t^{yw}(C, A) = (I_{t+1} \otimes C)\boldsymbol{G}_t^{xw}(A)$ and that $\boldsymbol{G}_t^{yu}$ can be decomposed in multiple ways:

$$\boldsymbol{G}_t^{yu} = (I_{t+1} \otimes C)\boldsymbol{G}_t^{xu} = \boldsymbol{G}_t^{yw}(I_{t+1} \otimes B) = \boldsymbol{G}_t^{yw}(\boldsymbol{I}_t - \boldsymbol{Z}_t^{+} \otimes A)\boldsymbol{G}_t^{xu}.$$

As a corollary of the previous lemma, we obtain bounds on the open loop maps for the outputs:

Lemma 27. *For any singular value $\sigma(\boldsymbol{G}_t^{yw})$, $\sigma(\boldsymbol{G}_t^{yu})$ of $\boldsymbol{G}_t^{yw}$ and $\boldsymbol{G}_t^{yu}$ holds:*

$$\frac{\sigma_{\min}(C)}{1 + \|A\|_2} \leqslant \sigma(\boldsymbol{G}_t^{yw}) \leqslant \sqrt{(t + 1)}\bar{\sigma}_t^o(C, A)$$
$$\frac{\sigma_{\min}(C)\sigma_{\min}(B)}{1 + \|A\|_2} \leqslant \sigma(\boldsymbol{G}_t^{yu}) \leqslant \|C\|_2\sqrt{t + 1}\bar{\sigma}_t^c(A, B)$$
$$\sigma(\boldsymbol{G}_t^{yu}) \leqslant \sqrt{t + 1}\bar{\sigma}_t^o(C, A)\|B\|_2.$$

Furthermore, by definition, it is easy to see that $\sigma_{\max}^2(\boldsymbol{G}_t^{zu}) \leqslant \sigma_{\max}^2(\boldsymbol{G}_t^{yu}) + \sigma_{\max}^2(D)$ and $\sigma_{\min}^2(\boldsymbol{G}_t^{yu}) + \sigma_{\min}^2(D) \leqslant \sigma_{\min}^2(\boldsymbol{G}_t^{zu})$. We can apply Lem. 27 and conclude the following range for the singular values of $\boldsymbol{G}_t^{zu}$:

$$\sigma^2(\boldsymbol{G}_t^{zu}) \geqslant \left(\frac{\sigma_{\min}(C)\sigma_{\min}(B)}{1 + \|A\|_2}\right)^2 + \sigma_{\min}^2(D) \tag{4.37a}$$
$$\sigma^2(\boldsymbol{G}_t^{zu}) \leqslant (t + 1)\|C\|_2\|B\|_2\bar{\sigma}_t^o(C, A)\bar{\sigma}_t^c(A, B) + \|D\|_2^2. \tag{4.37b}$$

Perturbation Inequalities in Operator Norm

Here we discuss how the feedforward operators $G^{(\cdot)}$ vary with changes in the system matrices (A, B), while assuming C and D are fixed. For the next lemmas, fix two sets of parameters A_1, B_1 and A_2, B_2 and denote $G^{xw,i}$, $G^{xu,i}$, $G^{yu,i}$, $i \in \{1, 2\}$ etc. the corresponding operators $G^{xw}(A_i)$, $G^{xu}(A_i, B_i)$, $G^{yu}(A_i, B_i, C)$, etc.. Furthermore, denote $\partial_{12}A := A_1 - A_2$, $\partial_{12}B := B_1 - B_2$, and correspondingly write $\partial_{12}G = G(A_1, B_1) - G(A_2, B_2)$ to denote the corresponding changes in the open-loop maps. Correspondingly, $\partial_{12}G_t^{xu} = G_t^{xu,1} - G_t^{xu,2}$, $\partial_{12}G_t^{xw} = G_t^{xw,1} - G_t^{xw,2}$, etc. First, we derive a useful decomposition of the terms $\partial_{12}G$. To that end, notice that for two invertible matrices A and B we can always write

$$A^{-1} - B^{-1} = A^{-1}(B - A)B^{-1} = B^{-1}(B - A)A^{-1}.$$

Now using the fact that $G^{xw,i}$ is always invertible and $G_t^{xw,i} = (I_t - Z_t^+ \otimes A_i)^{-1}$, we can rewrite the differences $\partial_{12}G_t^{xw} = G_t^{xw,1} - G_t^{xw,2}$ and $G_t^{xu,1} - G_t^{xu,2}$ as:

$$\partial_{12}G_t^{xw} = G_t^{xw,2}(Z_t^+ \otimes \partial_{21}A)G_t^{xw,1} \tag{4.38a}$$

$$\partial_{12}G_t^{xu} = (G_t^{xw,1} - G_t^{xw,2})(I_t \otimes B_1) + G_t^{xw,2}(I_t \otimes \partial_{12}B) \tag{4.38b}$$

$$= G_t^{xw,2}(Z_t^+ \otimes \partial_{21}A)G_t^{xu,1} + G_t^{xw,2}(I_t \otimes \partial_{12}B)$$

$$\partial_{12}G_t^{yw} = G_t^{yw,1}(Z_t^+ \otimes \partial_{21}A)G_t^{xw,2} \tag{4.38c}$$

$$\partial_{12}G_t^{yu} = G_t^{yw,2}(Z_t^+ \otimes \partial_{21}A)G_t^{xu,1} + G_t^{yw,2}(I_t \otimes \partial_{12}B). \tag{4.38d}$$

Applying the triangle inequality to these equations, we can directly obtain the following perturbation inequalities in the induced 2-norm:

$$\|\partial_{12}G_t^{xw}\|_2 \leqslant \|G^{xw,1}\|_2 \|G^{xw,2}\|_2 \|\partial_{12}A\|_2$$

$$\|\partial_{12}G_t^{xu}\|_2 \leqslant \|G^{xw,2}\|_2 \|G^{xu,1}\|_2 \|\partial_{12}A\|_2 + \|G_t^{xw,2}\|_2 \|\partial_{12}B\|_2$$

$$\|\partial_{12}G_t^{yw}\|_2 \leqslant \|G^{yw,1}\|_2 \|G^{xw,2}\|_2 \|\partial_{12}A\|_2$$

$$\|\partial_{12}G_t^{yu}\|_2 \leqslant \|G^{yw,2}\|_2 \|G^{xu,1}\|_2 \|\partial_{12}A\|_2 + \|G_t^{yw,2}\|_2 \|\partial_{12}B\|_2.$$

Applying the results of the previous section Lem. 26, we can further bound the above inequalities in terms of the singular values of the controllability grammians. For the following results, we will denote $\bar{\rho}_t^{c,i} = \bar{\rho}_t^c(A_i)$, $\bar{\sigma}_t^{c,i} = \bar{\sigma}_t^c(A_i, B_i)$, $\bar{\rho}_t^{o,i} = \bar{\rho}_t^o(C, A_i)$ corresponding to two sets of system parameters (A_1, B_1) and (A_2, B_2); for each variable, we will use $\bar{x}_t^{c/o*} := \max_i\{x_t^{c/oi}\}$ ($\underline{x}_t^{c/o*} := \min_i\{\underline{x}_t^{c/oi}\}$) to denote the maximum (minimum) of both cases, i.e., $\bar{\rho}_t^{c*} = \max\{\bar{\rho}_t^{c,1}, \bar{\rho}_t^{c,2}\}$, ($\underline{\varrho}_t^{c*} = \min\{\underline{\varrho}_t^{c,1}, \underline{\varrho}_t^{c,2}\}$) etc.

Lemma 28. *Let $G_{t,1}$ and $G_{t,2}$ be the open loop maps of two sets of system parameters (A_1, B_1) and (A_2, B_2). Then:*

$$\|\partial_{12}G_t^{xw}\|_2 \leqslant (t+1)\bar{\rho}_t^{c*}\|\partial_{12}A\|_2 \tag{4.39}$$

$$\|\partial_{12}G_t^{xu}\|_2 \leqslant (t+1)\bar{\sigma}_t^{c*}\bar{\rho}_t^{c*}\|\partial_{12}A\|_2 + \sqrt{t+1}\bar{\rho}_t^{c*}\|\partial_{12}B\|_2$$

$$\|\partial_{12}G_t^{yw}\|_2 \leqslant (t+1)\bar{\sigma}_t^{o*}\bar{\rho}_t^{c*}\|\partial_{12}A\|_2$$

$$\|\partial_{12}G_t^{yu}\|_2 \leqslant (t+1)\bar{\sigma}_t^{c*}\bar{\sigma}_t^{o*}\|\partial_{12}A\|_2 + \sqrt{t+1}\bar{\sigma}_t^{o*}\|\partial_{12}B\|_2.$$

We can utilize the above decomposition also for the controllability and observability matrices $P_t(A, B)$ and $Q_t(C, A)$. Notice that $P_k \in \mathbb{R}^{n \times (k+1)m} = E_t^\top G_t^{xu}$ is the last row of submatrices in G_k^{xu} and $Q_k(C, A) \in \mathbb{R}^{\ell(k+1) \times n} = G_t^{yw} E_0$ is the first column of submatrices in G_k^{yw}, hence from (4.38c) we obtain the decomposition:

$$Q_t(C, A_1) - Q_t(C, A_2) = G_t^{yw,1}(C, A_1)(Z_t^+ \otimes \partial_{21}A)G_t^{xw,2}(A_2)E_1$$
$$G_t^{yw,1}(C, A_1)(Z_t^+ \otimes \partial_{21}A)Q_t(I, A_2)$$

and from (4.38a) and (4.38b) we get:

$$P_t(A_1, B_1) - P_t(A_2, B_2) = P_t(A_2, I)\left((Z_t^+ \otimes \partial_{21}A)G_t^{xu,1} + I_t \otimes \partial_{12}B\right).$$

Applying the triangle-inequality to these decompositions and substituting the operator norm bounds derived in Lem. 27 and Lem. 26, we obtain the following perturbation inequality for the controllability and observability matrices:

Lemma. *Denote $\sigma^{u,i} = \lambda_{\max}(W_t^c(A_i, B_i))$ and $\sigma^{y,i} = \lambda_{\max}(W_t^o(C, A_i))$ corresponding to the two sets of system parameters (A_1, B_1) and (A_2, B_2). Then $\partial_{12}P_t$ and $\partial_{12}Q_t$ are bounded as:*

$$\|\partial_{12}P_t\|_2 \leqslant \bar{\rho}_t^{c,2}\left(\sqrt{(t+1)}\bar{\sigma}_t^{c,1}\|\partial_{12}A\|_2 + \|\partial_{12}B\|_2\right) \tag{4.40}$$

$$\|\partial_{12}Q_t\|_2 \leqslant \bar{\rho}_t^{o,2}\sqrt{(t+1)}\bar{\sigma}_t^{o,1}\|\partial_{12}A\|_2. \tag{4.41}$$

4.7 From $\mathcal{H}_2$-Optimal Control to Least Squares

Having the basic definitions and background, our next step is to reduce the $\mathcal{H}_2$-problem described by the equations (4.20) to a Least-Squares problem. Denote $\phi_k^{j,x} \in \mathbb{R}^n$, $\phi_k^{j,u} \in \mathbb{R}^m$ as the jth column of $R_k \in \mathbb{R}^{n \times n}$, $M_k \in \mathbb{R}^{m \times n}$ and e_j the unit vector in the j-th coordinate axis. From now on, we fix the finite horizon t. As described in [11], we can separate the problem by columns and can equivalently restate (4.20) in terms of each column $\phi_k^{j,x}$ and $\phi_k^{j,u}$:

$$S_j := \min \left\| \begin{bmatrix} C & 0 \\ 0 & D \end{bmatrix} \begin{bmatrix} \phi_1^{j,x} & \phi_2^{j,x} & \cdots & \phi_t^{j,x} \\ \phi_1^{j,u} & \phi_2^{j,u} & \cdots & \phi_t^{j,u} \end{bmatrix} \right\|_F^2 \tag{4.42}$$

$$\text{s.t.: } \phi_1^{j,x} = e_j$$
$$\phi_{k+1}^{j,x} = A\phi_k^{j,x} + B\phi_k^{j,u}, \quad \forall 1 \leqslant k \leqslant t$$
$$\phi_{t+1}^{j,x} = 0.$$

We rewrite (4.42) further and introduce new variables to avoid tedious notation. Define $u_k = \phi_k^{j,u}, \forall k : 0 \leqslant k \leqslant t - 1, \boldsymbol{u} = [u_1^\top, \ldots, u_t^\top]^\top$ and let $\boldsymbol{C}, \boldsymbol{D}$ denote the lifted weight matrices $\boldsymbol{C} = I_t \otimes C, \boldsymbol{D} = I_t \otimes D$. Now we rewrite the subproblem S_j as

$$S_j = \min_{\boldsymbol{u}} \left\| \begin{bmatrix} \boldsymbol{C}\boldsymbol{G}_t^{xu}(A, B) \\ \boldsymbol{D} \end{bmatrix} \boldsymbol{u} - \boldsymbol{\eta}_t(A) \right\|_2^2 + (C^\top C)_{jj} \tag{4.43a}$$

$$\text{s.t.: } \quad 0 = A^{t+1}e_j + P_t(A, B)\boldsymbol{u} \tag{4.43b}$$

where $\boldsymbol{\eta}_t^\top = -[(Q_t(C, A)Ae_j)^\top, 0]$ and $\boldsymbol{E}_1^\top = [I_n, 0_{n \times n}, \cdots, 0_{n \times n}]$. To simplify notation, we introduce the virtual outputs $z = [Cx, Du]^\top \in \mathbb{R}^{(n+m) \times m}, y = Cx$ and the operators $\boldsymbol{G}^{yu}, \boldsymbol{G}^{zu}$ as

$$\boldsymbol{G}^{yu}(A, B) = \boldsymbol{C}\boldsymbol{G}^{xu}(A, B) \qquad \boldsymbol{G}^{zu}(A, B) = \begin{bmatrix} \boldsymbol{C}\boldsymbol{G}^{xu}(A, B) \\ \boldsymbol{D} \end{bmatrix} \tag{4.44}$$

to rephrase the problem into (4.45), where we dropped the constant term $(C^\top C)_{jj}$ as it is not needed for analysis.

$$S_j = \min_{\boldsymbol{u}} \quad \|\boldsymbol{G}_t^{zu}(A, B)\boldsymbol{u} - \boldsymbol{\eta}_t(A)\|_2^2 \tag{4.45a}$$

$$\text{s.t.: } \quad 0 = A^{t+1}e_j + P_t(A, B)\boldsymbol{u}. \tag{4.45b}$$

4.8 Representation as a Least-Squares Problem

We now rewrite (4.43) as a least square problem. Define $\boldsymbol{u}_c^* := P_t^\top (P_t P_t^\top)^{-1} A^t e_j$, which is the solution to the optimization problem

$$\min_{\boldsymbol{u}} \quad \|\boldsymbol{u}\|_2^2$$

$$\text{s.t. } \quad -A^{t+1}e_j = P_t \boldsymbol{u}.$$

We can interpret $\boldsymbol{u}_c^*$ as the smallest control action, measured in ℓ_2, that drives the system from the origin to $-A^{t+1}e_j$ in t time-steps. This relates to controllability grammians as described in [49]. In terms of the Moore-Penrose Inverse, we can also write $\boldsymbol{u}_c^* := P_t^\dagger A^{t+1} e_j = P_t^\top (W_t^c)^\dagger A^{t+1} e_j$, where W_t^c is the t-th controllability grammian.

We drop the index j from ϕ^{*j} and reparameterize $\boldsymbol{u} = -\boldsymbol{u}_c^* + \boldsymbol{u}'$ where $\boldsymbol{u}' \in \text{null}(P_t)$ and describe (4.43) as the optimization problem:

$$S_j \quad := \min_{\boldsymbol{u}' \in \text{null}(P_t(A,B))} \left\| \boldsymbol{G}_t^{zu}(A,B)(\boldsymbol{u}' - \boldsymbol{u}_c^*(A,B)) - \boldsymbol{\eta}_t(A) \right\|_2^2. \tag{4.46}$$

Let $\boldsymbol{u}^*(A,B)$ be a minimizer of the above problem for fixed A, B, we are interested in the SLS solutions

$$\phi^*(A,B) := \begin{bmatrix} \boldsymbol{C}^{-1} & 0 \\ 0 & \boldsymbol{D}^{-1} \end{bmatrix} (\boldsymbol{G}_t^{zu}(A,B)(\boldsymbol{u}^* - \boldsymbol{u}_c^*(A,B)) - \boldsymbol{\eta}_t(A))$$

and how these solutions are perturbed with changes in A, B. We (over-)parameterize $\boldsymbol{u}$ as $\boldsymbol{u} = (I - P_t^\dagger P_t)\boldsymbol{\eta}$, to cast problem S_j into an unconstrained one:

$$\min_{\boldsymbol{\eta}} \left\| \underbrace{\boldsymbol{G}_t^{zu}(A,B)\boldsymbol{u}_c^*(A,B) + \boldsymbol{\eta}_t(A)}_{\boldsymbol{g}} - \underbrace{\boldsymbol{G}_t^{zu}(A,B)(I - P_t^\dagger P_t)}_{\boldsymbol{H}} \boldsymbol{\eta} \right\|_2^2. \tag{4.47}$$

The optimal value of problem S_j is $\|\nu^*\|_2^2$, where $\nu^* := (\boldsymbol{H}\boldsymbol{H}^\dagger - I)\boldsymbol{g}$ and is achieved at $\eta^* = \boldsymbol{H}^\dagger\boldsymbol{g}$. The corresponding optimal ϕ^* takes the form min-norm solution η^* to the above problem is $\eta^* = \boldsymbol{H}^\dagger\boldsymbol{g}$ and $S_j = \|\nu^*\|_2$ and therefore the optimal solution ϕ^* takes the form:

$$\phi^* = \begin{bmatrix} \boldsymbol{C}^{-1} & 0 \\ 0 & \boldsymbol{D}^{-1} \end{bmatrix} \nu^*, \text{ where } \nu^* = (\boldsymbol{H}\boldsymbol{H}^\dagger - I)\boldsymbol{g}. \tag{4.48}$$

Hence, up to the constant $\kappa_{CD} = \max\{\|C^{-1}\|_2, \|D^{-1}\|_2\}$, the sensitivity of ϕ^* scales linearly with the sensitivity of the corresponding optimal ν^*. Hence, if ϕ_1^*, ν_1^* and ϕ_2^*, ν_2^* are optimal solutions for two different sets of parameters (A_1, B_1) and (A_2, B_2), then $\|\phi_1^* - \phi_2^*\|_2$ is bounded as:

$$\|\phi_1^* - \phi_2^*\|_2 \leqslant \kappa_{CD}\|\nu_1^* - \nu_2^*\|_2.$$

The goal of the next section is to bound $\|\nu_1^* - \nu_2^*\|_2$ in terms of $\|A_1 - A_2\|_2$ and $\|B_1 - B_2\|_2$.

4.9 Perturbation of Least Squares

Consider two optimal solutions $\nu_1^* = (\boldsymbol{H}_1\boldsymbol{H}_1^\dagger - I)\boldsymbol{g}_1$ and $\nu_2^* = (\boldsymbol{H}_2\boldsymbol{H}_2^\dagger - I)\boldsymbol{g}_2$, of parameters (A_1, B_1) and (A_2, B_2). The difference $\nu_1^* - \nu_2^*$ can be written as

$$\nu_1^* - \nu_2^* = (\boldsymbol{H}_1\boldsymbol{H}_1^\dagger - I)(\boldsymbol{g}_1 - \boldsymbol{g}_2) + (\boldsymbol{H}_1\boldsymbol{H}_1^\dagger - \boldsymbol{H}_2\boldsymbol{H}_2^\dagger)\boldsymbol{g}_2,$$

and the term $H_1 H_1^\dagger - H_2 H_2^\dagger$ can be further broken down into:

$$(I - H_1 H_1^\dagger)(H_1 - H_2)H_2^\dagger + \left[(I - H_2 H_2^\dagger)(H_1 - H_2)H_1^\dagger\right]^\top.$$

Now, since the operators $(M M^\dagger - I)$ and $M M^\dagger$ are projections for any matrix M, we know that $\|H_i H_i^\dagger - I\|_2 = 1$ [5]. Therefore, we can bound the term $H_1 H_1^\dagger - H_2 H_2^\dagger$ in terms of the $\| \cdot \|_2$ norm as:

$$\|H_1 H_1^\dagger - H_2 H_2^\dagger\|_2 \leqslant \|H_1 - H_2\|_2(\|H_1^\dagger\|_2 + \|H_2^\dagger\|_2). \tag{4.49}$$

Applying the triangle-inequality on (4.49) and substituting the above bound yields:

$$\|v_1^* - v_2^*\|_2 \leqslant \|g_1 - g_2\|_2 + \|H_1 - H_2\|_2(\|H_1^\dagger\|_2 + \|H_2^\dagger\|_2)\|g_2\|_2. \tag{4.50}$$

In the following, we proceed to analyze each of the terms in the above inequality separately. To this end, we will make frequent use of the Lemmas. The facts stated in (29) can be easily verified. The proof of Lem. 30 is more involved and can be found in [129].[6]

Lemma 29. *For arbitrary matrices $X, Y \in \mathbb{R}^{n \times m}$ and $A, B \in \mathbb{R}^{n \times n}$, it holds that*

(i) $A_1^k - A_2^k = \sum_{j=0}^{k-1} A_1^{k-1-j}(A_1 - A_2)A_2^j.$

(ii) $XX^\dagger - YY^\dagger = (I - XX^\dagger)(X - Y)Y^\dagger + \left[(I - YY^\dagger)(X - Y)X^\dagger\right]^\top.$

(iii) $X^\dagger X - Y^\dagger Y = Y^\dagger(X - Y)(I - X^\dagger X) + \left[X^\dagger(X - Y)(I - Y^\dagger Y)\right]^\top.$

(iv) If A and B are invertible, then $A^{-1} - B^{-1} = A^{-1}(B - A)B^{-1}.$

The following is a corollary from Theorem 4.1 in [129]:

Lemma 30. *Let X and Y be matrices with equal rank, let $\| \cdot \|_2$ denote the induced 2-norm and $\| \cdot \|_F$ denote the Frobenius norm. The following inequalities hold:*

$$\|X^\dagger - Y^\dagger\|_2 \leqslant \varphi\|X^\dagger\|_2\|Y^\dagger\|_2\|X - Y\|_2$$
$$\|X^\dagger - Y^\dagger\|_F \leqslant \sqrt{2}\|X^\dagger\|_2\|Y^\dagger\|_2\|X - Y\|_F$$

where $\varphi = \frac{1+\sqrt{5}}{2}$ denotes the golden ratio constant.

---

[5] Unless $H_i = 0$.

[6] See Corollary of Thm.4.1 in [129].

Bounding $\|H_1 - H_2\|_2$

First, notice that we can express $H^\top$ as $(I - P_t^\dagger P_t)(G_t^{zu})^\top$ and that

$$H_1 - H_2 = G_t^{zu,1}(P_{t,2}^\dagger P_{t,2} - P_{t,1}^\dagger P_{t,1}) + (G_t^{zu,1} - G_t^{zu,2})(I - P_{t,2}^\dagger P_{t,2}).$$

Now using the previous lemma and noticing that $\|G_t^{zu,1} - G_t^{zu,2}\|_2 = \|G_t^{yu,1} - G_t^{yu,2}\|_2$ we obtain the following inequality

$$\|H_1 - H_2\|_2 \leqslant \|G_t^{yu,1} - G_t^{yu,2}\|_2 + \|G_t^{zu,1}\|_2 \left(\|P_{t,1}^\dagger\|_2 + \|P_{t,2}^\dagger\|_2\right) \|P_{t,2} - P_{t,1}\|_2$$

and arrive at:

$$\|H_1 - H_2\|_2(\|H_1^\dagger\|_2 + \|H_2^\dagger\|_2)\|g_2\|_2 \leqslant \mu_1 \|\partial_{12} G_t^{yu}\|_2 + \mu_2 \|\partial_{12} P_t\|_2$$

with the constants

$$\mu_1 = (\|H_1^\dagger\|_2 + \|H_2^\dagger\|_2)\|g_2\|_2 \tag{4.51a}$$

$$\mu_2 = \|G_t^{zu,1}\|_2(\|P_{t,1}^\dagger\|_2 + \|P_{t,2}^\dagger\|_2)(\|H_1^\dagger\|_2 + \|H_2^\dagger\|_2)\|g_2\|_2 \tag{4.51b}$$

Lemma 31. *Recall that D is invertible. Therefore,*

$$\|H^\dagger(A, B)\|_2 \leqslant \|G_t^{zu\dagger}\|_2 \leqslant (\sigma_{\min}(G_t^{zu}))^{-1}.$$

Proof. In order to bound $\|M^\dagger\|_2$ from above, we have to bound $\sigma_{-1}(M)$ from below. Notice that $G_t^{zu}(A, B)$ is full-column rank, since D is invertible and thus $\mathrm{rank}(G_t^{zu}(A, B)) = (t+1) \cdot n_u$. The projection $\Pi_{\mathcal{N}(P_t)} := (I - P_t^\dagger(A, B)P_t(A, B)) \in \mathbb{R}^{(t+1)n_u \times (t+1)n_u}$ has rank $r_H < (t+1)n_u = \mathrm{rank}(G_t^{zu}(A, B))$. Hence, $G_t^{zu}(A, B)$ and P_t have the same null space and therefore $H = G_t^{zu}(A, B)\Pi_{\mathcal{N}(P_t)}$ is of rank r_H. From these observations, we can equivalently say that $\sigma_{-1}(H)$ is the r_H-th largest singular eigenvalue of H. Using the Minimax principle, we can therefore represent $\sigma_{-1}(H)$ as the solution to the following max-min problem:

$$\sigma_{-1}(H) = \max_{\mathrm{proj.}\Pi,\, \mathrm{s.t.:\, rank}(\Pi)=r_H} \min_{x\, \mathrm{s.t.:}\, \|\Pi x\|=1} x^\top \Pi H^\top H \Pi x \tag{4.52}$$

$$= \max_{\mathrm{proj.}\Pi,\, \mathrm{s.t.:\, rank}(\Pi)=r_H} \min_{x\, \mathrm{s.t.:}\, \|\Pi x\|=1} x^\top \Pi \Pi_{\mathcal{N}(P_t)} G_t^{zu\top} G_t^{zu} \Pi_{\mathcal{N}(P_t)} \Pi x. \tag{4.53}$$

Now recall that $\Pi_{\mathcal{N}(P_t)}$ is of rank r_H, hence it is a feasible choice for the variable Π of the outer optimization problem. Furthermore, this leads us to the bound

$$\sigma_{-1}(H) \geqslant \min_{x\, \mathrm{s.t.:}\, \|\Pi_{\mathcal{N}(P_t)}x\|=1} x^\top \Pi_{\mathcal{N}(P_t)} G_t^{zu\top} G_t^{zu} \Pi_{\mathcal{N}(P_t)} x \tag{4.54}$$

$$= \min_{x\, \mathrm{s.t.:}\, z\in\mathcal{N}(P_t),\, \|z\|_2=1} z^\top G_t^{zu\top} G_t^{zu} z \tag{4.55}$$

$$\geqslant \min_{z\, \mathrm{s.t.:}\, \|z\|=1} z^\top G_t^{zu\top} G_t^{zu} z = \sqrt{\lambda_{\min}(G_t^{zu\top} G_t^{zu})}. \tag{4.56}$$

Finally, the desired statement follows, since $\sqrt{\lambda_{\min}(G_t^{zu\top}G_t^{zu})} = \sigma_{\min}(G_t^{zu})$ and by inverting the inequality above. $\qquad\square$

To bound the worst-case of $\|g\|_2$ recall its definition:

$$g = G_t^{zu}(A,B)u_c^*(A,B) + \eta_t(A)$$
$$= G_t^{zu}(A,B)P_t^\dagger(A,B)A^{t+1}e_j - Q_t(C,A)Ae_j.$$

Definition 4.4. *For a pair (A_i, B_i) define the constants $\alpha_{t,i}$, and $\bar{\sigma}_t^{zu,i}, \underline{\sigma}_t^{zu,i}$ as:*

$$\alpha_t := \max_{0 \leqslant k \leqslant t+1} \|A^k\|_2 \quad \bar{\sigma}_t^{zu,i} := \sigma_{\max}(G_t^{zu}(A_i, B_i)) \quad \underline{\sigma}_t^{zu,i} := \sigma_{\min}(G_t^{zu}(A_i, B_i)).$$

We can obtain an upper bound for $\|g\|_2$, as a corollary of Lem. 27 and Lem. 24

$$\|g\|_2 \leqslant \left(\|C\|_2\sqrt{t}\bar{\sigma}_t^c + \|D\|_2\right)(\underline{\sigma}_t^c)^{-1}\alpha_t + \bar{\sigma}_t^o\|A\|_2, \tag{4.57}$$

and using the constants defined in Definition (4.4) we can bound the constant μ_1 as:

$$\mu_1 \leqslant 2\frac{\bar{\sigma}_t^{zu*}}{\underline{\sigma}_t^{zu*}\underline{\sigma}_t^{c*}}\alpha_t + \bar{\sigma}_t^{o*}\|A_*\|_2 \tag{4.58}$$

$$\mu_2 \leqslant \frac{4}{\underline{\sigma}_t^{zu*}}\left(\frac{\bar{\sigma}_t^{zu*}}{\underline{\sigma}_t^{c*}}\right)^2\alpha_t + 2\frac{\bar{\sigma}_t^{zu*}}{\underline{\sigma}_t^{c*}}\bar{\sigma}_t^{o*}\|A_*\|_2 \tag{4.59}$$

where the expression with superscript $\underline{\sigma}_t^{c*} = \min\{\underline{\sigma}_t^{c,1}, \underline{\sigma}_t^{c,2}\}$, $\bar{\sigma}_t^{o*} = \max\{\bar{\sigma}_t^{o,1}, \bar{\sigma}_t^{o,2}\}$ represents the worst-case pick of $i = 1, 2$ for the corresponding expression. We can obtain explicit bounds by substituting the bounds of Lem. 27

Bounding $\|g_1 - g_2\|_2$

Recall that $g_1 - g_2$ takes the form:

$$g_1 - g_2 = G_t^{zu,1}P_{t,1}^\dagger A_1^{t+1}e_j - G_t^{zu,2}P_{t,2}^\dagger A_2^{t+1}e_j + \eta_t(A_1) - \eta_t(A_2). \tag{4.60}$$

We split $g_1 - g_2 = \Delta_1 + \Delta_2 + \Delta_3 + \Delta_4$ into the four telescoping terms $\Delta_1, \ldots, \Delta_4$ below and proceed to bound each of them individually.

$$\Delta_1 = \left(G_t^{zu,1} - G_t^{zu,2}\right)P_{t,1}^\dagger A_1^{t+1}e_j \qquad \Delta_2 = G_t^{zu,2}\left(P_{t,1}^\dagger - P_{t,2}^\dagger\right)A_1^{t+1}e_j$$

$$\Delta_3 = G_t^{zu,2}P_{t,2}^\dagger\left(A_1^{t+1}e_j - A_2^{t+1}e_j\right) \qquad \Delta_4 = \eta_t(A_1) - \eta_t(A_2)$$

1. $\underline{\Delta_1}$: $\|\Delta_1\|_2 \leqslant \|G_t^{yu,1} - G_t^{yu,2}\|_2\|P_{t,1}^\dagger\|_2\|A_1^{t+1}e_j\|_2.$
2. $\underline{\Delta_2}$: Applying the $\|\cdot\|_2$-bound from Lem. 30 yields:

$$\|\Delta_2\|_2 \leqslant \varphi\|P_{t,1}^\dagger\|_2\|P_{t,2}^\dagger\|_2\|G_t^{zu,2}\|_2\|A^{t+1}e_j\|_2\|P_{t,1} - P_{t,2}\|_2.$$

3. $\underline{\Delta_3}$: We apply Lem. 29 to obtain $A_1^{t+1} - A_2^{t+1} = \sum_{j=0}^{t} A_1^{t-j}(A_1 - A_2)A_2^j$. Notice we can rewrite that as

$$A_1^{t+1} - A_2^{t+1} = P_t(A_1, I)(A_1 - A_2)Q_t(I, A_2)$$

and apply the bounds of Lem. 24 to conclude:

$$\|\Delta_3\|_2 \leqslant \sqrt{\lambda_{\max}(W_t^c(A_1, I))\lambda_{\max}(W_t^o(I, A_2))}\|\boldsymbol{G}_t^{zu,2}\|_2\|P_{t,2}^\dagger\|_2\|A_1 - A_2\|_2$$
$$\leqslant \bar{\rho}_t^{o,2}\bar{\rho}_t^{c,1}\|\boldsymbol{G}_t^{zu,2}\|_2\|P_{t,2}^\dagger\|_2\|A_1 - A_2\|_2.$$

4. $\underline{\Delta_4}$: We can bound $\|\Delta_4\|_2$ as

$$\|\Delta_4\|_2 \leqslant \|\partial_{12}Q_t\|_2\|A_1e_j\|_2 + \|Q_{t,2}\|_2\|(A_1 - A_2)e_j\|_2.$$

Combining all the previous inequalities gives us:

$$\|\boldsymbol{g}_1 - \boldsymbol{g}_2\|_2 \leqslant \gamma_1\|\partial_{12}\boldsymbol{G}_t^{yu}\|_2 + \gamma_2\|\partial_{12}P_t\|_2 + \gamma_3\|\partial_{12}Q_t\|_2 + \gamma_4\|\partial_{12}A\|_2$$

with the γ_i defined as:

$$\gamma_1 = \|P_{t,1}^\dagger\|_2\|A_1^{t+1}e_j\|_2 \tag{4.61a}$$
$$\gamma_2 = \varphi\|P_{t,1}^\dagger\|_2\|P_{t,2}^\dagger\|_2\|\boldsymbol{G}_t^{zu,2}\|_2\|A^{t+1}e_j\|_2 \tag{4.61b}$$
$$\gamma_3 = \|A_1e_j\|_2 \tag{4.61c}$$
$$\gamma_4 = \|Q_{t,2}\|_2 + \bar{\rho}_t^{o,2}\bar{\rho}_t^{c,1}\|\boldsymbol{G}_t^{zu,2}\|_2\|P_{t,2}^\dagger\|_2 \tag{4.61d}$$

with abbrevations $\bar{\rho}_t^{c,i} = \lambda_{\max}(W_t^c(A_i, I))$ and $\bar{\rho}_t^{o,i} = \lambda_{\max}(W_t^o(I, A_i))$. Combining the inequalities we proved in the last two subsections, we obtain a first bound on the optimal solutions $\|\phi_1^* - \phi_2^*\|_2 \leqslant \kappa_{CD}\|\nu_1^* - \nu_2^*\|_2$:

$$\kappa_{CD}^{-1}\|\phi_1^* - \phi_2^*\|_2 \leqslant \gamma_1\mu_1\|\partial_{12}\boldsymbol{G}_t^{yu}\|_2 + \gamma_2\mu_2\|\partial_{12}P_t\|_2 + \gamma_3\|\partial_{12}Q_t\|_2 + \gamma_4\|\partial_{12}A\|_2.$$

This inequality informs a natural interpretation: The sensitivity to parameter changes of the optimal solutions ϕ^* can be reduced to analyzing the change of four system-theoretic quantities: the open-loop map from control input to output $\|\partial_{12}\boldsymbol{G}_t^{yu}\|_2$, the controllability matrix $\partial_{12}P_t$, observability $\partial_{12}Q_t$, and the system matrix $\partial_{12}A$.

Lemma 32. *Let C, D be fixed invertible matrices and let $[A_1, B_1]$ and $[A_2, B_2]$ be two t-controllable pairs of system and input matrices. Let ϕ_1^* and ϕ_2^* be the optimal solutions of the $\mathcal{H}_2$-problem (4.45), corresponding to both sets of parameters (A_1, B_1) and (A_2, B_2). Then, for ϕ_1^* and ϕ_2^* holds the inequality:*

$$\kappa_{CD}^{-1}\|\phi_1^* - \phi_2^*\|_2 \leqslant \delta_1\|\partial_{12}\boldsymbol{G}_t^{yu}\|_2 + \delta_2\|\partial_{12}P_t\|_2 + \delta_3\|\partial_{12}Q_t\|_2 + \delta_4\|\partial_{12}A\|_2$$

where $\delta_1, \ldots, \delta_4$ and μ_1 are the constants defined below:

$$\mu_1 \leqslant \frac{2}{\underline{\sigma}_t^{zu*}} \frac{\bar{\sigma}_t^{zu*}}{\underline{\sigma}_t^{c*}} \alpha_t + \bar{\sigma}_t^{o*} \|A_*\|_2 \tag{4.62a}$$

$$\mu_2 \leqslant \frac{4}{\underline{\sigma}_t^{zu*}} \left(\frac{\bar{\sigma}_t^{zu*}}{\underline{\sigma}_t^{c*}}\right)^2 \alpha_t + 2\frac{\bar{\sigma}_t^{zu*}}{\underline{\sigma}_t^{c*}} \bar{\sigma}_t^{o*} \|A_*\|_2 \tag{4.62b}$$

$$\delta_1 = \|P_{t,1}^\dagger\|_2 \|A_1^{t+1} e_j\|_2 + \mu_1 \tag{4.62c}$$

$$\delta_2 = \varphi \|P_{t,1}^\dagger\|_2 \|P_{t,2}^\dagger\|_2 \bar{\sigma}_t^{zu*} \|A^{t+1} e_j\|_2 + \mu_2 \tag{4.62d}$$

$$\delta_3 = \|A_1 e_j\|_2 \tag{4.62e}$$

$$\delta_4 = \|Q_{t,2}\|_2 + \bar{\rho}_t^{o,2} \bar{\rho}_t^{c,1} \bar{\sigma}_t^{zu*} \|P_{t,2}^\dagger\|_2. \tag{4.62f}$$

Final Perturbation Bound

Our final step is to apply the results of section Section 4.6 to bound the ∂-terms on the right-hand side of the perturbation inequality Lem. 32 in terms of $\partial_{12}A$ and $\partial_{12}B$:

$$\|\partial_{12} G_t^{yu}\|_2 \leqslant (t+1)\bar{\sigma}_t^{c*}\bar{\sigma}_t^{o*}\|\partial_{12}A\|_2 + \sqrt{t+1}\bar{\sigma}_t^{o*}\|\partial_{12}B\|_2$$

$$\|\partial_{12} P_t\|_2 \leqslant \bar{\rho}_t^{c,2}\left(\sqrt{(t+1)}\bar{\sigma}_t^{c,1}\|\partial_{12}A\|_2 + \|\partial_{12}B\|_2\right)$$

$$\|\partial_{12} Q_t\|_2 \leqslant \bar{\rho}_t^{o,2}\sqrt{(t+1)}\bar{\sigma}_t^{o,1}\|\partial_{12}A\|_2.$$

We recall also the lower and upper bounds of the singular values of G_t^{zu}

$$\sigma^2(G_t^{zu}) \geqslant \left(\frac{\sigma_{\min}(C)\sigma_{\min}(B)}{1+\|A\|_2}\right)^2 + \sigma_{\min}^2(D) =: (\underline{\sigma}_t^{zu})^2$$

$$\sigma^2(G_t^{zu}) \leqslant (t+1)\|C\|_2\|B\|_2\bar{\sigma}_t^o(C,A)\bar{\sigma}_t^c(A,B) + \|D\|_2^2 =: (\bar{\sigma}_t^{zu})^2$$

and the relationship between controllability/observability matrices and grammians:

$$\|P_t\|_2 = \bar{\sigma}_t^c \qquad \|P_t^\dagger\|_2 = (\underline{\sigma}_t^c)^{-1} \qquad \|Q_t\|_2 = \bar{\sigma}_t^o \qquad \|Q_t^\dagger\|_2 = (\underline{\sigma}_t^o)^{-1}.$$

We formulate the constants from (4.62) in terms of the controllability and observability singular values of the system:

$$\delta_1 \leqslant (\underline{\sigma}_t^{c*})^{-1}\alpha_{t*} + 2(\underline{\sigma}_t^{zu*})^{-1}\bar{\sigma}_t^{zu*}(\underline{\sigma}_t^{c*})^{-1}\alpha_t + \bar{\sigma}_t^{o*}\|A_*\|_2$$

$$\delta_2 \leqslant \varphi(\underline{\sigma}_t^{c*})^{-2}\bar{\sigma}_t^{zu*}\alpha_{t*} + 4(\underline{\sigma}_t^{zu*})^{-1}(\bar{\sigma}_t^{zu*})^2(\underline{\sigma}_t^{c*})^{-2}\alpha_t + 2\bar{\sigma}_t^{zu*}(\underline{\sigma}_t^{c*})^{-1}\bar{\sigma}_t^{o*}\|A_*\|_2$$

$$\delta_3 \leqslant \|A_*\|_2$$

$$\delta_4 \leqslant \bar{\sigma}_t^{o*} + \bar{\rho}_t^{o*}\bar{\rho}_t^{c*}\bar{\sigma}_t^{zu*}(\underline{\sigma}_t^{c*})^{-1}, \qquad \alpha_{t*} := \max_{0\leqslant k\leqslant t+1} \|A_1^k\|_2 \vee \|A_2^k\|_2.$$

After a bit of bookkeeping, we obtain a first bound quantifying the sensitivity of the solutions with respect to perturbations in the parameters A and B:

Lemma 33. *Recall the setup of Lem. 32. Then the following inequality holds:*

$$\kappa_{CD}^{-1}\|\phi_1^* - \phi_2^*\|_2 \leqslant \Gamma_a\|\partial_{12}A\|_2 + \Gamma_b\|\partial_{12}B\|_2$$

where Γ_a, Γ_b are defined as:

$$\Gamma_a = (t+1)\bar{\sigma}_t^{c*}\bar{\sigma}_t^{o*}\delta_1 + \delta_2(\bar{\rho}_t^{c*}\sqrt{(t+1)\bar{\sigma}_t^{c*}}) + \delta_3\bar{\rho}_t^{o*}\sqrt{(t+1)\bar{\sigma}_t^{o*}} + \delta_4$$
$$\Gamma_b = \delta_1\sqrt{t+1}\,\bar{\sigma}_t^{o*} + \delta_2\bar{\rho}_t^{c*}.$$

4.10 Lipshitzness of $\mathcal{H}_2$-Optimal CLM Parameterization

With Lem. 33, we have derived analytic bounds that quantify the difference between the two optimal CLMs $\Phi^*(A_1, B_1, t)$ and $\Phi^*(A_2, B_2, t)$ in terms of the difference in system parameters $\partial_{12}A = A_1 - A_2$, $\partial_{12}B = B_1 - B_2$, and control-theoretic properties of the linear dynamics corresponding to each set of parameters. The latter is measured in terms of singular eigenvalues measuring controllability and observability (w.r.t. C) of the individual pairs of system matrices $[A_1, B_1]$ and $[A_2, B_2]$. Our original motivation for this analysis was to investigate the continuity of the mapping Φ^* as a parametrization of optimal CLMs over a compact set S of system matrices $\{[A[\omega], B[\omega]] \mid \omega \in \Omega\}$ corresponding to the continuous parametrization functions $A : \Omega \to \mathbb{R}^{n\times n}$, $B : \Omega \to \mathbb{R}^{m\times n}$, and parameter space (Ω, d) used in Algorithm 1. With the help of Lem. 33, we will now derive that the mapping $\Phi^*(\cdot|t) : S \ni [A', B'] \mapsto \Phi^*(A', B', t) \in \mathcal{C}(\ell^{\mathcal{X}}, \ell^{\mathcal{X}\times\mathcal{U}})$ is indeed Lipshitz over a compact set $S \subset \mathbb{R}^{n\times(n+m)}$ of matrices, as long as S consists all of t-controllable pairs $[A, B]$. To this end, in the next lemma, we verify that all constants used in Lem. 33, such as $\delta_i^{c/o}, \mu, \rho_i^{c/o}$, etc., which depend on (A_i, B_i) can be uniformly bounded above for the described sets S.

Lemma 34. *Let S be a compact[7] subset of t-controllable matrix pairs $(A, B) \subset \mathbb{R}^{n\times n} \times \mathbb{R}^{n\times m}$ with full-column rank B, and let $C \in \mathbb{R}^{n\times n}$ and $D \in \mathbb{R}^{m\times m}$ be invertible matrices. Then, there exist positive constants $0 < \underline{\sigma}_S^c \leqslant \overline{\sigma}_S^c$, $0 < \underline{\sigma}_S^o \leqslant \overline{\sigma}_S^o$, $0 < \underline{\rho}_S^c \leqslant \overline{\rho}_S^c$, $0 < \underline{\rho}_S^o \leqslant \overline{\rho}_S^o$, $0 < \underline{b}_S \leqslant \overline{b}_S$, and some $0 \leqslant \overline{\alpha}_S^{(1)} \leqslant \ldots \overline{\alpha}_S^{(k)} \ldots$ such that for all $(A, B) \in S$ and $i \in \{1, \ldots, n\}$:*

$$\underline{\sigma}_S^c \leqslant \sigma_i(P_t(A, B)) \leqslant \overline{\sigma}_S^c \qquad \underline{\sigma}_S^o \leqslant \sigma_i(Q_t(C, A)) \leqslant \overline{\sigma}_S^o$$
$$\underline{\rho}_S^c \leqslant \sigma_i(P_t(A, I)) \leqslant \overline{\rho}_S^c \qquad \underline{\rho}_S^o \leqslant \sigma_i(Q_t(I, A)) \leqslant \overline{\rho}_S^o$$
$$\max_{1\leqslant j\leqslant k}\|A^j\|_2 \leqslant \overline{\alpha}_S^{(k)}, \forall k \geqslant 1 \qquad \underline{b}_S \leqslant \sigma_i(B) \leqslant \overline{b}_S.$$

---

[7]Take the usual norm for the product-space is defined as $|(A, B)| := |A| + |B|$.

Proof. All bounds follow from the simple fact that continuous scalar functions over compact sets achieve a maximum and a minimum, and by letting the corresponding constant be defined by that maximum/minimum value. Thus we need to verify the continuity of the corresponding functions. It is obvious that $\|A^j\|_2$ and P_t and Q_t are continuous functions of A and B. It is well-known [25], that the k-th largest singular value $\sigma_k : \mathbb{R}^{p \times q} \to \mathbb{R}^+$ is a continuous function over the space of matrices; a quick way to convince ourselves is by writing σ_k as a difference of the k-th and the $k - 1$-th Ky-Fan-Norm [25] and using continuity of norms. Lastly, we know that the minimum singular values $\underline{\sigma}_S^c$, $\underline{\sigma}_S^o$, $\underline{\rho}_S^c$, $\underline{\rho}_S^o$, $\underline{b}_S$ are all positive, because they are achieved at respective pairs $\{(A_i, B_i)\}$ all of which are t-controllable, i.e., $\sigma_{\min}(P_t(A_i, B_i)) > 0$, and B_i is full column rank, i.e., $\sigma_{\min}(B_i) > 0$. $\qquad\square$

The concluding theorem of our perturbation analysis follows as a corollary of the previous two lemmas, and establishes the Lipshitzness of the LQ-optimal CLM parametrization map $\boldsymbol{\Phi}^*$ over compact sets of t-controllable LTI systems:

Theorem (Lipshitzness of LQ-optimal CLM oracle). *Let $t \in \mathbb{N}$ be fixed, $C \in \mathbb{R}^{n \times n}$, $D \in \mathbb{R}^{m \times m}$ be fixed invertible matrices, and $\boldsymbol{\Phi}^*(A, B, t)$ represent the unique optimum of the optimal control problem $\mathrm{LQ}(A, B, t)$ described by (4.20), (4.21). Assume that $\mathcal{S}_{ab}$ is a compact subset of $\mathbb{R}^{n \times n} \times \mathbb{R}^{n \times m}$ such that for all $(A, B) \in \mathcal{S}_{ab}$, $[A, B]$ is t-controllable and B is full column rank. Then, there exist fixed constants $L_a, L_b \in \mathbb{R}^+$ such that for all $(A_1, B_1), (A_2, B_2) \in \mathcal{S}_{ab}$, the following inequality holds:*

$$\||\boldsymbol{\Phi}^*(A_1, B_1, t) - \boldsymbol{\Phi}^*(A_2, B_2, t)\||_\infty \leqslant L_a|\partial_{12}A| + L_b|\partial_{12}B|$$

where $\partial_{12}A := A_1 - A_2$ and $\partial_{12}B := B_1 - B_2$.

Proof. Using Lem. 33 and the constants defined in Lem. 34 for $\mathcal{S} = \mathcal{S}_{ab}$, we can see that there are constants $\overline{\Gamma}_a$ and $\overline{\Gamma}_b$ which are functions of the constants $t, \overline{\sigma}_S^c, \underline{\sigma}_S^c, \overline{\sigma}_S^o, \underline{\sigma}_S^o, \overline{\rho}_S^c, \underline{\rho}_S^c, \overline{\rho}_S^o, \underline{\rho}_S^o, \underline{b}_S, \overline{b}_S, \overline{\alpha}_S^{(k)}$ such that the difference between any two vectorized solutions $\phi_1^* := \phi^*(A_1, B_1, t)$, $\phi_2^* := \phi^*(A_2, B_2, t)$, where (A_1, B_1) and (A_2, B_2) belong to $\mathcal{S}_{ab}$, is bounded in $|\cdot|_2$ as $|\phi_1^* - \phi_2^*|_2 \leqslant \overline{\Gamma}_a|\partial_{12}A|_2 + \overline{\Gamma}_b|\partial_{12}B|_2$. Thus, ϕ^* is a Lipshitz-continuous map over $\mathcal{S}_{ab}$. This means that the matrices $M_k^{\Phi^*}$ associated with component functions of $\Phi_k^* = U^{*[k]}\boldsymbol{\Phi}^*P^k$ are $|\cdot|_F$-Lipshitz continuous and, by equivalence of norms, also $|\cdot|_\infty$-Lipshitz continuous. Finally, because $\boldsymbol{\Phi}^*(A_i, B_i, t)$ is always t-FIR and time-invariant, there are fixed L_a and

L_b such that for any component Φ_k^*, $k \in \mathbb{N}$, and perturbations $\partial_{12}A$ and $\partial_{12}B$, the following inequality holds:

$$\sup_{w_{k:0}} \frac{|\partial_{12}\Phi_k^*(w_{k:0})|}{\max_{j \leqslant k} |w_j|} \leqslant L_a|\partial_{12}A| + L_b|\partial_{12}B|.$$

The desired conclusion follows, since $\|\Phi^*(A_1, B_1, t) - \Phi^*(A_2, B_2, t)\|_\infty$ is the supremum over k of the left-hand side of the above inequality. $\square$

Analytic expressions of L_a and L_b in terms of the constants can be easily derived, however, doing so is very tedious. The asymptotic behavior in terms t and controllability/observability singular values is given below:

$$L_a = \mathcal{O}\left(\left(\frac{t\overline{\sigma}_{\mathcal{S}}^c\overline{\sigma}_{\mathcal{S}}^o}{\underline{\sigma}_{\mathcal{S}}^c}\right)^{\frac{3}{2}}\left(\frac{\overline{\sigma}_{\mathcal{S}}^o}{\underline{\sigma}_{\mathcal{S}}^c}\right)^{\frac{1}{2}}\overline{\alpha}_{\mathcal{S}}^{(t)}\right) \quad L_b = \mathcal{O}\left(\left(\frac{t\overline{\sigma}_{\mathcal{S}}^c\overline{\sigma}_{\mathcal{S}}^o}{\underline{\sigma}_{\mathcal{S}}^c}\right)\left(\frac{\overline{\sigma}_{\mathcal{S}}^o}{\underline{\sigma}_{\mathcal{S}}^c}\right)\overline{\alpha}_{\mathcal{S}}^{(t)}\right).$$

Part II

Learning to Control Unknown Systems

OVERVIEW

In regards to robustness to model uncertainty, the previous part primarily focuses on problem settings where the system dynamics are uncertain but can be narrowed down to a set of possible models. In this setting, all the models can be simultaneously stabilized by a fixed controller using methods of robust control theory. On the other hand, this part of the book focuses on the more general "large uncertainty" scenario, where the set of possible models is bounded but can be arbitrarily large. In practical terms, Part 2 of the book assumes that the dynamics of the system are almost entirely unknown. In this second part, we introduce new theory and algorithms for a general framework of learning-to-control with worst-case safety and performance guarantees, even in settings where dynamic uncertainty is very large. We approach this problem from two complementary perspectives: Chapter 5 pursues a model-free, data-driven approach, while Chapter 6 follows a model-based learning and control approach.

The results presented in Chapter 5 are based on the work published in [5] and develop new methods for stability analysis and control design without the need for a model. In particular, we demonstrate the first instance of an all model-free formulation of controller, closed-loop dynamics, and robust stability analysis. We present a simple model-free control algorithm that can robustly learn and stabilize an unknown discrete-time linear system with full control and state feedback, subject to arbitrary bounded disturbance and noise sequences. The controller does not require any prior knowledge of the system dynamics, disturbances, or noise, yet it can guarantee robust stability and provides asymptotic and worst-case bounds on the state and input trajectories. To the best of our knowledge, this is the first model-free algorithm to come with such robust stability guarantees without the need to make any prior assumptions about the system. Simulation results also show that despite its generality and simplicity, the controller demonstrates good closed-loop performance, including fast convergence, small learning transients, and nearly optimal asymptotic gain.

In Chapter 6, we approach the problem of learning-to-control unknown systems from a model-based perspective. In this case, we are given a compact parameterization of all possible system dynamics, which can be arbitrarily large. The results presented are based on the work published in [5] and introduce a new modular framework for model-based learning-to-control. This framework provides robust safety and cost-performance guarantees for closed loop under worst-case scenarios of realized disturbances, noise, or other environmental conditions. Our approach involves decomposing the problem into two subproblems: online learning, referred to as "consistent model chasing," and the underlying control problem in the absence of model uncertainty, known as "oracle design." Each subproblem can be addressed separately, and its solutions (a control oracle and model chaser) are used to instantiate a certainty-equivalent learning-to-control scheme. This scheme inherits both control- and learning-theoretic guarantees, certifying robustness of the closed-loop, even for large model uncertainty in the system dynamics. We discuss how the control oracle is implicitly given by standard nominal control design, provided that this procedure satisfies certain regularity properties over the space of models. Designing the corresponding model chaser represents a problem which we term "consistent model chasing" and discuss how to solve it with existing techniques from online learning.

Chapter 5

ROBUST MODEL FREE LEARNING AND CONTROL OF LINEAR SYSTEMS

We present a simple model-free control algorithm that is able to robustly learn and stabilize an unknown discrete-time linear system with full control and state feedback subject to arbitrary bounded disturbance and noise sequences. The controller does not require any prior knowledge of the system dynamics, disturbances, or noise, yet it can guarantee robust stability and provides asymptotic and worst-case bounds on the state and input trajectories. To the best of our knowledge, this is the first model-free algorithm that comes with such robust stability guarantees without the need to make any prior assumptions about the system. We would like to highlight the new convex geometry-based approach taken towards robust stability analysis, which served as a key enabler in our results. We will conclude with simulation results that show that despite generality and simplicity, the controller demonstrates good closed-loop performance.

5.1 Introduction

Motivation and Problem Statement

Learning to stabilize unknown dynamical systems from online data has been an active research area in the control community since the 1950s [74] and has recently attracted the attention of the machine learning community, in particular in the context of reinforcement learning. Although extensive research has been conducted on this topic, very few of the algorithms developed have reached the level of adoption in real-world applications, as one would expect. In particular in areas where frequent interaction with the physical world is necessary, system failure is costly and the deployment of control algorithms is only possible if the algorithm can guarantee that minimal safety and performance specifications will be met during operation. Although there have been previous research [132],[18] and recent research efforts [57],[43], [2], [24], [38],[44] to address this problem, very few algorithms have come with the necessary performance and safety guarantees to be deployed in real-world applications so far. Motivated by this, we revisit the basic problem of learning to stabilize a linear system and aim to find learning and control strategies with the least restrictive assumptions that can still give robust stability bounds for the closed loop.

In this chapter, we focus on the problem of adaptively stabilizing a linear discrete-time system

$$z_{k+1} = A_0 z_k + u_k + d_k \tag{5.1a}$$

$$x_k = z_k + n_k \tag{5.1b}$$

with state z_k, bounded disturbance d_k, bounded noise n_k, and control action u_k that is only allowed to depend on noisy state measurements until time k, i.e., $x_0, \ldots, x_k$. We are interested in finding controllers that can stabilize (in the sense of BIBO- or input-to-state stability guarantees) without requiring any additional assumptions about the unknown system matrix A_0 and the disturbance/noise sequences (d_k), (n_k). Although the system (5.1) admits a very restrictive class of linear systems (full state feedback and control), nearly all available learning and control approaches need to make some prior assumptions about this system in order to state stability and performance guarantees. Most commonly, these assumptions come in the form of a priori bounds on d_k, n_k, and/or A_0.

Related Work

We will review the relevant literature in the context of our problem setting. Classical control approaches are found in the literature on adaptive control with [72], [71], [112] focusing on the deterministic setting and [132] on the stochastic setting. The self-tuning regulator [18] and its variations come with asymptotic optimality [60], yet robust stability guarantees without restrictive assumptions are few and can only be made in the probabilistic sense. On the deterministic side, [72], [71] point out that instabilities can occur with traditional adaptive schemes and provide improved versions of adaptive controllers that come with robust stability and performance guarantees. However, the desired guarantees depend on knowing some limits of the system parameters and disturbance signals. Other challenges associated with classical adaptive control approaches are discussed in [8], [10]. Methods in safe reinforcement learning [57], [24], [5], [53] have made great progress toward methods that guarantee robust safety properties for classes of non-linear systems, yet the synthesis procedures involved are computationally expensive, and require knowledge of an initially robust stabilizing controller, even in the case of a simple linear system (5.1). Recent work [2], [42], [43], [92], [38], [44] has made significant progress in providing algorithms with robust finite-time performance guarantees for the adaptive linear quadratic Gaussian regulator problem. However, in the context of our simple linear problem setting, all methods require that the uncertainty in the system dynamics (i.e., A_0) is small enough at the outset to provide stability guarantees of the closed loop.

Main Contribution and Overview of the Chapter

In this work, we present a simple controller that can adaptively stabilize (5.1) without any additional assumptions about disturbance, noise, or the system matrix A_0. The presented algorithm performs tractable computations (solving a linear program each time step) and provides both uniform asymptotic and worst-case guarantees on the state and input trajectories. An additional surprising feature of the presented algorithm is that it is not based on the certainty-equivalence principle and has a completely model-free formulation. To the best of our knowledge, this is the first model-free adaptive controller that can give our robust stability guarantees without requiring prior knowledge about the unknown system (5.1).

Our core theoretical contribution is a novel approach towards stability analysis. We first show that in any closed-loop trajectory (x_t), there are only a finite number of time instances t_i at which x_{t_i+1} is significantly larger than x_{t_i}. We term those time-instances as "unstable transitions" and our first main theorem shows an upper-bound on the occurrence of these unstable transitions in the closed loop state trajectory. Then, our stability and performance guarantees follow as corollaries of this result. We develop a new technique based on convex geometry to bound the occurrence of unstable transitions in the closed loop. Vaguely speaking, our main idea is to show that if an unstable transition occurs at some time t', our proposed adaptive controller learns enough from this observation to prevent similar unstable transitions from occurring in the future. Mathematically, we formulate this idea in two steps: 1. We define a distance function d between unstable transitions and show that, w.r.t. to d, we can identify the set of unstable transitions with a bounded separated set P of equal cardinality. 2. We bound the cardinality of P by a metric-entropy type of quantity, which leads to an upper bound on the maximum number of times that unstable transitions can occur. We discuss the convex geometry-based techniques in detail, to emphasize their potential use for robust design and analysis of learning and control algorithms, particularly in the model-free setting. The chapter is organized as follows. We formulate our problem in Section 5.2 and give a brief overview of our main results in Section 5.3. In Section 5.5, we derive the model-free closed-loop equation and explain the intuition behind the proposed control law. In Section 5.6, we present and discuss our main results in detail. Section 5.7 and Section 5.8 highlight the main techniques and ideas used to prove our results.Section 5.9 highlights a parallel between the role of metric entropy in the context of our results and in the context of statistical learning theory, which could serve as an interesting intersection for future research. We conclude with some experimental results in Section 5.10.

5.2 Problem Setup

For our discussion, we transform the system (5.1) w.r.t to the measurements x_t and obtain the equivalent[1] system

$$x_{t+1} = A_0 x_t + u_t + w_t \tag{5.2}$$
$$w_t := d_t + n_{t+1} - A_0 n_t,$$

where w_t represents the lumped bounded disturbance at time t which summarizes the influence on the system of the original noise and disturbance. A causal controller can be represented as a collection $\boldsymbol{K} = (K_0, K_1, \dots)$ of control laws $K_t : (x_t, \dots, x_0) \mapsto u_t$. The closed loop of $\boldsymbol{K}$ and (5.2) is then described by the equation

$$x_{t+1} = A_0 x_t + K_t(x_t, \dots, x_0) + w_t. \tag{5.3}$$

Our goal is now to design $\boldsymbol{K}$ such that the closed-loop (5.3) is bounded-input bounded-output stable for any A_0 and any bounded sequence[2] (w_t). More specifically, we want to design $\boldsymbol{K}$ such that any closed-loop trajectory (x_t) satisfies bounds of the form

$$\sup_t \|x_t\| \leq f_1(A_0, \sup_t \|w_t\|)$$
$$\limsup_{t \to \infty} \|x_t\| \leq f_2(A_0, \sup_t \|w_t\|)$$

for some fixed functions f_1 and f_2.

5.3 Preview of Main Result

We will start by describing the implementation of our proposed controller and a summary of its performance guarantees in a closed-loop with system (5.2).

Proposed control strategy

For adaptive stabilization of (5.2) we propose a dynamic controller $\boldsymbol{K}^{cc} \in \mathcal{C}(\ell^{\mathcal{X}}, \ell^{\mathcal{U}})$, which at every time step t computes the input as

$$u_t = K_t^{cc}(x_t, \dots, x_0, u_{t-1}, \dots, u_0)$$

---

[1] Since $x_t = z_t + n_t$, controlling the system state z_t is equivalent to controlling the noisy measurement x_t.

[2] We use bracket notation to distinguish a sequence (w_t) from its element w_t at time t.

based on all previous measurements $x_t, \ldots, x_0$ and previously taken actions $u_{t-1}, \ldots, u_0$. The controller $\boldsymbol{K}^{\mathrm{cc}}$ computes the input u_t at time t as

$$K_t^{\mathrm{cc}}(x_t, \ldots, x_0, u_{t-1}, \ldots, u_0) := \left(U_{t-1} - X_{t-1}^+ \right) \lambda_{t-1}(x_t) \tag{5.4}$$

where $\lambda_{t-1}(x_t)$ is defined as the solution of the convex optimization problem

$$\begin{aligned} \min_{\lambda} \quad & \|\lambda\|_1 \\ \text{s.t.} \quad & X_{t-1}\lambda = x_t \end{aligned} \tag{5.5}$$

and where the matrices U_t, X_t and X_t^+ are composed of state x_t and input u_t measurements up until time t as

$$X_t := [x_t, x_{t-1}, \ldots, x_0, X_{-1}] \tag{5.6a}$$

$$U_t := [u_t, u_{t-1}, \ldots, u_0, U_{-1}] \tag{5.6b}$$

$$X_{t-1}^+ := [x_t, x_{t-1}, \ldots, x_1, X_{-1}^+] \tag{5.6c}$$

with fixed chosen matrices X_{-1}, U_{-1}, $X_{-1}^+ \in \mathbb{R}^{n \times n_0}$ such that $n_0 > n$ and $\mathrm{rank}(X_{-1}) = n$. The matrices X_{-1}, U_{-1}, X_{-1}^+ with columns $\hat{x}_i$, $\hat{u}_i$ and $\hat{x}_i^+$ defined as

$$\begin{aligned} X_{-1} := [\hat{x}_1, \ldots, \hat{x}_{n_0}], \quad & U_{-1} := [\hat{u}_1, \ldots, \hat{u}_{n_0}], \\ X_{-1}^+ := [\hat{x}_1^+&, \ldots, \hat{x}_{n_0}^+] \end{aligned} \tag{5.7}$$

serve to initialize the controller $\boldsymbol{K}^{\mathrm{cc}}$. Depending on the application scenario, the matrices can be chosen as follows:

(I1) **No prior knowledge:** choose $\hat{u}_i = 0$, $\hat{x}_i^+ = 0$, $\hat{x}_i = \varepsilon e_i$ for $1 \leq i \leq n$ where e_i is the ith Cartesian unit vector and $\varepsilon > 0$ is some positive scalar. The parameter ε can be viewed as an initial guess on $\sup_t \|w_t\|_1$, the supremum of the disturbance sequence in 1-norm.

(I2) **Prior data available:** Assume we had noisy data available x_{-j}, x_{-j}^+, u_{-j}, $1 \leq j \leq k$ collected from the system (5.2) before $t = 0$. i.e., the data satisfies

$$x_{-j}^+ = A_0 x_{-j} + u_{-j} + w_{-j} \tag{5.8}$$

with w_{-j} denoting the corresponding lumped disturbances. Then, in addition to the initialization (I1) $\hat{u}_i = 0$, $\hat{x}_i^+ = 0$, $\hat{x}_i = \varepsilon e_i$ for $1 \leq i \leq n$, we can incorporate the data x_{-j}, x_{-j}^+, u_{-j} by appending additional columns as

$$\hat{u}_{n+i} := u_{-i}, \quad \hat{x}_{n+i} := x_{-i}, \quad \hat{x}_{n+i}^+ := x_{-i}^+, \; 1 \leq i \leq k.$$

Closed-Loop Guarantees

In this chapter, we will show that the controller K^{cc} stabilizes system (5.1) without requiring any knowledge of A_0 or (w_t). We will term the controller K^{cc} defined by (5.4), (5.5) the *causal cancellation controller*, as it can be interpreted at time t to cancel out the part of the dynamics that can be inferred from all previously collected observations $x_t, \ldots, x_0$ and actions $u_{t-1}, \ldots, u_0$.

As presented in detail in Section 5.6, for **any** initialization X_{-1}, U_{-1}, X_{-1}^+ (only assuming $\mathrm{rank}(X_{-1}) = n$), the controller (5.5) **always** ensures a closed loop for which:

   (i) the state (x_t) and input (u_t) are uniformly bounded.

  (ii) an analytic upper-bound can be derived for the worst-case state-deviation.

 (iii) after some finite time, (x_t) and (u_t) converge exponentially to a bounded limit set.

The above guarantees will be phrased w.r.t. a norm $\| \cdot \|_\mathsf{W}$ which measures (x_t) and (u_t) relative to the size of the disturbance (w_t) that produced them. Moreover, as described in (I2), we can incorporate prior data into the initialization of the controller K^{cc}. In the case where the provided data is "more informative" than the default initialization (I1), the closed loop guarantees and bounds tighten. Hence, the proposed control scheme K^{cc} does not need prior knowledge to give closed-loop stability guarantees, but if prior knowledge is available, it can be leveraged through the initialization (I2) to improve closed-loop guarantees.

5.4   Preliminaries of Convex Geometry

Convex Bodies and Norms

It is a well-known fact in convex geometry that there is a one-to-one relationship between symmetric convex bodies (see Def. 5.1) and norms in $\mathbb{R}^n$. Here we will discuss this equivalence and how it relates to the properties of the $\| \cdot \|_\mathsf{S}$-norms which we used in this chapter. The following discussion is adapted from chapter 1.7 of the standard text [113], to which we refer for more detail.

Definition 5.1 (Symmetric Convex Body). *A set* $\mathsf{B} \subset \mathbb{R}^n$ *is a symmetric convex body if* B *is a closed, bounded, convex set with non-empty interior and* $z \in \mathsf{B} \Leftrightarrow -z \in \mathsf{B}$.

Symmetric convex bodies and norms are equivalent in $\mathbb{R}^n$, in the sense that any norm on $\mathbb{R}^n$ is uniquely defined by its corresponding unit norm ball, and the set of all possible unit norm balls in $\mathbb{R}^n$ is precisely the set of symmetric convex bodies in $\mathbb{R}^n$. We summarize this in the following lemma:

Lemma 35. *For any norm $\|\cdot\|$ in $\mathbb{R}^n$, the corresponding norm ball $\{x\,|\,\|x\| \leqslant 1\}$ is a symmetric convex set in $\mathbb{R}^n$. On the contrary, for any symmetric convex body B in $\mathbb{R}^n$, the function $g(\mathsf{B}, \,\cdot\,) : \mathbb{R}^n \mapsto \mathbb{R}^+$ defined by*

$$g(\mathsf{B}, x) := \min\{r \geqslant 0\,|\,x \in r\mathsf{B}\}, \quad \forall x \in \mathbb{R}^n \tag{5.9}$$

is a norm on $\mathbb{R}^n$.

In convex geometry, the function $g(\mathsf{B}, \cdot)$ is often called the *gauge function* or the *Minkowski functional* of B and describes a concrete way to evaluate a norm based on knowing its unit ball. For our purposes, it will be convenient to extend the above definition to derive norms from general bounded sets $\mathsf{S} \subset \mathbb{R}^n$ in the following way: Given an arbitrarily bounded set S, we will refer to $\|\cdot\|_\mathsf{S}$ as the norm $g(\mathrm{c}(\mathsf{S}), \,\cdot\,)$, obtained by the Minkowski functional of the *absolute convex hull* $\mathrm{c}(\mathsf{S})$ of the set S. We define this formally in Def. 5.4 and Def. 5.5:

Definition. *Let S be a set in $\mathbb{R}^n$, then the set of all finite linear combinations $\sum_{i=1}^{N} \lambda_i x_i$ of elements x_i in S with $\sum_{i=1}^{N} |\lambda_i| \leqslant 1$ is called the absolute convex hull of S, and we will refer to its closure as $\mathrm{c}(\mathsf{S})$:*

$$\mathrm{c}(\mathsf{S}) := \mathrm{cl}\left\{ \sum_{i=1}^{N} \lambda_i x_i \,\middle|\, \{x_i\}_{i=1}^{N} \subset \mathsf{S}, \ \sum_{i=1}^{N} |\lambda_i| \leqslant 1 \right\}. \tag{5.10}$$

Remark. *Equivalently, $\mathrm{c}(\mathsf{S})$ is the closure of the convex hull of the set $(-\mathsf{S}) \cup \mathsf{S}$.*

Definition. *For a fixed bounded set $\mathsf{S} \subset \mathbb{R}^n$, let $\|\cdot\|_\mathsf{S} : \mathbb{R}^n \mapsto \mathbb{R}_{\geqslant 0}$ be the norm defined for all $x \in \mathbb{R}^n$ as*

$$\|x\|_\mathsf{S} := \begin{cases} \min\{r \geqslant 0\,|\,x \in r\mathrm{c}(\mathsf{S})\}, & \text{for } x \in \mathrm{span}(\mathsf{S}) \\ \infty, & \text{else} \end{cases}$$

and for sets $\mathsf{S}' \subset \mathbb{R}^n$, define $\|\mathsf{S}'\|_\mathsf{S}$ as the quantity

$$\|\mathsf{S}'\|_\mathsf{S} := \max_{z \in \mathrm{c}(\mathsf{S}')} \|z\|_\mathsf{S}. \tag{5.11}$$

The definition is overloading the common notation for the norm $\|\cdot\|_S$ as used in [113] where S is required to be a symmetric convex body. Applying Def. 5.5 to the disturbance set W, (5.26) defines the previously introduced norm $\|\cdot\|_W$ that we use to formulate the stability analysis. Figure 5.1 illustrates an example of how the set $c(W)$ and the norm $\|x\|_W$ are related in two dimensions. The following lemma summarizes some key properties following from the above definitions. Note property (iii) and (iv), which show that we can verify the set-membership and set inclusions of symmetric convex bodies in terms of the norm. Furthermore, property (ii) describes a practical evaluation of $\|\cdot\|_S$ for finite sets S and shows

$$\|\lambda_{t-1}(x)\|_1 = \min_{\lambda} \ \|\lambda\|_1 \qquad\qquad = \|x\|_{X_{t-1}} \tag{5.12}$$
$$\text{s.t.} \quad X_{t-1}\lambda = x.$$

Lemma 36. *Norms according to Definition 5.5 satisfy:*

(i) *For any set S in Def. 5.5, $c(S)$ is a symmetric convex body in $\mathbb{R}^n$. Moreover, $c(S)$ is the unit norm ball of $\|\cdot\|_S$, so we can equiv. write $\|\cdot\|_S = \|\cdot\|_{c(S)}$.*

(ii) *If S is a finite set $S = \{p_1, \ldots, p_N\}$, then for any $x \in \mathbb{R}^n$, $\|x\|_S$ can be computed as:*

$$\|x\|_S = \min\left\{ \sum_{i=1}^N |\lambda_i| \ \bigg| \ \lambda_1, \ldots, \lambda_N \ s.t. \ \sum_{i=1}^N \lambda_i p_i = x \right\}.$$

(iii) *for all $x \in \mathbb{R}^n$ holds $x \in c(S) \Leftrightarrow \|x\|_S \leq 1$.*

(iv) *$c(S_1) \subset c(S_2)$ holds if and only if for all $x \in \mathbb{R}^n$ holds $\|x\|_{S_1} \geq \|x\|_{S_2}$.*

(v) *for all $\gamma > 0$, holds $\|\cdot\|_{\frac{1}{\gamma}S} = \gamma\|\cdot\|_S$.*

Proof. The statements of Lem. 36 are easy to verify: (i), (iii), (iv) follow directly from Lem. 35 and (ii) follows by using the description of the set shown in (5.10) to rewrite the definition of $\|\cdot\|_S$. $\qquad\square$

We use the definition (5.11) of $\|S_1\|_{S_2}$ to measure the size of a set S_1 w.r.t. the norm $\|\cdot\|_{S_2}$ of another set S_2. The following properties can be easily verified:

Lemma 37. *Let S_1, S_2 be some bounded sets in $\mathbb{R}^n$ and recall the definitions Def. 5.5. Then we have the following.*

(i) *$\|S_1\|_{S_2} := \min\{ t \mid S_1 \subset t c(S_2), \ t \geq 0\}$.*

(ii) equivalence of norms in $\mathbb{R}^n$: $\frac{1}{\|S_1\|_{S_2}} \| \cdot \|_{S_2} \leqslant \| \cdot \|_{S_1} \leqslant \|S_2\|_{S_1} \| \cdot \|_{S_2}$.

(iii) $\|S_1\|_{S_2} \leqslant 1 \Leftrightarrow S_1 \subset c(S_2)$.

(iv) $S_1 \subset \|S_1\|_{S_2} c(S_2)$.

Property (i) states that we can equivalently define $\|S_1\|_{S_2}$ as the smallest factor t such that S_1 is contained in $tc(S_2)$. This definition is visualized in the middle plot of Figure 5.1 for some exemplary sets W and X_1 in $\mathbb{R}^2$. The other properties can be derived as immediate consequences of property (i): (i) $\Rightarrow$ (iv) $\Rightarrow$ (iii), (ii).

Distance Between Norms

For bounded sets S_1, S_2 we define $d(S_1, S_2)$ as a multiplicative distance $d(S_1, S_2)$ between the two norms $\| \cdot \|_{S_1}$ and $\| \cdot \|_{S_2}$:

Definition 5.2. *Let* $S_1, S_2 \subset \mathbb{R}^n$ *be sets with norms* $\| \cdot \|_{S_1}$, $\| \cdot \|_{S_2}$ *defined as in Def. 5.5. Then, define* $d(S_1, S_2)$ *as*

$$d(S_1, S_2) := \max\{\|S_1\|_{S_2}, \|S_2\|_{S_1}\}. \tag{5.13}$$

Lemma 38. *The definitions 5.2 imply the following.*

(i) $d(S, S) = 1$.

(ii) $d(S_1, S_2) = d(S_2, S_1)$.

(iii) $d(S_1, S_2) \leqslant d(S_1, S')d(S', S_2)$.

Proof. Statement (i) and (ii) are trivial. Part (iii) follows using (iv) and (i) of Lem. 37: Note that

$$S_1 \subset d(S_1, S')c(S') \subset d(S_1, S')d(S', S_2)c(S_2) \tag{5.14}$$
$$S_2 \subset d(S_2, S')c(S') \subset d(S_2, S')d(S', S_1)c(S_1) \tag{5.15}$$

leads to $\max\{\|S_1\|_{S_2}, \|S_2\|_{S_1}\} \leqslant d(S_2, S')d(S', S_1)$. $\qquad\square$

Lem. 38 shows that the map $d(\,\cdot\,,\,\cdot\,)$ can be viewed as a multiplicative distance between sets: These properties imply that $\log d(\,\cdot\,,\,\cdot\,)$ is a pseudometric over the space of bounded sets in $\mathbb{R}^n$.

5.5 The Model-Free Adaptive Controller and Closed-Loop Equations

We will start by first discussing the key idea and intuition behind the causal cancelation controller $\boldsymbol{K}^{\mathrm{cc}}$. Section 5.5 is deriving that all (X_t^+, X_t, U_t) satisfies the open loop equation of the unknown system for some appropriately defined disturbance matrix W_t. This is used to show in Section 5.5 that at time t and state x_t, the causal cancelation control law K_t^{cc} approximates the ideal deadbeat control action $u_t^* = -A_0 x_t$ directly from online data $(X_{t-1}^+, X_{t-1}, U_{t-1})$ without requiring an explicit estimate A_0. This relation leads to a model-free form of the closed loop equation, shown in Section 5.5, which is used for the later stability analysis.

Open Loop Equation for Data Matrices

Recall from (5.6), that (X_t^+, X_t, U_t) are constructed from some fixed initialization $(X_{-1}^+, X_{-1}, U_{-1})$ and some state (x_t) and input (u_t) sequences of the system (5.2) with respect to some fixed lumped disturbance (w_t). Define the disturbance matrix $W_t \in \mathbb{R}^{n \times (t+1+n_0)}$ as the matrix

$$W_t := [w_t, w_{t-1}, \ldots, w_0, W_{-1}] \tag{5.16a}$$

$$W_{-1} := [\hat{w}_1, \ldots, \hat{w}_{n_0}] := X_{-1}^+ - A_0 X_{-1} - U_{-1} \tag{5.16b}$$

of lumped disturbances $w_t, \ldots, w_0$ and the matrix W_{-1} which is composed of the columns $\hat{w}_1, \ldots, \hat{w}_{n_0}$. With the above auxiliary definition, we can easily see that the matrices U_t, X_t, X_t^+, and W_t satisfy the linear equation

$$X_{t-1}^+ = A_0 X_{t-1} + U_{t-1} + W_{t-1}, \tag{5.17}$$

which resembles the open-loop dynamics of the unknown system. We will term $\hat{w}_i$ as "virtual" disturbances, which are defined to account for errors introduced through the initial guesses X_{-1}, U_{-1}, X_{-1}^+. If we take $\hat{x}_i, \hat{u}_i, \hat{x}_i^+$ to be the ith columns of the initialization matrices X_{-1}, U_{-1}, X_{-1}^+ and W_{-1}, we can rewrite the definition (5.16b) columnwise in the form

$$\hat{x}_i^+ = A_0 \hat{x}_i + \hat{u}_i + \hat{w}_i, \quad 1 \leqslant i \leqslant n_0 \tag{5.18}$$

to see that each pair of $(\hat{x}_i, \hat{u}_i)$ and $\hat{x}_i^+$ can be posed as a transition of the true unknown system (5.2), w.r.t. the virtual disturbance $\hat{w}_i$.

Example 2. *If we initialize according to procedure (I1), then* $\hat{w}_i = -\varepsilon A_0 e_i$ *and* $W_{-1} = -\varepsilon A_0$.

Model-Free Approximation of Deadbeat Control Action

In a compact form, the components $\{K_t^{cc}\}$ of the causal cancellation controller are mappings $(x_t, X_{t-1}^+, X_{t-1}, U_{t-1}) \mapsto u_t$ such that:

$$u_t = \left(U_{t-1} - X_{t-1}^+\right) \lambda_{t-1}(x_t) \tag{5.19}$$

$$\text{where } \lambda_{t-1}(x_t) := \underset{\lambda \text{ s.t. } X_{t-1}\lambda=x_t}{\arg\min} \|\lambda\|_1 .$$

Remark. *The technical issue that a minimizer of (5.5) might not be unique is not relevant for the analysis and for simplicity will be ignored.*

The function $\lambda_{t-1}(x_t)$ is defined to always satisfy

$$X_{t-1}\lambda_{t-1}(x_t) = x_t \tag{5.20}$$

and represents a decomposition of the state x_t as a linear combination of the columns of X_{t-1}.

Rewriting equation (5.17) as

$$U_{t-1} - X_{t-1}^+ = -A_0 X_{t-1} - W_{t-1} \tag{5.21}$$

and substituting the right hand side of equation (5.21) into (5.19) and using (5.20) allows us to rewrite the controller equivalently as

$$u_t = \left(U_{t-1} - X_{t-1}^+\right) \lambda_{t-1}(x_t) \overset{(5.20)}{=} -A_0 x_t - W_{t-1}\lambda_{t-1}(x_t). \tag{5.22}$$

The above shows that the control law (5.19) is a direct way to approximate the ideal deadbeat control action $-A_0 x_t$ from the online data matrices $U_{t-1}, X_{t-1}^+, X_{t-1}$. The additional term $-W_{t-1}\lambda_{t-1}(x_t)$ is the corresponding approximation error at time t. As will become clear later, the optimization step in (5.19) is minimizing an upper bound of $-W_{t-1}\lambda_{t-1}(x_t)$ relative to the norm $\| \cdot \|_{\mathsf{W}_{t-1}}$ (see definition in Section 5.6).

The Model-Free Closed-Loop equations

Setting $K = K^{cc}$ and using (5.22), the closed loop equations (5.3) take the form of

$$x_{t+1} = -W_{t-1}\lambda_{t-1}(x_t) + w_t \tag{5.23a}$$

$$u_t = \left(U_{t-1} - X_{t-1}^+\right) \lambda_{t-1}(x_t) \tag{5.23b}$$

which will serve as the basis for our stability analysis. The above equation says that the closed-loop dynamics is entirely determined by W_{t-1} (containing virtual and past lumped disturbances), as well as how we choose to decompose x_t as a linear combination $X_{t-1}\lambda_{t-1}(x_t)$ of past data. Suitably, we could call equation (5.23a) to be a model-free description of the closed loop, since the dynamics are formulated independent of the underlying true unknown system A_0.

5.6   Main Results

Theorem 27 and Theorem 29 are the main theoretical results. Theorem 27 states that any state trajectory (x_t) of the closed loop has finitely many "unstable transitions" (defined in Def. 5.6). Theorem 29 is a consequence of Theorem 27 and presents our main stability bounds for state and input trajectories of the closed loop. To formulate our results, we first introduce the necessary notation and definitions.

Notation and Definitions

Definition 5.3. *If $M \in \mathbb{R}^{n \times N}$ is a matrix with N columns m_i, then define the corresponding variable M in sans serif font to denote the set $M := \{m_1, \ldots, m_N\}$.*

Definition 5.4. *Let S be a set in $\mathbb{R}^n$, then the set of all finite linear combinations $\sum_{i=1}^N \lambda_i x_i$ of elements x_i in S with $\sum_{i=1}^N |\lambda_i| \leq 1$ is called the absolute convex hull of S and we will refer to its closure as $\mathrm{c}(S)$:*

$$\mathrm{c}(S) := \mathrm{cl}\left\{ \sum_{i=1}^N \lambda_i x_i \;\middle|\; \{x_i\}_{i=1}^N \subset S, \;\; \sum_{i=1}^N |\lambda_i| \leq 1 \right\}. \tag{5.24}$$

Definition 5.5. *For a fixed bounded set $S \subset \mathbb{R}^n$, let $\|\cdot\|_S : \mathbb{R}^n \mapsto \mathbb{R}_{\geq 0}$ be the norm defined for all $x \in \mathbb{R}^n$ as*

$$\|x\|_S := \begin{cases} \min\{r \geq 0 \,|\, x \in r\mathrm{c}(S)\}, & \text{for } x \in \mathrm{span}(S) \\ \infty, & \text{else} \end{cases} \tag{5.25}$$

and for sets $S' \subset \mathbb{R}^n$, define $\|S'\|_S$ as the quantity

$$\|S'\|_S := \max_{z \in \mathrm{c}(S')} \|z\|_S .$$

Key properties of the above norm and relevant concepts from convex geometry are discussed in the appendix Section 5.4. For a fixed disturbance (w_t) and virtual disturbance $\hat{w}^i$, define W as the corresponding fixed set.

$$W := \{w_t | t \in \mathbb{N}\} \cup \{\hat{w}_i | 1 \leq i \leq n_0\}. \tag{5.26}$$

Let $\|\cdot\|_W$ and $\|\cdot\|_{X_t}$ denote the norms constructed from the fixed disturbances and

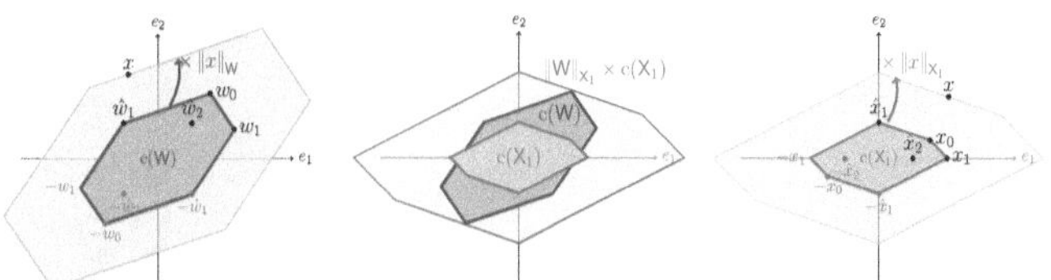

Figure 5.1: Examples in $\mathbb{R}^2$. *Left*: $c(W)$ and $\|x\|_W$ for $W = \{\hat{w}_1, \hat{w}_2\} \cup \{w_t | r \in \mathbb{N}\}$ and $(w_t) = (w_0, w_1, 0, 0, \dots)$. *Right*: $c(X_1)$ and $\|\cdot\|_{X_1}$ for $X_1 = [x_1, x_0, \hat{x}_2, \hat{x}_1]$. *Middle*: $\kappa_2 = \|W\|_{X_1}$ is the smallest factor r such that $c(W) \subset rc(X_1)$.

data matrix X_t according to Def. 5.5 and Def. 5.3. For a fixed trajectory (x_t), W and fixed initial time τ, the constant κ_τ refers to the quantity

$$\kappa_\tau := \|W\|_{X_{\tau-1}}. \tag{5.27}$$

Figure 5.1 shows an example in $\mathbb{R}^2$ that illustrates the geometric relationship between the sets $c(W)$, $c(X_t)$ and the evaluation of their respective norms at some point x. The arrows indicate that one set is a scaled copy of the other set. The middle picture in Figure 5.1 shows a geometric interpretation of the corresponding constant κ_τ for $\tau = 2 : \kappa_\tau$ is the smallest scaling factor r such that the set $r \times c(X_{\tau-1})$ contains the set $c(W)$.

Finite Occurrence of Unstable Transitions

Our approach is to analyze the behavior of the closed loop by quantifying how many "unstable transitions" can occur in the future time window $[\tau, \infty)$ of a closed-loop trajectory (x_t), given $(X_\tau, X_{\tau-1}^+, U_{\tau-1})$, which represents the data collected up to time τ. For a fixed $0 < \mu < 1$ and a trajectory (x_t), we define the occurrence of a μ-unstable transition as follows:

Definition 5.6 (μ-unstable transition). *The trajectory (x_t) has a μ-unstable transition at time t if the pair of consecutive states (x_{t+1}, x_t) satisfies*

$$\|x_{t+1}\|_W > \max\left\{\tfrac{1}{1-\mu}, \mu\|x_t\|_W + 1\right\}. \tag{5.28}$$

In other words, $(x_{t+1}, x_t) \in \mathcal{U}_\mu$, where $\mathcal{U}_\mu$ denotes the set of all pairs $(x, x^+) \in \mathbb{R}^n \times \mathbb{R}^n$ that satisfy the inequality (5.28):

$$\mathcal{U}_\mu := \left\{(x^+, x) |\, \|x^+\|_W > \max\left\{\tfrac{1}{1-\mu}, \mu\|x\|_W + 1\right\}\right\}. \tag{5.29}$$

The condition (5.28) represents a growth condition on a transition (x_t, x_{t+1}) on the trajectory (x_t). For each trajectory (x_t), we define a corresponding set $\mathcal{X}_\mu$ that collects all states x_t at which (x_{t+1}, x_t) belongs to $\mathcal{U}_\mu$:

Definition 5.7. *Given a trajectory* (x_t), *an initial time* τ *and some* $0 < \mu < 1$, *define* $\mathcal{X}_\mu((x_t); \tau) \subset \mathbb{R}^n$ *as*

$$\mathcal{X}_\mu((x_t); \tau) := \{\, x_t \mid (x_t, x_{t+1}) \in \mathcal{U}_\mu, t \geqslant \tau \,\}. \tag{5.30}$$

Remark. *Note that if* $\mu < \mu'$, *then* $\mathcal{X}_\mu((x_t), \tau) \supset \mathcal{X}_{\mu'}((x_t), \tau)$.

The core technical contribution of our chapter is Theorem 27, which places an upper bound on the number of μ-unstable transitions that can occur in the closed loop trajectory (x_t):

Theorem 27. *For any trajectory* (x_t) *of the closed loop* (5.23a) *and any* $\tau \geqslant 0$, *the set* $\mathcal{X}_\mu((x_t); \tau)$ *is a finite set for any* $\mu \in \mathcal{I}_{\kappa_\tau}$, *where* $\mathcal{I}_{\kappa_\tau}$ *is the open interval.*

$$\mathcal{I}_{\kappa_\tau} := \left(\left(\sqrt{\tfrac{1}{4} + \tfrac{1}{\kappa_\tau}} + \tfrac{1}{2} \right)^{-1}, 1 \right). \tag{5.31}$$

Moreover, the cardinality is bounded above as $|\mathcal{X}_\mu((x_t); \tau)| \leqslant N(\mu; \kappa_\tau)$, *where* $N : \mathbb{R} \times \mathbb{R} \mapsto \mathbb{R}$ *stands for the function*

$$N(\mu; \kappa_\tau) := \tfrac{1}{2} \left(\frac{\mu}{\mu - \sqrt{\kappa_\tau(1-\mu)}} \right)^n \max\{1, \tfrac{\mu}{1-\mu}\}^n \tag{5.32}$$

and κ_τ *is a constant computed from* $\mathsf{X}_{\tau-1}$ *as:*

$$\kappa_\tau = \|\mathsf{W}\|_{\mathsf{X}_{\tau-1}}. \tag{5.33}$$

Remark 28. *Recall, that we initialize* X_{-1} *such that* $\mathrm{rank}(X_{-1}) = n$; *This guarantees* $\mathrm{rank}(X_{\tau-1}) = n$, *assures* $\kappa_\tau < \infty$ *and that the interval* $\mathcal{I}_{\kappa_\tau}$ *is always non-empty. In addition, it can be verified that* $N(\mu; \kappa_\tau) < \infty$ *for any feasible* μ.

Theorem 27 states that for the suitably chosen μ, the set $\mathcal{X}_\mu((x_t); \tau)$ is finite for any closed loop trajectory (x_t). The constant κ_τ controls the interval of feasible μ as well as the total number of unstable transitions $\mathcal{U}_\mu$ that can occur in the time interval $[\tau, \infty)$. As κ_τ decreases, the bound $N(\mu; \kappa_\tau)$ tightens ($N(\mu; \kappa_\tau) \leqslant N(\mu; \kappa_\tau')$ for $\kappa_\tau \leqslant \kappa_\tau'$) and the interval (5.31) widens. Geometrically, κ_τ describes the size of the disturbance set W relative to the set $\mathrm{c}(\mathsf{X}_{\tau-1})$ (see Figure 5.1 for an example in $\mathbb{R}^2$) and the result states that we have fewer unstable transitions if the collected observations are larger in size than the disturbance. Therefore, we can view κ_τ as a constant that quantifies how informative the data $\mathsf{X}_{\tau-1}$ observed before τ are to control the system for time $t \geqslant \tau$.

The proof of Theorem 27 is postponed to Section 5.8.

Closed Loop Stability Bounds

As a consequence of Theorem 27, we obtain our main closed loop stability bounds presented in Theorem 29. The result gives bounds on the trajectories (x_t) and (u_t) in terms of the fixed disturbance (w_t) and virtual disturbance $\hat{w}_i$.

Remark. *Recall from (5.16b) that instead of analyzing the closed loop dynamics for fixed A_0, X_{-1}^+, X_{-1}, U_{-1}, we can equivalently analyze the closed loop dynamics for fixed $\hat{w}_i$.*

Theorem 29. *Let (x_t), (u_t) be the trajectories of the closed loop (5.23a) for some fixed (w_t) and $\hat{w}_i$ with the corresponding set W defined as (5.26). Let τ be some fixed time and let $\kappa_\tau := \|\mathsf{W}\|_{X_{\tau-1}}$. Then, for any $\mu \in \mathcal{I}_{\kappa_\tau}$, where $\mathcal{I}_{\kappa_\tau}$ is the interval*

$$\mathcal{I}_{\kappa_\tau} := \left(\left(\sqrt{\tfrac{1}{4} + \tfrac{1}{\kappa_\tau}} + \tfrac{1}{2} \right)^{-1}, 1 \right), \tag{5.34}$$

the trajectories (x_t) and (u_t) satisfy the bounds (i), (ii) and (iii):

(i) $\limsup_{t\to\infty} \|x_t\|_{\mathsf{W}} \leq \frac{1}{1-\mu}$,

$\limsup_{t\to\infty} \|u_t\|_{\mathsf{W}} \leq (\|A_0\|_{\mathsf{W}} + \kappa_\tau)\frac{1}{1-\mu}$.

(ii) there exists an $T' > 0$ such that for all $k > 0$ holds

$$V_1(x_{T'+k}) \leq \mu^k V_1(x_{T'}) \tag{5.35}$$

where $V_1(x) := \max\{0, \|x\|_{\mathsf{W}} - \frac{1}{1-\mu}\}$.

(iii) the worst-case norm of (x_t) and (u_t) is bounded above as [3]

$$\sup_{t\geq\tau} \|x_t\|_{\mathsf{W}} \leq f(\kappa_\tau, \mu, \|x_\tau\|_{\mathsf{W}}) + g(\mu, \kappa_\tau) \tag{5.36}$$

$$\sup_{t\geq\tau} \|u_t\|_{\mathsf{W}} \leq (\|A_0\|_{\mathsf{W}} + \kappa_\tau)\sup_{t\geq\tau} \|x_t\|_{\mathsf{W}}$$

where $N(\mu; \kappa_\tau)$ is defined as the function

$$N(\mu; \kappa_\tau) := \tfrac{1}{2} \left(\frac{\mu}{\mu - \sqrt{\kappa_\tau(1-\mu)}} \right)^n \max\{1, \tfrac{\mu}{1-\mu}\}^n, \tag{5.37}$$

$\|A_0\|_{\mathsf{W}} := \max_{x\in\mathsf{W}} \|A_0 x\|_{\mathsf{W}}$ is a constant and f and g abbreviate the functions

$$f(\kappa_\tau, \mu, \|x_\tau\|_{\mathsf{W}}) = \max\{1, \kappa_\tau^{N(\mu;\kappa_\tau)}\} \max\{\tfrac{1}{1-\mu}, \|x_\tau\|_{\mathsf{W}}\}$$

$$g(\kappa_\tau, \mu) = \frac{1 - \kappa_\tau^{N(\mu;\kappa_\tau)}}{1 - \kappa_\tau}. \tag{5.38}$$

---

[3] For $\tau = 0$ and $x_0 \notin \mathrm{span}(\mathsf{W})$, replace $\|x_0\|_{\mathsf{W}}$ with $\|A_0 x_0\|_{\mathsf{W}}$ in (5.36).

The bounds in Theorem 29 are phrased w.r.t. to the norm $\|\cdot\|_W$ that is constructed from the set W (see Figure 5.1 as an example of $\|\cdot\|_W$ in $\mathbb{R}^2$). The set W captures disturbances due to (w_t) and due to $\hat{w}_i$, where $\hat{w}_i$ describes the mismatch between the initial guess matrices X_{-1}^+, X_{-1}, U_{-1} and the true system matrix A_0. $\|x_t\|_W$ measures x_t relative to the underlying set of (lumped and virtual) disturbances W that realized it.

The result also quantifies how the bound guarantees improvement with online data: Given some initial time τ, the above result gives stability bounds on the future trajectories of x_t, u_t, $t \geq \tau$ which depend on the total states observed X_τ before time τ, the constant κ_τ and $\mu \in \mathcal{I}_{\kappa_\tau}$, which acts as a free variable. The constant κ_τ can be interpreted as a signal-to-noise ratio between state observations X_τ and the disturbance set W (see Figure 5.1 for an example in $\mathbb{R}^2$). A smaller κ_τ indicates that the data $X_{\tau-1}^+$, $X_{\tau-1}$, $U_{\tau-1}$ collected before time τ are more informative about how to stabilize the system for future time-steps $t \geq \tau$. κ_τ is always nonincreasing in τ and the bounds (iii), (i) of Theorem 29 tighten as τ increases. The bounds in Theorem 29 depend on a free variable μ which can be chosen in the interval $\mathcal{I}_{\kappa_\tau}$. We can tighten the bounds (i) and (iii) by minimizing the right-hand side over $\mu \in \mathcal{I}_{\kappa_\tau}$. For bound (i), the choice

$$\mu^* = \left(\sqrt{\tfrac{1}{4} + \tfrac{1}{\kappa_\tau}} + \tfrac{1}{2} \right)^{-1} \tag{5.39}$$

minimizes $\frac{1}{1-\mu}$ over $\mu \in \mathcal{I}_{\kappa_\tau}$ and achieves a minimal value which is almost linear in κ_τ:

$$\frac{1}{1-\mu^*} = \kappa_\tau \left(\tfrac{1}{2} + \sqrt{\tfrac{1}{4} + \tfrac{1}{\kappa_\tau}} \right) + 1 \quad (\leq \kappa_\tau + 2). \tag{5.40}$$

For $\tau = 0$ we get the following improved asymptotic upper bound for the state trajectory:

Corollary 39. *If (x_t) satisfies (5.23a) then*

$$\limsup_{t\to\infty} \|x_t\|_W \leq \kappa_0 \left(\tfrac{1}{2} + \sqrt{\tfrac{1}{4} + \tfrac{1}{\kappa_0}} \right) + 1 =: m(\kappa_0).$$

Example

Assume $n = 1$ and the scalar system $x_{t+1} = a_0 x_t + u_t + w_t$. Pick $X_{-1} = \varepsilon$ with some $\varepsilon > 0$ and X_{-1}^+, $U_{-1} = 0$. Let (w_t) be some fixed bounded scalar disturbance with $\|(w_t)\|_\infty = 1$. Then $W = c(-a_0\varepsilon \cup \{w_t | t \in \mathbb{N}\})$ and $\|x\|_W = \frac{|x|}{\max\{|a_0|\varepsilon, 1\}}$. The constant κ_0 takes the value

$$\kappa_0 = \|W\|_{X_{-1}} = \max\{\varepsilon^{-1}, |a_0|\}.$$

If we substitute this into Cor. 39, and rewrite it in terms of $|x_t|$ we obtain the bound

$$\limsup_{t \to \infty} |x_t| \leqslant \varepsilon \kappa_0^2 \left(\tfrac{1}{2} + \sqrt{\tfrac{1}{4} + \tfrac{1}{\kappa_0}} \right) + \varepsilon \kappa_0. \tag{5.41}$$

5.7 Proving Closed Loop Stability

In this section, we derive the closed loop stability bounds presented in Theorem 29 from the results of Theorem 27. The derivation of Theorem 27 is postponed to Section 5.8. First, we will derive some useful inequalities that are used frequently in the derivations.

Bounding One-Time Step Closed Loop Transitions

Recall the closed loop equation (5.23a) and the definition of the norm Def. 5.5 and the sets W and X_t. In the Appendix, Lem. 36 summarizes some important properties of the norms $\| \cdot \|_\mathsf{S}$. We use these to obtain the following bounds on the one time-step growth of the state:

Lemma 40. *Consider a state trajectory* (x_t) *of the closed loop for a fixed* W, *then at each time step* $t > \tau$ *holds:*

$$\|x_{t+1}\|_\mathsf{W} \leqslant \left\| W_{t-1} \tfrac{\lambda_{t-1}}{\|\lambda_{t-1}(x_t)\|_1} \right\|_\mathsf{W} \|\lambda_{t-1}(x_t)\|_1 + 1 \tag{5.42a}$$

$$\leqslant \|\lambda_{t-1}(x_t)\|_1 + 1 \tag{5.42b}$$

$$\leqslant \|x_t\|_{\mathsf{X}_{t-1}} + 1 \tag{5.42c}$$

$$\leqslant \|\mathsf{W}\|_{\mathsf{X}_{t-1}} \|x_t\|_\mathsf{W} + 1 \tag{5.42d}$$

$$\leqslant \kappa_\tau \|x_t\|_\mathsf{W} + 1. \tag{5.42e}$$

Recall that the vector $\lambda_{t-1}(x_t)$ poses as a linear decomposition of x_t in terms of the previous observations X_{t-1}, which is obtained through the minimization in (5.19). The right-hand side of the inequality (5.42c) and (5.42b) are equivalent. This follows from the equivalence relation

$$\|\lambda_{t-1}(\cdot)\|_1 = \| \cdot \|_{\mathsf{X}_{t-1}}, \tag{5.43}$$

which follows from property (ii) of Lem. 36 and is discussed in the appendix. The inequality (5.42c) offers valuable insight into the closed loop behavior: The smaller x_t is relative to the absolute convex hull of all previous observations X_{t-1}, the tighter the bound is on $\|x_{t+1}\|_\mathsf{W}$. Hence, $\|x_t\|_{\mathsf{X}_{t-1}}$ captures how well we can control a certain state x_t given the observations made up until time t. If we rewrite $\|x_t\|_{\mathsf{X}_{t-1}}$ as $\|x_t\|_\mathsf{W} \|x_t/\|x_t\|_\mathsf{W}\|_{\mathsf{X}_{t-1}}$ and use the fact that the normalized vector $x_t/\|x_t\|_\mathsf{W}$ lies in the set W, we obtain the looser upper-bound (5.42d). Finally, (5.42e) is obtained by recalling that, per definition $\|W\|_{X_t}$ is non-increasing in t and therefore for all $t > \tau$ it holds $\|W\|_{X_{t-1}} \leqslant \|W\|_{X_{\tau-1}} = \kappa_\tau$.

Obtaining Bounds on Closed Loop Trajectories

Recall the definition of a μ-unstable transition in Def. 5.6 and consider Lem. 41: If a μ-unstable transition does not occur, (5.44) and (5.45) show that the quantities $V_1(x_t; \mu)$ and $V_2(x_t; \mu)$ do not increase for that time-step; On the other hand, (5.46) provides a bound on the increase of $V_2(x_t; \mu)$ if a μ-unstable transition does occur.

Lemma 41. *Let (x_t) be a trajectory of (5.23a) with $t \geqslant 0$ and define the scalar functions $V_1(x; \mu) := \max\{0, \|x\|_W - \frac{1}{1-\mu}\}$ and $V_2(x; \mu) := \max\{\|x\|_W, \frac{1}{1-\mu}\}$. Then*

(i) if $(x_{t+1}, x_t) \notin \mathcal{U}_\mu$, then

$$V_1(x_{t+1}; \mu) \leqslant \mu V_1(x_t; \mu) \tag{5.44}$$

$$V_2(x_{t+1}; \mu) \leqslant V_2(x_t; \mu). \tag{5.45}$$

(ii) if $(x_{t+1}, x_t) \in \mathcal{U}_\mu$, then

$$V_2(x_{t+1}; \mu) \leqslant \kappa_t V_2(x_t; \mu) + 1 \tag{5.46}$$

$$V_1(x_{t+1}; \mu) > \mu V_1(x_t; \mu). \tag{5.47}$$

Proof. See Appendix. $\square$

The bounds of Theorem 29 follow by combining the results of Theorem 27 with the above lemma. To highlight the main proof techniques, we focus only on the derivation of (i) and (ii) of Theorem 29 and refer to the Appendix for a detailed proof of the remaining statements.

Consider some arbitrary closed-loop trajectory (x_t), fix $\tau = 0$, and choose some $\mu \in \mathcal{I}_{\kappa_0}$, where κ_τ depends on the set W and the initial guess matrix $X_{-1} = [\hat{x}_1, \ldots, \hat{x}_{n_0}]$. Recall that κ_τ measures the relative size between the disturbance set W and the set $c(X_{-1})$. According to Theorem 27, the trajectory (x_t) is guaranteed to have at most $N(\mu; \kappa_0)$-many μ-unstable transitions. Hence, there is some finite time, call it $T'((x_t))$, such that for all time $t > T'((x_t))$ it holds $(x_{t+1}, x_t) \notin \mathcal{U}_\mu$, and therefore the reverse inequality of (5.28) holds, that is:

$$\|x_{t+1}\|_W \leqslant \max\left\{\tfrac{1}{1-\mu}, \mu\|x_t\|_W + 1\right\}, \quad \forall t > T'((x_t)). \tag{5.48}$$

Now, apply the statement (i) of Lem. 41, to conclude that for all $t > T'((x_t))$ holds $V_1(x_{t+1}; \mu) \leqslant \mu V_1(x_t; \mu)$. Therefore, we get the convergence bound.

$$V_1(x_{T'((x_t))+k}; \mu) \leqslant \mu^k V_1(x_{T'((x_t))}; \mu), \quad k \geqslant 0 \tag{5.49}$$

which proves that the trajectory (x_t) has to be bounded. We also conclude that $\lim_{t\to\infty} V_1(x_t; \mu) = 0$, which leads to the asymptotic bound.

$$\limsup_{t\to\infty} \|x_t\|_W \leqslant \limsup_{t\to\infty}(V_1(x_t; \mu) + \tfrac{1}{1-\mu}) = \tfrac{1}{1-\mu}. \tag{5.50}$$

Similar type of arguments are used to derive the other statements of Theorem 29 and are presented in the Appendix.

5.8 Proving Finite Occurrence of Unstable Transitions

Here, we will discuss the key steps in proving Theorem 27. The general idea will be to first argue that if an unstable transition occurred at time t' and state $x_{t'}$, (i.e., $(x_{t'+1}, x_{t'}) \in \mathcal{U}_\mu$) then any future unstable transitions $(x_{t+1}, x_t) \in \mathcal{U}_\mu$, $t > t'$ must originate from some state x_t which is significantly different from $x_{t'}$; in a second step, we then prove that there is a finite upper bound on how many significantly "different" unstable transitions can occur in the same trajectory, which leads to the result presented in Theorem 27. In the following derivations we will make use of various simple facts from convex geometry, which are summarized in the appendix, Section 5.4. Matching the presentation of the theorem, in the derivations we will use the constant $\kappa_\tau := \|W\|_{X_{\tau-1}}$ corresponding to some fixed set W, the trajectory (x_t) of the closed loop (5.23a) and the initial time τ. Throughout the discussion, μ will represent some fixed value in the open interval.

$$\mathcal{I}_{\kappa_\tau} := \left(\left(\sqrt{\tfrac{1}{4} + \tfrac{1}{\kappa_\tau}} + \tfrac{1}{2}\right)^{-1}, 1\right) \tag{5.51}$$

and δ will refer to the corresponding transformed variable $\delta := \frac{\mu^2}{1-\mu}\frac{1}{\kappa_\tau}$, which always satisfies $\delta > 1$. The following one-to-one relationship between both constants μ and δ will be frequently used and can be easily verified:

$$\delta = \tfrac{\mu^2}{1-\mu}\tfrac{1}{\kappa_\tau}, \quad \text{for } \mu \in \left(\left(\sqrt{\tfrac{1}{4} + \tfrac{1}{\kappa_\tau}} + \tfrac{1}{2}\right)^{-1}, 1\right) \tag{5.52a}$$

$$\Leftrightarrow \quad \mu = \left(\sqrt{\tfrac{1}{4} + \tfrac{1}{\delta\kappa_\tau}} + \tfrac{1}{2}\right)^{-1}, \quad \text{for } \delta \in (1, \infty). \tag{5.52b}$$

Our argument can be structured into the following three statements, which we prove separately in the next sections:

(a) We can radially project the set $\mathcal{X}_\mu$ onto the ball $\frac{\delta}{\mu}W$ and show that the resulting set, called $\mathcal{P}_\mu$, has the same cardinality as $\mathcal{X}_\mu$.

(b) The set $\mathcal{P}_\mu$ forms a δ-separated subset of $\frac{\delta}{\mu}W$ with respect to a particularly chosen distance function $d(\,\cdot\,,\,\cdot\,;\mathsf{X}_{\tau-1})$.

(c) There are some constants c and C, such that for any δ separated subset P of $\frac{\delta}{\mu}W$ we can construct a superset $\mathsf{P} \subset \mathcal{N}(\mathsf{P})$ in $\mathbb{R}^n$ whose volume can be bounded above and below as $|\mathsf{P}|c_{\text{in}} \leqslant \text{Vol}(\mathcal{N}(\mathsf{P})) \leqslant C_{\text{out}}$; hence, the cardinality of any δ separated set, included $\mathcal{P}_\mu$, is bounded above by $\frac{C_{\text{out}}}{c_{\text{in}}}$.

Projection Onto the Ball $\frac{\delta}{\mu}W$

Define the projection $\Pi_\mu : \mathbb{R}^n \mapsto \frac{\delta}{\mu}W$ as $\Pi_\mu(p) := \frac{\delta}{\mu\|p\|_W}p$ and define $\mathcal{P}_\mu((x_t);\tau)$ as the set resulting from applying Π_μ to every point in $\mathcal{X}_\mu((x_t);\tau)$:

$$\mathcal{P}_\mu((x_t);\tau) := \{\, \Pi_\mu(x_t) \mid x_t \in \mathcal{X}_\mu((x_t);\tau) \,\}. \tag{5.53}$$

Remark. *To limit the notational burden, we will state the explicit dependency on the trajectory (x_t) and τ only in lemmas and theorems. For the derivations, we will simply write $\mathcal{X}_\mu$, $\mathcal{P}_\mu$ instead of $\mathcal{P}_\mu((x_t);\tau)$, $\mathcal{X}_\mu((x_t);\tau)$.*

Per construction, for every point $p \in \mathcal{P}_\mu$ holds $\|p\|_W = \frac{\delta}{\mu}$ and therefore each $p \in \mathcal{P}_\mu$ lies on the surface of the ball $\frac{\delta}{\mu}W$. Recall that for a time instance t, where $x_t \in \mathcal{X}_\mu$ holds.

$$\|x_{t+1}\|_W > \max\left\{\tfrac{1}{1-\mu}, \mu\|x_t\|_W + 1\right\} \tag{5.54a}$$

$$\|x_{t+1}\|_W \leqslant \|x_t\|_{\mathsf{X}_{t-1}} + 1,$$

$$\leqslant \|W\|_{\mathsf{X}_{t-1}} \|x_t\|_W + 1 \tag{5.54b}$$

where (5.54a) is due to the definition of the set $\mathcal{X}_\mu$ and (5.54b) follows from Lem. 40. Combining the above inequalities, we can further establish that any $x_t \in \mathcal{X}_\mu$ also satisfies the inequalities (5.55a):

Lemma 42.

$$\mu\|x_t\|_W > \tfrac{\mu^2}{(1-\mu)} \tfrac{1}{\|W\|_{\mathsf{X}_{t-1}}} \tag{5.55a}$$

$$\left\|\tfrac{1}{\mu\|x_t\|_W}x_t\right\|_{\mathsf{X}_{t-1}} > 1. \tag{5.55b}$$

Proof. Combining the lower-bound (5.54a) and the upper-bound (5.54b) yields

$$\|x_t\|_{X_{t-1}} + 1 > \mu \|x_t\|_W + 1$$
$$\|W\|_{X_{t-1}} \|x_t\|_W > \tfrac{1}{1-\mu} - 1 = \mu\tfrac{1}{1-\mu}.$$

$\square$

Using the above inequalities we can show in Lem. 43 that $\mathcal{P}_\mu$ has the same cardinality as $\mathcal{X}_\mu$. Hence, instead of reasoning about the size of $\mathcal{X}_\mu$ directly, we can equivalently study the size of the set $\mathcal{P}_\mu$. As will become apparent in the following sections, the main advantage of analyzing the projected set $\mathcal{P}_\mu$ rather than $\mathcal{X}_\mu$ is that we can leverage $\mathcal{P}_\mu$ as a subset of $\tfrac{\delta}{\mu}W$.

Lemma 43. $|\mathcal{X}_\mu| = |\mathcal{P}_\mu|$.

Proof. From the definition of $\mathcal{P}_\mu$ it is clear that $\mathcal{P}_\mu$ has at most as many elements as $\mathcal{X}_\mu$, hence trivially we have $|\mathcal{P}_\mu| \leqslant |\mathcal{X}_\mu|$. To establish $|\mathcal{P}_\mu| \geqslant |\mathcal{X}_\mu|$, we have to show that there are no two time instances $t_1 \neq t_2$ for which $x_{t_1}, x_{t_2} \in \mathcal{X}_\mu$ gets mapped to the same point $p \in \mathcal{P}_\mu$. For the sake of proof by contradiction, assume for some $x_{t_1}, x_{t_2} \in \mathcal{X}_\mu$ where w.l.o.g. $t_1 < t_2$, holds $\delta(\mu \|x_{t_1}\|_W)^{-1} x_{t_1} = \delta(\mu \|x_{t_2}\|_W)^{-1} x_{t_2}$. Then, using Lem. 42 it follows:

$$\left\| \tfrac{1}{\mu\|x_{t_1}\|_W} x_{t_1} \right\|_{X_{t_2-1}} = \left\| \tfrac{1}{\mu\|x_{t_2}\|_W} x_{t_2} \right\|_{X_{t_2-1}} \overset{(5.55b)}{>} 1$$

$$\Rightarrow \|x_{t_1}\|_{X_{t_2-1}} > \mu \|x_{t_1}\|_W \overset{(5.55a)}{>} \tfrac{\mu^2}{(1-\mu)} \tfrac{1}{\|W\|_{X_{t_1-1}}}. \tag{5.56}$$

Now, since $t_2 > t_1$, it is clear that $x_{t_1} \in c(X_{t_2-1})$ and therefore $\|x_{t_1}\|_{X_{t_2-1}} \leqslant 1$. Furthermore, with (5.56) and since μ is in the interval $\mathcal{I}_{\kappa_\tau}$, we are forced to conclude the following

$$\|W\|_{X_{t_1-1}} > \frac{\mu^2}{(1-\mu)} \geqslant \|W\|_{X_{\tau-1}} \tag{5.57}$$

which is a contradiction, since $t_1 \geqslant \tau$ and we know that $\|W\|_{X_t}$ is non-increasing in t.

$\square$

Separateness of the Set $\mathcal{P}_\mu$

The previous section established, that the bounded set $\mathcal{P}_\mu \subset \tfrac{\delta}{\mu}W$ has equal number of elements as $\mathcal{X}_\mu$. Here, we will show that the points in the set $\mathcal{P}_\mu$ are "evenly spread" across the surface of $\tfrac{\delta}{\mu}W$. Formally, we will term $\mathcal{P}_\mu$ to be a δ-*separated* subset of $\frac{\delta}{\mu}W$. This property will ultimately lead to the cardinality bound derived in the next Section 5.8. The next lemma shows that any two points $p, p' \in \mathcal{P}_\mu$, $p \neq p'$ respect the inequality (5.58).

Lemma 44. *Let $\mathcal{P}_\mu((x_t); \tau)$ be the projected set (5.53) and recall the definitions of the variables δ, μ and κ_τ in (5.52). Then, for any two distinct points $p_1, p_2 \in \mathcal{P}_\mu((x_t); \tau)$, $p_1 \neq p_2$ holds:*

$$\max\{\|p_2\|_{X_{\tau-1}\cup p_1}, \|p_1\|_{X_{\tau-1}\cup p_2}\} > \delta. \tag{5.58}$$

Proof. Fix two arbitrary and distinct points $p_1, p_2 \in \mathcal{P}_\mu$, $p_1 \neq p_2$, then according to the definition of $\mathcal{P}_\mu$ there are two corresponding elements $x_{t_1}, x_{t_2} \in \mathcal{X}_\mu((x_t); \tau)$ with $t_1 \neq t_2$ such that $p_1 = \delta(\mu \|x_{t_1}\|_W)^{-1} x_{t_1}$ and $p_2 = \delta(\mu \|x_{t_2}\|_W)^{-1} x_{t_2}$. We will prove the desired statement, by showing that depending on which unstable transition occurred first, i.e., $t_2 > t_1$ or $t_1 < t_2$, either $\|p_2\|_{X_{\tau-1}\cup p_1} > \delta$ or $\|p_1\|_{X_{\tau-1}\cup p_2} > \delta$ has to be satisfied. The inequality (5.58) then follows taking the maximum of both cases. Therefore, to complete our argument, we will assume the case $t_2 > t_1$ and proceed to prove $\|p_2\|_{X_{\tau-1}\cup p_1} > \delta$; The case $t_1 < t_2$ then follows by interchanging t_1 and t_2: First, notice that since $\|W\|_{X_{t_1-1}} \leq \|W\|_{X_{\tau-1}}$, we can conclude that for any $x_t \in \mathcal{X}_\mu((x_t); \tau)$, the following inequality is satisfied:

$$\frac{\delta}{\mu \|x_t\|_W} \leq \frac{\mu^2}{(1-\mu)} \frac{1}{\|W\|_{X_{\tau-1}}} \frac{1}{\mu \|x_t\|_W} < 1. \tag{5.59}$$

Now, for x_{t_2} recall from (5.55b) that

$$\|p_2\|_{X_{t_2-1}} = \left\| \frac{\delta}{\mu \|x_{t_2}\|_W} x_{t_2} \right\|_{X_{t_2-1}} > \delta. \tag{5.60}$$

We will now use repeatedly the property (iv) of Lem. 36, to bound the left-hand side of (5.60) from above. To this end, consider first the following chain of inclusions:

$$c(X_{t_2-1}) \overset{a)}{\supseteq} c(X_{t_1}) \overset{b)}{\supseteq} c(X_{\tau-1} \cup x_{t_1}) \dots$$
$$\dots \overset{c)}{\supseteq} c(X_{\tau-1} \cup p_1). \tag{5.61}$$

The inclusions $a)$, $b)$ follow directly from the definition of X_t. For inclusion $c)$, observe that $p_1 = \delta(\mu \|x_t\|_W)^{-1} x_t$ and recall from (5.59) that the scalar constant $\delta(\mu \|x_t\|_W)^{-1}$ is less than one. Now, since $c(X_\tau \cup x_{t_1})$ is a symmetric convex body, we know that it contains 0. Hence, we can view p_1 as a convex combination of 0 and x_t, which proves that $p_1 \in c(X_{\tau-1} \cup x_t)$ and therefore the set inclusion $c)$. Now, using property (iv) of Lem. 36 we can translate the inclusion (5.61) into a chain of corresponding inequalities to bound the left hand side of (5.60) and ultimately obtain:

$$\|p_2\|_{\mathsf{X}_{\tau-1}\cup\mathsf{p}_1} > \delta.$$

$\square$

The term on the left hand side of inequality (5.58) can be seen as a binary operation $d(\,\cdot\,,\,\cdot\,;\mathsf{X}_{\tau-1})$ on the points p_1 and p_2 which measures a particular notion of distance characterized by the symmetric convex body $c(\mathsf{X}_{\tau-1})$. We will define this operation more generally for some set B below and can use it to restate inequality (5.58) as

$$d(p_1, p_2; \mathsf{X}_{\tau-1}) > \delta.$$

Definition 5.8. *Let* B *be some bounded set in* $\mathbb{R}^n$ *and define the map* $d(\,\cdot\,,\,\cdot\,;\mathsf{B}) :$ $\mathbb{R}^n \times \mathbb{R}^n \mapsto \mathbb{R}_{\geqslant 0}$ *for each* $x, y \in \mathbb{R}^n$ *as*

$$d(x, y; \mathsf{B}) := \min\left\{\, r \,\left|\, \begin{array}{l} c(\mathsf{B}\cup\mathsf{x}) \subset rc(\mathsf{B}\cup\mathsf{y}) \\ c(\mathsf{B}\cup\mathsf{y}) \subset rc(\mathsf{B}\cup\mathsf{x}) \end{array} \right.\right\} \tag{5.62}$$

or equivalently as

$$d(x, y; \mathsf{B}) := \max\{\|\mathsf{B}\cup\mathsf{x}\|_{\mathsf{B}\cup\mathsf{y}}, \|\mathsf{B}\cup\mathsf{y}\|_{\mathsf{B}\cup\mathsf{x}}\}. \tag{5.63}$$

The definition of $d(\,\cdot\,,\,\cdot\,;\mathsf{B})$ in the form of equation (5.62) gives a geometric intuition as to why the value $d(x, y; \mathsf{B})$ can be viewed as a notion of distance between x and y. As an example, consider in Figure 5.2 the two points $x, y \in \mathbb{R}^2$ that satisfy $d(x, y; \mathsf{B}) > r$ and where B is taken as the box $[-1, 1] \times [-1, 1]$ in $\mathbb{R}^2$; Equation (5.62) then implies that x lies outside the set $rc(\mathsf{B}\cup\mathsf{y})$ and y lies outside of $rc(\mathsf{B}\cup\mathsf{x})$. This scenario is presented in Figure 5.2 for $r = 1.3$ and illustrates how condition $d(x, y; \mathsf{B}) > r$ enforces a separation between x and y. We will call x and y to be $(r; \mathsf{B})$-separated. More generally, we will introduce the following terminology:

Definition 5.9. *A set* $\mathsf{P} \subset \mathbb{R}^n$ *is* $(\varepsilon; \mathsf{B})$-*separated (with suitable set* B*), if* $d(p, p'; \mathsf{B}) > \varepsilon$ *holds for any* $p, p' \in \mathsf{P}$, $p \neq p'$.

In terms of the above definition, Lem. 44 states that $\mathcal{P}_\mu$ is a $(\delta; \mathsf{X}_{\tau-1})$-*separated* subset of $\frac{\delta}{\mu}\mathsf{W}$.

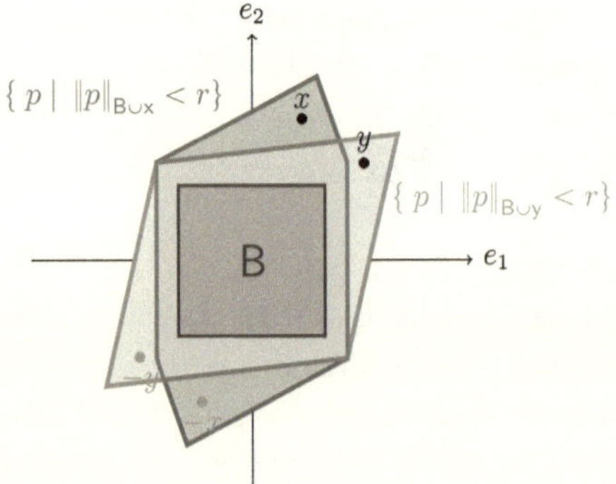

Figure 5.2: Geometry of the distance function $d(\,\cdot\,,\,\cdot\,;\mathsf{B})$: The points x and y satisfy the inequality $d(x, y; \mathsf{B}) > r$ in $n = 2$, where B is taken as the two dimensional cube $\mathsf{B} := \{(x_1, x_2) \mid |x_i| \leqslant 1\}$ and $r = 1.3$.

Bounding the Cardinality of $\mathcal{P}_\mu$ Through Volume Bounds

In this section we will complete the proof of Theorem 27, by showing that any $(\varepsilon, \mathsf{B})$-separated subset of some bounded set S has to be a finite set. This argument then leads to the results in Theorem 27, since $\mathcal{P}_\mu$ is a $(\delta; \mathsf{X}_{\tau-1})$-*separated* subset of $\frac{\delta}{\mu}\mathsf{W}$.

To illustrate the general idea, assume that we would like to construct a $(\varepsilon; \mathsf{B})$ separated set $\mathsf{P} = \{p_1, p_2, \ldots, \}$, contained within some larger bounded set $\mathsf{S} \in \mathbb{R}^2$. In particular, assume that we start with some $p_1 \in \mathsf{S}$, pick $p_2 \in \mathsf{S}$ such that $d(p_1, p_2; \mathsf{B}) > \varepsilon$ and proceed to select each p_n st. $d(p_n, p_k; \mathsf{B}) > \varepsilon$ holds for all previous $k < n$. As illustrated in Figure 5.2, it becomes intuitively clear that any constructed $(\varepsilon; \mathsf{B})$-separated subset P in S has to have finite cardinality, as it becomes increasingly harder to find "enough" room for a new point $p_n \in \mathsf{S}$ that respects the separation condition w.r.t. previous points $d(p_n, p_k; \mathsf{B}) > \varepsilon$, $k < n$. In the next section, we will show by means of a volumetric argument that this intuition extends to n-dimensions and leads to a cardinality bound on the set $\mathcal{P}_\mu$. Denote P to represent some $(\delta; \mathsf{X}_{\tau-1})$-*separated* subset of $\frac{\delta}{\mu}\mathsf{W}$, i.e., not necessarily $\mathcal{P}_\mu$. We will bound $|\mathsf{P}|$ first by constructing a corresponding cover set $\mathcal{N}(\mathsf{P}) \supset \mathsf{P}$ and then showing that the volume of $\mathcal{N}(\mathsf{P})$ is bounded below and above as

$$|\mathsf{P}|c_{\mathrm{in}} \leqslant \mathrm{Vol}(\mathcal{N}(\mathsf{P})) \leqslant C_{\mathrm{out}},$$

with some constants c_{in}, C_{out} independent of P. The desired cardinality bound then takes the form $|\mathsf{P}| \leqslant \frac{C_{\text{out}}}{c_{\text{in}}}$. The following sections will discuss: 1) the set $\mathcal{N}(\mathsf{P})$, 2) establishing the lower bound $|\mathsf{P}|c_{\text{in}}$, 3) proving the upper bound C_{out}, and 4) formulating the statement of Theorem 27.

Covering Set $\mathcal{N}(\mathsf{P})$

For some point $p \in \mathbb{R}^n$ and $\varepsilon > 1$, define $\mathsf{N}(p; \varepsilon, \mathsf{B})$ to stand for the set

$$\mathsf{N}(p; \varepsilon, \mathsf{B}) := \{p' \in \mathbb{R}^n \,|\, d(p, p'; \mathsf{B}) \leqslant \varepsilon\}. \tag{5.64}$$

See Figure 5.3 as an example of the geometry of the set $\mathsf{N}(p; \varepsilon, \mathsf{B})$ in $\mathbb{R}^2$ with $\mathsf{B} = [-1, 1] \times [-1, 1]$. For a set P, correspondingly define the set $\mathcal{N}(\mathsf{P})$ as the following union of sets.

$$\mathcal{N}(\mathsf{P}) := \bigcup_{p \in \mathsf{P}} \mathsf{N}(p; \delta^{\frac{1}{2}}, \mathsf{X}_{\tau-1}). \tag{5.65}$$

$\mathcal{N}(\mathsf{P})$ is a cover of P, since it can be easily verified that $\mathcal{N}(\mathsf{P}) \subset \mathsf{P}$. It can be easily seen that the map $d(\,\cdot\,,\,\cdot\,; \mathsf{B})$ inherits the following properties from Lem. 38:

Lemma 45. *For all $x, y, z \in \mathbb{R}^n$ holds:*

(i) $d(x, x; \mathsf{B}) = 1$.

(ii) $d(x, y; \mathsf{B}) = d(y, x; \mathsf{B}) = d(y, -x; \mathsf{B})$.

(iii) $d(x, y; \mathsf{B}) \leqslant d(x, z; \mathsf{B})d(z, y; \mathsf{B})$.

As shown in Corollary 46, the property (iii) of Lemma 45 can be used to show that for (δ, B)-separated sets P, the sets in the union (5.65) are pairwise disjoint and we can therefore evaluate the volume $\text{Vol}(\mathcal{N}(\mathsf{P}))$ as the sum:

$$\text{Vol}(\mathcal{N}(\mathsf{P})) = \sum_{p \in \mathsf{P}} \text{Vol}\big(\mathsf{N}(p; \delta^{\frac{1}{2}}, \mathsf{X}_{\tau-1})\big). \tag{5.66}$$

Corollary 46 (of Lem. 45). *If for some $x, y \in \mathbb{R}^n$ holds $d(x, y; \mathsf{B}) > \varepsilon$, then $\mathsf{N}(x; \varepsilon^{\frac{1}{2}}, \mathsf{B}) \cap \mathsf{N}(y; \varepsilon^{\frac{1}{2}}, \mathsf{B}) = \varnothing$.*

Proof. For the sake of proving the statement through contradiction, assume that there was some point $z \in \mathsf{N}(x; \varepsilon^{\frac{1}{2}}, \mathsf{B}) \cap \mathsf{N}(y; \varepsilon^{\frac{1}{2}}, \mathsf{B})$. Then we know that z satisfies both $d(x, z; \mathsf{B}) \leqslant \varepsilon^{\frac{1}{2}}$ and $d(y, z; \mathsf{B}) \leqslant \varepsilon^{\frac{1}{2}}$. But from the property (iii) of Lem. 45, we also have to conclude

$$d(x, y; \mathsf{B}) \leqslant d(x, z; \mathsf{B})d(z, y; \mathsf{B}) \leqslant \varepsilon$$

which leads to the intended contradiction. $\qquad\square$

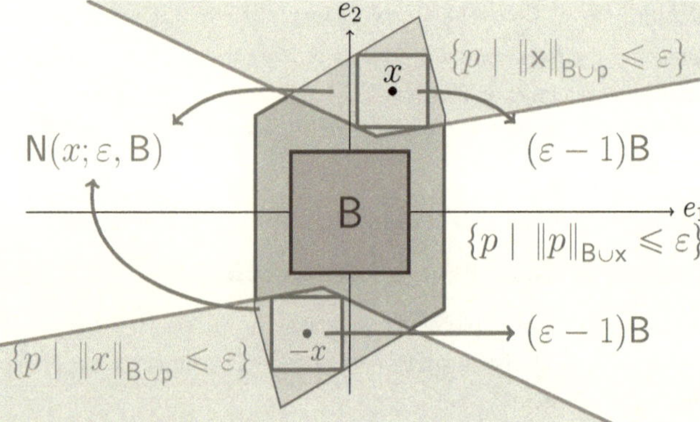

Figure 5.3: $N(x; \varepsilon, B)$ is the intersection of $\{p| \ \|x\|_{B \cup p} \leqslant \varepsilon\}$ and $\{p| \ \|p\|_{B \cup x} \leqslant \varepsilon\}$ and contains two translates of the set $(\varepsilon - 1)B$. In the picture, B is taken as the two dimensional cube $B := \{(x_1, x_2)| \ |x_i| \leqslant 1\}$ and $\varepsilon = 1.6$.

Lower Bound on Volume of $\mathcal{N}(P)$

To lower-bound the quantity (5.66), we will make use of the following lemma:

Lemma 47. *Let $x, y \in \mathbb{R}^n$, then if $y = x + (\varepsilon - 1)p$ for some $p \in B$ and $\varepsilon > 1$, then it holds $d(x, y; B) \leqslant \varepsilon$.*

Proof. We need to prove $\|x\|_{B \cup y} \leqslant \varepsilon$ and $\|y\|_{B \cup x} \leqslant \varepsilon$. $\underline{\|y\|_{B \cup x} \leqslant \varepsilon}$: From the triangle inequality, we obtain

$$\|y\|_{B \cup x} = \|x + (\varepsilon - 1)p\|_{B \cup x} \leqslant \|x\|_{B \cup x} + (\varepsilon - 1) \|p\|_{B \cup x}$$

and using the fact that $x, p \in c(B \cup x)$ by the norm definition (5.25) we get $\|y\|_{B \cup x} \leqslant 1 + (\varepsilon - 1) = \varepsilon$.

$\underline{\|x\|_{B \cup y} \leqslant \varepsilon}$: Rewrite x as $x = \varepsilon \left(\frac{1}{\varepsilon}(y) + \frac{\varepsilon - 1}{\varepsilon}(-p)\right)$ and notice that $-p, y \in c(B \cup y)$, which shows that $x \in \varepsilon c(B \cup y)$. Hence, via the norm definition (5.25) we conclude $\|x\|_{B \cup y} \leqslant \varepsilon$. $\qquad\square$

If we use the property (ii) of Lem. 45, then Lem. 47 tells us that each set $N(p; \varepsilon, B)$ contains the sets $x \oplus (\varepsilon - 1)B$ and $-x \oplus (\varepsilon - 1)B$, where the operator $\oplus$ denotes the Minkowski sum of two sets. For $\mathbb{R}^2$, Figure 5.3 illustrates the geometric relationship between the set $N(x; \varepsilon, B)$ and the set B which is taken again to be the ∞-norm unit ball. We can see that the set $N(p; \varepsilon, B)$ is a union of two symmetrical polytopes and contains two non-overlapping translations (by the vector x and $-x$) of the set $(\varepsilon - 1)\mathsf{B}$. Now, using the fact that n-dimensional volume $\mathrm{Vol}(\,\cdot\,)$ is a homogenous function of degree n, we can obtain the following lower bound on the volume of any set $\mathsf{N}(p; \varepsilon, \mathsf{B})$:

$$\mathrm{Vol}(\mathsf{N}(p; \varepsilon, \mathsf{B})) \geqslant 2(\varepsilon - 1)^n \mathrm{Vol}(\mathsf{B}). \tag{5.67}$$

Combining this observation with our previous finding (5.66), we obtain the following lower bound on the volume of $\mathrm{Vol}(\mathcal{N}(\mathsf{P}))$:

Lemma 48. *Let $\mathcal{N}(\mathsf{P})$ be the collection (5.65) corresponding to a $(\delta, \mathsf{X}_{\tau-1})$-separated $(\delta > 1)$ set P, then the volume $\mathrm{Vol}(\mathcal{N}(\mathsf{P}))$ is bounded below by*

$$\mathrm{Vol}(\mathcal{N}(\mathsf{P})) \geqslant 2(\delta^{\frac{1}{2}} - 1)^n \mathrm{Vol}(\mathrm{c}(\mathsf{X}_{\tau-1}))|\mathsf{P}|, \tag{5.68}$$

where $|\mathsf{P}|$ denotes the cardinality of the set P.

Proof. Apply (5.67) to every term in the sum (5.66). $\qquad\square$

Upper Bound on Volume of $\mathcal{N}(\mathsf{P})$

Consider some arbitrary point $q \in \mathsf{N}(p; \delta^{\frac{1}{2}}, \mathsf{X}_{\tau-1})$ for some p in the δ-separated set P and recall that $p \in \frac{\delta}{\mu}\mathsf{W}$. Then, from the construction of the sets N as (5.64) we can conclude that

$$\|q\|_{\mathsf{X}_{\tau-1} \cup p} \leqslant d(q, p; \mathsf{X}_{\tau-1}) \leqslant \delta^{\frac{1}{2}}. \tag{5.69}$$

Moreover, since we can upper bound W as $\mathsf{W} \subset \kappa_\tau \mathrm{c}(\mathsf{X}_{\tau-1})$, we also obtain $\mathsf{X}_{\tau-1} \cup p \subset \max\{1, \frac{\kappa_\tau \delta}{\mu}\}\mathrm{c}(\mathsf{X}_{\tau-1})$ and therefore the point q satisfies

$$\|q\|_{\max\{1, \frac{\kappa_\tau \delta}{\mu}\}\mathsf{X}_{\tau-1}} \leqslant \|q\|_{\mathsf{X}_{\tau-1} \cup p} \leqslant \delta^{\frac{1}{2}}$$

$$\Leftrightarrow \qquad q \in \delta^{\frac{1}{2}} \max\{1, \tfrac{\kappa_\tau \delta}{\mu}\}\mathrm{c}(\mathsf{X}_{\tau-1}). \tag{5.70}$$

Hence, (5.70) shows that the collection $\mathcal{N}(\mathsf{P})$ is a subset of $\delta^{\frac{1}{2}} \max\{1, \frac{\kappa_\tau \delta}{\mu}\}\mathrm{c}(\mathsf{X}_{\tau-1})$ which proves the following upperbound on the volume $\mathrm{Vol}(\mathcal{N}(\mathsf{P}))$:

Lemma 49. *Let $\mathcal{N}(\mathsf{P})$ be the collection (5.65) corresponding to a $(\delta, \mathsf{X}_{\tau-1})$-separated $(\delta > 1)$ set P, then the volume $\mathrm{Vol}(\mathcal{N}(\mathsf{P}))$ is bounded above by*

$$\mathrm{Vol}(\mathcal{N}(\mathsf{P})) \leqslant \delta^{\frac{n}{2}} \max\{1, \tfrac{\kappa_\tau \delta}{\mu}\}^n \mathrm{Vol}(\mathrm{c}(\mathsf{X}_{\tau-1})). \tag{5.71}$$

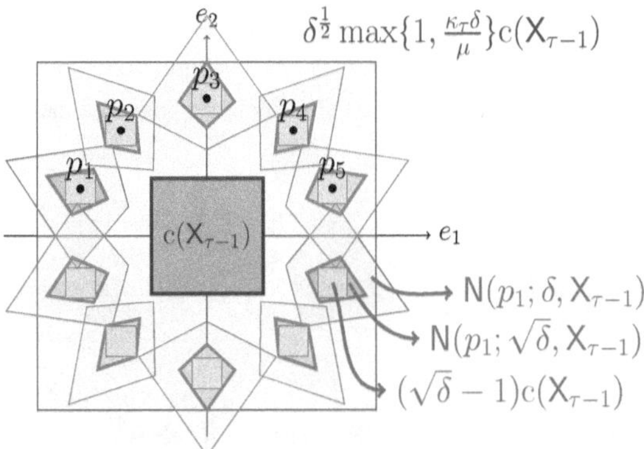

Figure 5.4: The geometry of a $(\delta, c(X_{\tau-1}))$-separated set $P = \{p_1,\ldots,p_5\}$. All sets are geometrically accurate, assuming $c(X_{\tau-1})$ is the box $[-1,1] \times [-1,1]$ and $\delta = 1.6$.

Cardinality Bound for $(\delta, X_{\tau-1})$-Separated Sets P

Finally, the lower bound (5.68) and upper bound (5.71) imply the following bound on the cardinality of any $(\delta, X_{\tau-1})$-separated set $P \subset \frac{\delta}{\mu}W$ with $\delta > 1$:

$$|P| \leq \frac{1}{2}\left(\frac{\sqrt{\delta}}{\sqrt{\delta}-1}\right)^n \max\{1,\tfrac{\delta\kappa_\tau}{\mu}\}^n. \tag{5.72}$$

Figure 5.4 shows a pictorial summary of our derivation of the above inequality in $\mathbb{R}^2$. A $(\delta, c(X_{\tau-1}))$-separated set $P = \{p_1,\ldots,p_5\}$ is defined to satisfy $p_i \notin N(p_j,\delta,X_{\tau-1})$, for all $i \neq j$ and as a consequence, we showed in Cor. 46 that the sets in the cover $\mathcal{N}(P) = \cup_j N(p_j,\sqrt{\delta},X_{\tau-1})$, are all disjoint. Then, Lem. 47 helped us establish that $\mathcal{N}(P)$ contains $2|P|$ many translations of the set $(\sqrt{\delta}-1)c(X_{\tau-1})$, which lead to the volume bound (5.67). We obtain the upperbound (5.71) by showing that $\mathcal{N}(P)$ has to be contained in the bigger box $\delta^{\frac{1}{2}}\max\{1,\frac{\kappa_\tau\delta}{\mu}\}c(X_{\tau-1})$. So, in the context of the picture Figure 5.4, we obtained our final cardinality bound (5.72) by dividing the volume of the outer larger box by the volume of the smaller boxes.

Recalling the relationship between the variables δ, μ and κ_τ in (5.52), we can express μ in terms of some $\delta > 1$ and κ_τ as

$$\mu = \delta\kappa_\tau\left(\sqrt{\tfrac{1}{4}+\tfrac{1}{\delta\kappa_\tau}}-\tfrac{1}{2}\right) = \left(\sqrt{\tfrac{1}{4}+\tfrac{1}{\delta\kappa_\tau}}+\tfrac{1}{2}\right)^{-1} \tag{5.73}$$

and can equivalently rewrite (5.72) in terms of constant κ_τ and $\delta > 1$ as a free variable :

$$|\mathsf{P}| \leqslant \tfrac{1}{2} \left(\tfrac{\sqrt{\delta}}{\sqrt{\delta}-1} \right)^{n} \max\{\tfrac{1}{\delta\kappa_\tau}, \sqrt{\tfrac{1}{4} + \tfrac{1}{\delta\kappa_\tau}} + \tfrac{1}{2}\}^{n}(\delta\kappa_\tau)^{n}. \tag{5.74}$$

In summary, (5.74) establishes a bound on the cardinality of $|\mathsf{P}|$ which serves as an upper-bound on $|\mathcal{P}_\mu((x_t), \tau)| = |\mathcal{X}_\mu((x_t), \tau)|$, thus the total number of unstable transitions $\mathcal{U}_\mu$ that can occur in the interval $[\tau, \infty)$ of any closed loop trajectory (x_t). We conclude by restating the results Theorem 27 again in terms of δ:

Theorem. *For any trajectory (x_t) of the closed loop* (5.23a) *and any $\tau \geqslant 0$, the cardinality $|\mathcal{X}_\mu((x_t); \tau)|$ of the set $\mathcal{X}_\mu((x_t); \tau)$ is finite for any μ chosen as*

$$\mu = \left(\sqrt{\tfrac{1}{4} + \tfrac{1}{\delta\kappa_\tau}} + \tfrac{1}{2} \right)^{-1}, \delta > 1 \tag{5.75}$$

for some $\delta > 1$ and bounded above as $|\mathcal{X}_\mu((x_t); \tau)| \leqslant N(\delta; \kappa_\tau)$, where N stands for the function

$$N(\delta; \kappa_\tau) := \tfrac{1}{2} \left(\tfrac{\sqrt{\delta}}{\sqrt{\delta}-1} \right)^{n} \max\{\tfrac{1}{\delta\kappa_\tau}, \sqrt{\tfrac{1}{4} + \tfrac{1}{\delta\kappa_\tau}} + \tfrac{1}{2}\}^{n}(\delta\kappa_\tau)^{n} \tag{5.76}$$

and κ_τ is a constant computed from $\mathsf{X}_{\tau-1}$ as:

$$\kappa_\tau = \|\mathsf{W}\|_{\mathsf{X}_{\tau-1}} := \max_{z \in \mathsf{W}} \|z\|_{\mathsf{X}_{\tau-1}}. \tag{5.77}$$

5.9 A Connection Between Metric Entropy Bounds and Model-Free Stability Analysis

The notion of metric entropy[4] dates back to early work of A.N. Kolmogorov [80] in 1959 and more recently has been proven useful for studying stochastic processes in the field of high-dimensional statistics. As an example, Chap. 5 of [125] discusses how bounds on the metric entropy of a metric space can be leveraged to obtain probabilistic bounds on the supremum of sub-Gaussian processes over that same metric space; Particularly in machine learning applications, these mathematical results can then be used to derive learning theoretic guarantees of algorithms.

Reexamining the line of arguments that lead to our theoretical guarantees suggests that there might be a possibly fruitful connection between metric entropy bounds and worst-case performance bounds in the context of learning and control problems. In retrospect, the main technique for stability analysis can be described as representing the collection $\mathcal{X}_\mu$ of unstable transitions as a packing set P_μ in the *totally bounded*

---

[4]This is to be distinguished from the Kolmogorov-Sinai entropy of a dynamical system introduced in [79].

metric space $(\frac{\delta}{\mu}\mathsf{W}, d(\,\cdot\,,\,\cdot\,;\mathsf{X}_{\tau-1}))$; The bound on the cardinality $\mathcal{X}_\mu$ presented in Theorem 27 and derived in Section 5.8 can be viewed as the corresponding metric entropy bound. In hindsight, this inspires a new potential approach to algorithm design for learning and control: Synthesizing a control law for which unstable transitions (potentially as broader defined than $\mathcal{U}_\mu$ considered here) form a packing in some totally bounded metric space. Similar to our presented result, we could also hope that smaller metric entropy translates to improved closed loop performance guarantees.

We will proceed by introducing the metric entropy and related concepts based on [125] and [47]. In the next section, we will draw the connection to our stability analysis presented in Section 5.8.

Metric Entropy of Pseudo-Metric Spaces

A pseudometric space (S, d) consists of a set S and a pseudometric $d : \mathsf{S} \times \mathsf{S} \mapsto \mathbb{R}_{\geqslant 0}$, which satisfies the following properties:

(i) $d(x, x) = 0$ for any $x \in \mathsf{S}$.

(ii) $d(x, y) = d(y, x)$ for any $x, y \in \mathsf{S}$.

(iii) $d(x, y) \leqslant d(x, y) + d(y, z)$ for any $x, y, z \in \mathsf{S}$.

If in addition $d(x, y) = 0$ holds only if $x = y$, then d is called a metric and correspondingly (S, d) is a metric space. The ε-packing of S w.r.t to d is a set $\mathsf{P} \subset \mathsf{S}$ such that for each two distinct points $p_1, p_2 \in \mathsf{P}$, $p_1 \neq p_2$ holds $d(p_1, p_2) > \varepsilon$. Correspondingly, the ε-packing number of S is the cardinality of the largest ε-packing set P of S. Formally, this is defined in Def. 5.10

Definition 5.10. *Let* (S, d) *be a metric (or pseudo-metric) space. Then the ε-packing number $D(\mathsf{S}, \varepsilon)$ (or $D(\mathsf{S}, \varepsilon, d)$) of S is defined as*

$$D(\mathsf{S}, \varepsilon) := \sup \left\{ m \,\middle|\, \begin{array}{l} \text{for some } p_1, \ldots, p_m \in \mathsf{S}, \\ d(p_i, p_j) > \varepsilon \text{ for } 1 \leqslant i < j \leqslant m \end{array} \right\}. \tag{5.78}$$

If $D(\mathsf{S}, \varepsilon)$ is finite for any $\varepsilon > 0$, then (S, d) is often called *totally bounded*. In this case, we define the quantity $\log(D(\mathsf{S}, \varepsilon))$ as the *metric entropy* of the set S w.r.t. the metric d.

Remark 30. *An alternative definition of metric entropy as in [125], is* $\log(N(\mathsf{S}, \varepsilon))$ *where* $N(\mathsf{S}, \varepsilon)$ *is the covering number of the set* S. *The distinction between both (and other) definitions is of conventional matter, as it is well-known that packing numbers and covering numbers behave in equivalent manners. For our purpose, we will use the more fitting definition (5.10), which is for example used in the works of R.M. Dudley [47].*

Bounding Occurrence of Unstable Transitions Through Metric Entropy

The key result behind our analysis in Section 5.8 was to show that any $(\delta, \mathsf{X}_{\tau-1})$-separated subset $\mathsf{P} \subset \frac{\delta}{\mu}\mathsf{W}$ respects the cardinality bound (5.74). From Lem. 45 we can directly see that the operation $\log(d(\,\cdot\,,\,\cdot\,; \mathsf{B})$ satisfies the properties of a pseudo-metric Def. 5.9 and therefore $(\frac{\delta}{\mu}\mathsf{W}, \log(d(\,\cdot\,,\,\cdot\,; \mathsf{X}_{\tau-1}))$ is a pseudo-metric space. Correspondingly, the set $\mathsf{P} \subset \frac{\delta}{\mu}\mathsf{W}$ is $\log(\delta)$-packing in that same pseudo-metric space and our cardinality bound can be seen as an upper bound on the packing-number $D(\frac{\delta}{\mu}\mathsf{W}, \log(\delta))$ of the set $\frac{\delta}{\mu}\mathsf{W}$ w.r.t. to the pseudometric $\log(d(\,\cdot\,,\,\cdot\,; \mathsf{X}_{\tau-1}))$. Moreover, since we established the bound (5.74) for every $\delta > 1$ (or $\log(\delta) > 0$), the space $(\frac{\delta}{\mu}\mathsf{W}, \log(d(\,\cdot\,,\,\cdot\,; \mathsf{X}_{\tau-1}))$ is a totally bounded pseudometric space with inequality (5.74) implying a particular metric entropy bound. Hence in hindsight, our approach to stability analysis relied on mapping the set of unstable transitions $\mathcal{X}_\mu$ onto a fitting pseudometric space in which the metric entropy imposes a direct bound on the cardinality of the set $\mathcal{X}_\mu$. An interesting topic of further research is whether this general principle could be leveraged for model-free stability analysis and controller synthesis in broader learning and control problem settings.

5.10   Simulation

We conducted $N = 1000$ simulations of the causal cancelation controller $\boldsymbol{K}^{\mathrm{cc}}$ defined in (5.19). For the kth experiment, the trajectories (x_t^k), (u_t^k) are produced by the closed loop equations

$$x_{t+1}^k = A_0^k x_t^k + K_t^{\mathrm{cc}}(x_t^k, X_{t-1}^{k+}, X_{t-1}^k, U_{t-1}^k) + w_t^k, \tag{5.79}$$

and the system matrix $A_0^k \in \mathbb{R}^{3\times3}$, initial condition $x_0^k \in \mathbb{R}^3$ and disturbance w_t^k is picked at random.   All entries of A_0^k and x_0^k are picked i.i.d. from the standard Gaussian distribution $\mathcal{N}(0, 1)$. In each experiment, the causal cancellation controller (5.19) is initialized as $X_{-1} = \varepsilon I$, $X_{-1}^+ = 0$, $U_{-1} = 0$ with fixed choice $\varepsilon = 0.1$.

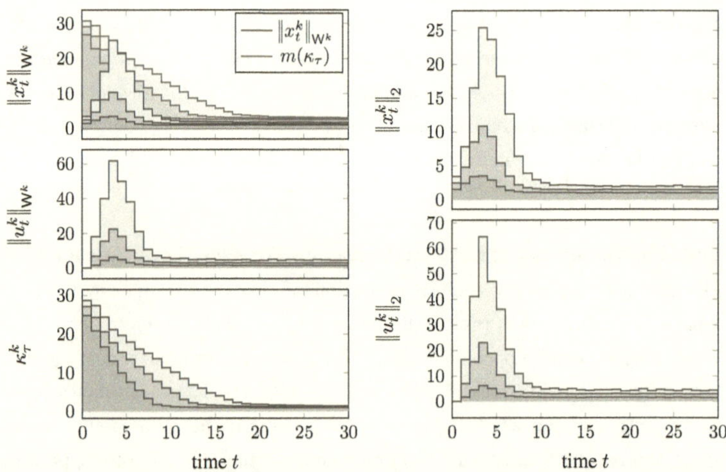

Figure 5.5: Results of 1000 closed-loop simulations with random A_0^k, x_0^k and disturbances w_t^k drawn from $[-1, 1]^3$. The plots on the left show the largest 1%, 10%, 50% percentile values of $\|x_t^k\|_{\mathsf{W}^k}$, $\|u_t^k\|_{\mathsf{W}^k}$, κ_τ^k, $m(\kappa_\tau)$. The right plot shows the same percentile values for the state x_t^k and input u_t^k measured in 2-norm.

Figure 5.7 shows the simulation results of a single experiment where A_0 is chosen as

$$A_0 = \begin{bmatrix} 1.4 & 0.2 & 1 \\ 0.2 & 1.3 & 1 \\ 0.5 & 0.3 & 2 \end{bmatrix} \qquad \lambda(A_0) = \begin{bmatrix} 2.7 \\ 1.13 \\ 0.86 \end{bmatrix}. \tag{5.80}$$

and has a large unstable eigenvalue $\lambda_i(A_0)$. Figure 5.5 and Figure 5.6 summarize the N closed-loop experiments for two scenarios of disturbances. The graphs show, as a function of t, the highest 1%, 10%, 50% percentiles of the values $\|x_t^k\|_{\mathsf{W}^k}$, $\|u_t^k\|_{\mathsf{W}^k}$, $\kappa_t^k = \|\mathsf{W}^k\|_{\mathsf{X}_{t-1}}$ and $m(\kappa_\tau)$ among the N experiments; the quantity $m(\kappa_\tau) := \kappa_\tau\left(\frac{1}{2} + \sqrt{\frac{1}{4} + \frac{1}{\kappa_\tau}}\right) + 1$ represents the updated theoretical asymptotic bound given the data collected up until time τ. In experiment k, the set W^k is constructed according to equation (5.26) from the disturbance sequence[5] $(w_t^k)_{t=0}^{T-1}$ and the virtual disturbances $\hat{w}_i^k$. For our initialization of X_{-1}, X_{-1}^+, U_{-1}, the vectors $\hat{w}_i^k$ take the values $-\varepsilon A_0^k e_i$, $1 \leqslant i \leqslant n$, where e_i denotes the ith axis of the standard basis in $\mathbb{R}^n$:

$$\mathsf{W}^k = \{w_t^k \mid 0 \leqslant t < T\} \cup \{-\varepsilon A_0^k e_i \mid 1 \leqslant i \leqslant n\}. \tag{5.81}$$

---

[5] We assume that after $t > T$ the disturbance w_t^k stays in the set $\mathrm{c}(\mathsf{W}^k)$.

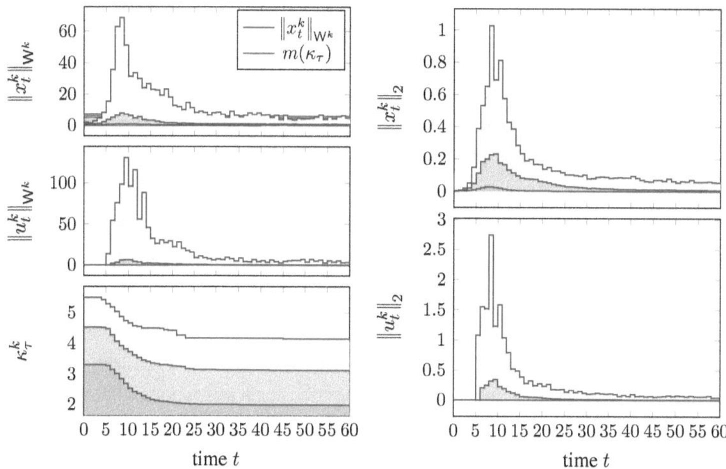

Figure 5.6: Results of 1000 closed-loop simulations with random A_0^k, x_0^k and $w_t^k = 0$. The plots on the left show the largest 1%, 10%, 50%* percentile values of $\left\|x_t^k\right\|_{\mathsf{W}^k}$, $\left\|u_t^k\right\|_{\mathsf{W}^k}$, κ_τ^k, $m(\kappa_\tau)$. (* this percentile is too small to visualize for $\left\|x_t^k\right\|_{\mathsf{W}^k}$ and $\left\|u_t^k\right\|_{\mathsf{W}^k}$) The right plot shows the same percentile values for the state and input measured in 2-norm.

For Figure 5.5, the disturbance sequence (w_t^k) of each experiment is picked i.i.d. uniformly from the interval $[-1, 1]^3$. For Figure 5.6, the initial condition $x_{0,i}$ is chosen i.i.d. according to the Gaussian distribution $\mathcal{N}(0, \sigma^2)$, $\sigma = 10^{-3}$ and $w_t^k = 0$. Since we have no disturbance, for this case, W^k is simply the set of virtual disturbances $\{-\varepsilon A_0^k e_i \mid 1 \leqslant i \leqslant n\}$. We discussed that, as a corollary of our main result (see (5.40)), the causal cancelation controller $\boldsymbol{K}^{\mathrm{cc}}$ guarantees for each experiment the asymptotic bound

$$\limsup_{t \to \infty} \left\|x_t^k\right\|_{\mathsf{W}^k} \leqslant m(\kappa_t^k), \quad \forall t \tag{5.82}$$

where we take the function $m(\,\cdot\,)$ to abbreviate the expression

$$m(s) := s\left(\tfrac{1}{2} + \sqrt{\tfrac{1}{4} + \tfrac{1}{s}}\right) + 1. \tag{5.83}$$

In Figure 5.5 and Figure 5.6, we overlayed the percentiles of $\left\|x_t^k\right\|_{\mathsf{W}^k}$ (blue) and $m(\kappa_t^k)$ (red) to show that qualitatively the experiments match the theoretical guarantee above. In each experiment, the controller eventually learns to stabilize the unknown system (consistently after 10 time-steps) and eventually (see $t > 20$ in Figure 5.5) is bounded above by the asymptotic bound $m(\kappa_t^k)$. Note also that as more online data is observed, the asymptotic bound $m(\kappa_t^k)$ tightens. Figure 5.6 is showing the closed loop performance in the no disturbance regime. This is to investigate how the controller $\boldsymbol{K}^{\mathrm{cc}}$ performs in the absence of excitation by the disturbance. We see in Figure 5.6 that the controller $\boldsymbol{K}^{\mathrm{cc}}$ stabilizes the system in all experiments, but compared to (5.5), we have a longer transient of learning. Note that in Figure 5.6, the percentiles of the constant κ_t^k do not decrease over time as much as in the experiments of Figure 5.5. Recall that for time t, the constant κ_t^k can be seen to approximate the remaining uncertainty of the unknown system A_0^k. Therefore, Figure 5.6 shows that despite the remaining uncertainty in the system, the controller still manages to stabilize the system. This reflects that $\boldsymbol{K}^{\mathrm{cc}}$ does not primarily care about identifying the unknown matrix A_0^k, but rather collects *only* enough data about the matrix A_0^k to be able to stabilize the closed loop.

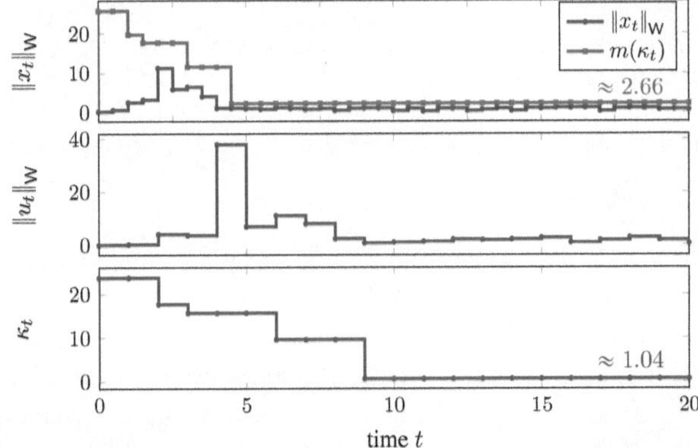

Figure 5.7: $\|x_t\|_{\mathrm{W}}$ and $\|u_t\|_{\mathrm{W}}$ trajectories for closed loop with uniform disturbance $w_{i,t}$ in $[-1,1]$ and $x_0 = [0.2, 0, 0.1]^T$.

5.11 Conclusion

In this chapter we derive a simple model-free controller that can adaptively and robustly stabilize a linear system with full actuation without any additional knowledge on disturbance, noise or parameter bounds. The controller comes with uniform asymptotic and worst-case guarantees on the state-deviation. The control design and stability analysis is enabled by a novel approach inspired by convex geometry, and simulations show that the controller is able to simultaneously learn and control the system in an efficient manner, even when applied to an open loop system with large unstable eigenvalues. Future work will further explore how this new perspective on adaptive control can provide more learning and control algorithms with robustness guarantees and non-restrictive assumptions in a more general setting. In addition, we will investigate how the presented ideas can help in providing robustness and performance bounds for present methods in adaptive control and reinforcement learning.

5.A   Proofs

Theorem 29

Proof. The proof follows by applying Lem. 41. The corresponding bounds for (u_t) are then obtained using the equation (5.22).

According to the setting of the theorem, consider some fixed trajectories (x_t), (u_t), reference time τ, $\mu \in \mathcal{I}_{\kappa_\tau}$ with $\kappa_\tau = \|W\|_{X_{\tau-1}}$. Then, as discussed before, a direct consequence of Theorem 27 is that there is some trajectory-dependent finite time $T' < \infty$, such that in the time interval $[0, T']$ there are at most $N(\mu; \kappa_\tau)$-many time instances $\mathcal{T} := \{t'_1, \ldots, t'_M\}$, where

$$\tau \leqslant t'_1 < t'_2 < \cdots < t'_M \leqslant T', \quad M < N(\mu; \kappa_\tau) \tag{5.84}$$

at which μ-unstable transitions occur and for all other time-instances $t \neq t'_i$ holds the opposite inequality of (5.28). Thus, depending on whether t belongs to $\mathcal{T}$, the transitions (x_{t+1}, x_t) of the trajectory (x_t) satisfy

$$\|x_{t+1}\|_W > \max\left\{\tfrac{1}{1-\mu}, \mu\|x_t\|_W + 1\right\}, \quad \forall t \in \mathcal{T} \tag{5.85a}$$

$$\|x_{t+1}\|_W \leqslant \max\left\{\tfrac{1}{1-\mu}, \mu\|x_t\|_W + 1\right\}, \quad \forall t \notin \mathcal{T}. \tag{5.85b}$$

Moreover, combining Lem. 41 with the above, we find that, w.r.t. the function $V_1(x; \mu) := \max\{0, \|x\|_W - \tfrac{1}{1-\mu}\}$, the transitions (x_{t+1}, x_t) respect the inequality

$$V_1(x_{t+1}; \mu) > \mu V_1(x_t; \mu), \quad \forall t \in \mathcal{T} \tag{5.86a}$$

$$V_1(x_{t+1}; \mu) \leqslant \mu V_1(x_t; \mu), \quad \forall t \notin \mathcal{T} \tag{5.86b}$$

and for function $V_2(x; \mu) := \max\{\|x\|_W, \tfrac{1}{1-\mu}\}$, the transitions (x_{t+1}, x_t) satisfy

$$V_2(x_{t+1}; \mu) \leqslant \|W\|_{X_{t-1}} V_2(x_t; \mu) + 1, \qquad \forall t \in \mathcal{T} \tag{5.87a}$$

$$V_2(x_{t+1}; \mu) \leqslant V_2(x_t; \mu), \qquad \forall t \notin \mathcal{T}. \tag{5.87b}$$

The bounds on (x_t) in part (i) and (ii) were derived before from (5.86), (see (5.49) and (5.50)). For (iii), notice that (5.87) implies that for any t, $V_2(x_t; \mu)$ can be bounded above by $V_2(x_{t'_M+1}; \mu)$, since apart from the time intervals $[t'_i, t'_{i+1}]$, the quantity $V_2(x_t; \mu)$ is guaranteed to be nonincreasing. Moreover (5.87) also shows that $V_2(x_{t'_M+1}; \mu)$ can be bounded above as

$$V_2(x_{t'_M+1}; \mu) \leq \alpha V_2(x_\tau; \mu) + \beta \qquad (5.88)$$

$$\alpha := \prod_{k=1}^{M} \|W\|_{X_{t'_k-1}}, \quad \beta := \sum_{k=0}^{M-1} \prod_{j=1}^{k} \|W\|_{X_{t'_{M-j}-1}}.$$

Recall that $\|W\|_{X_t}$ is not increasing (hence $\|W\|_{X_{t'_k-1}} \leq \kappa_\tau := \|W\|_{X_{\tau-1}}$) and the bound $M \leq N(\mu; \kappa_\tau)$, to see that the constants α and β are bounded above as

$$\alpha \leq \max\{1, \kappa_\tau{}^{N(\mu;\kappa_\tau)}\} \qquad \beta \leq \frac{1 - \kappa_\tau{}^{N(\mu;\kappa_\tau)}}{1 - \kappa_\tau}. \qquad (5.89)$$

We then obtain the final inequality (5.36) by substituting the above bounds into (5.88) and observing that

$$\sup_{t \geq \tau} \|x_t\|_W \leq \sup_{t \geq \tau} V_2(x_t; \mu) \leq V_2(x_{t'_M+1}; \mu). \qquad (5.90)$$

To obtain the corresponding bounds for the input (u_t), recall that u_t can be rewritten as

$$u_t = (U_{t-1} - X_{t:1}) \lambda_{t-1}(x_t)$$
$$= (-A_0 X_{t-1} - W_{t-1}) \lambda_{t-1}(x_t)$$
$$= -A_0 x_t - W_{t-1} \lambda_{t-1}(x_t)$$

and that $\|\lambda_{t-1}\|_1 = \|x_t\|_{X_{t-1}}$. This allows us to upper-bound $\|u_t\|_W$ by

$$\|u_t\|_W \leq \left\| A_0 \frac{x_t}{\|x_t\|_W} \right\|_W \|x_t\|_W + \left\| \frac{x_t}{\|x_t\|_W} \right\|_{X_{t-1}} \|x_t\|_W$$
$$\leq \left(\max_{x \in W} \|A_0 x\|_W + \|W\|_{X_{\tau-1}} \right) \|x_t\|_W$$
$$\leq \left(\|A_0\|_W + \kappa_\tau \right) \|x_t\|_W$$

and obtain desired bounds for (u_t) by adding the bounds already derived for (x_t). $\quad\square$

Lemma 41

Proof. Part (i) and (5.47): We can expand the inequality as

$$\|x_{t+1}\|_W \leq \max\left\{ \tfrac{1}{1-\mu}, \mu \|x_t\|_W + (1-\mu)\tfrac{1}{1-\mu} \right\} \qquad (5.91)$$

and can subtract $\frac{1}{1-\mu}$ on both sides to obtain

$$\|x_{t+1}\|_{\mathsf{W}} - \tfrac{1}{1-\mu} \leqslant \max\left\{0, \mu(\|x_t\|_{\mathsf{W}} - \tfrac{1}{1-\mu})\right\}$$
$$\Leftrightarrow \max\{0, \|x_{t+1}\|_{\mathsf{W}} - \tfrac{1}{1-\mu}\} \leqslant \mu \max\left\{0, \|x_t\|_{\mathsf{W}} - \tfrac{1}{1-\mu}\right\}.$$

Similarly, noticing that the second term on the right hand sight of (5.91) is a convex combination of $\|x_t\|_{\mathsf{W}}$ and $\frac{1}{1-\mu}$, we can conclude

$$\max\{\|x_{t+1}\|_{\mathsf{W}}, \tfrac{1}{1-\mu}\} \leqslant \max\{\|x_t\|_{\mathsf{W}}, \tfrac{1}{1-\mu}\}.$$

<u>Part (ii)</u>: We previously derived that the inequality (5.42d) holds for all time t:

$$\|x_{t+1}\|_{\mathsf{W}} \leqslant \|W\|_{\mathsf{X}_{t-1}} \|x_t\|_{\mathsf{W}} + 1.$$

Now, if in addition inequality (5.28) holds, then we obtain

$$\max\left\{\tfrac{1}{1-\mu}, \mu\|x_t\|_{\mathsf{W}} + 1\right\} < \|W\|_{\mathsf{X}_{t-1}} \|x_t\|_{\mathsf{W}} + 1.$$

Combining both the previous inequalities, we get the following result.

$$\max\left\{\tfrac{1}{1-\mu}, \|x_{t+1}\|_{\mathsf{W}}\right\} \leqslant \|W\|_{\mathsf{X}_{t-1}} \|x_t\|_{\mathsf{W}} + 1$$
$$\leqslant \|W\|_{\mathsf{X}_{t-1}} \max\left\{\tfrac{1}{1-\mu}, \|x_t\|_{\mathsf{W}}\right\} + 1.$$

$\square$

Chapter 6

ROBUST MODEL-BASED LEARNING AND CONTROL OF UNKNOWN SYSTEMS

In this chapter, we introduce a new framework "PixSel" for adaptive control, and, more generally, for one-shot control design of nonlinear discrete-time systems. Most notably, the theory and design methods can provide worst-case closed-loop guarantees on safety- and cost-performance even in the presence of arbitrarily large model uncertainty, and allow for problem settings with nonlinear time-varying dynamics in both system and controller. Our framework reveals a promising connection between online learning and robust control theory, which enables systematic and modular design of robust learning and control algorithms with provided safety and performance guarantees in the large uncertainty setting. To the best of our knowledge, this is the first time that a fundamental connection between the fields of online learning and control theory has ever been discovered in this context.

Our approach is based on decomposing one-shot control design into two separate sub-problems: Designing a "robust oracle" π, which encapsulates application specific nominal control design and desired guarantees, and designing a "consistent model chaser" SEL, a pure online learning problem which embodies the issue of stable and efficient adaptation. If each individual problem can be solved, we can use the resulting subroutines π and SEL to instantiate a CE[1]-based adaptive controller $\mathcal{A}_{\pi \times \text{SEL}}$ that inherits worst-case guarantees from nominal control design, which surprisingly still hold for arbitrarily large model uncertainty.

Our discussion will begin with studying the one-shot control design problem "Online Control with Mistake Guarantees" (OC-MG) first introduced in our work [5], which will serve as a motivation and introductory case study of the general "PixSel"-design framework (which ultimately has a far broader scope than the problem setting of OC-MG). Some of the broader implications of the "PixSel"-framework are discussed in Section 6.7 and Section 6.8, others are presented in recent [9] and topic of ongoing work.

---

[1]Certainty-Equivalence principle.

Figure 6.1: PixSel Framework for adaptive control.

6.1 Introduction

We study the problem of online control for nonlinear systems with large model uncertainty, under the requirement to provide upfront control-theoretic guarantees for the worst-case online performance; by large uncertainty, we mean to say that we are given an arbitrarily large set of potential models, of which an *unknown few* are exact descriptions of the true system dynamics. Algorithms with such capabilities can enable us to (at least partially) sidestep undertaking laborious system identification tasks prior to robust controller design. Motivated by real-world control applications, we formulate a class of problems which allow us a unified way to address common control problems such as stabilization, tracking, disturbance rejection, robust set invariance, etc. We introduce this as *online control* with *mistake guarantees* (OC-MG): We define a problem instance by specifying a desired system behavior and search for online control algorithms which can quantify, in terms of number of *mistakes*, how often the online controlled system could deviate from this behavior in the worst-case (i.e: worst possible scenario of true system dynamics, disturbances, noise, etc.). We propose a modular framework for OC-MG: Use robust control to design a *robust oracle* π, use online learning to design an algorithm SEL which *chases consistent models*, and fuse them together via a simple meta-algorithm; the end result is Algorithm 2, which we refer to as $\mathcal{A}_\pi(\text{SEL})$. Our approach is based on decomposing the original problem into the two independent sub-problems "robust oracle design" (ROD) and "consistent models chasing" (CMC) which for many problem instances can be readily addressed with existing tools from control theory (see Section 6.3) and online learning (see Section 6.5). We demonstrate in Section 6.10, that for general robotic systems, we can solve CMC through competitive convex body chasing [12, 13, 33, 114] and ROD using well-known robust control methods [54, 96, 118, 119, 141].

Once suitable subroutines π and SEL are selected, we can provide online performance guarantees for the resulting control algorithm $\mathcal{A}_\pi(\text{SEL})$ that holds in the large uncertainty setting:

- *Mistake guarantee:* A worst case bound on the total number of times desired system behavior is violated. See Theorem 38, 39, 40.
- *Safety guarantee:* A worst-case norm bound on the state-trajectory. See Theorem 36.

To provide the above guarantees, π and SEL have to be solutions to a corresponding ROD and CMC sub-problem. In Section 6.3 and Section 6.5 we discuss that in many problem settings this is not a restrictive assumption. In particular, assuming that ROD can be solved merely ensures that the overall OC-MG problem is well-posed: The underlying control problem (i.e., assuming no uncertainty) has to be tractably solvable with robust control; it is clear that this is a bare minimum requirement to state a meaningful OC-MG problem. In Section 6.5 we discuss different versions of the CMC sub-problem and present a reduction to nested convex body chasing [33] which is applicable for a large class of systems.

In Section 6.11, we follow our approach to design a high-performing control algorithm for a difficult nonlinear adaptive control problem: swinging-up a cartpole with large parametric uncertainty *and* state constraints. We benchmark the performance of the online algorithm $\mathcal{A}_\pi(\text{SEL})$ against the offline optimal algorithm over 900 problem settings (adversarial chosen system parameters, noise, disturbances) and show that $\mathcal{A}_\pi(\text{SEL})$ performs only marginally worse than the optimal offline controller, which has access to the true system model.

Problem Statement

Consider controlling a discrete-time *nonlinear dynamical system* with system equations:

$$x_{t+1} = f^*(t, x_t, u_t), \quad f^* \in \mathcal{F}, \tag{6.1}$$

where $x_t \in \mathcal{X}$ and $u_t \in \mathcal{U}$ denote the system state and control input at time step t and $\mathcal{X} \times \mathcal{U}$ denotes the state-action space. We assume that f^* is an *unknown* function and that we only know of an *uncertainty set* $\mathcal{F}$ that contains the true f^*.

Large Uncertainty Setting. We impose no further assumptions on $\mathcal{F}$ and explicitly allow $\mathcal{F}$ to represent arbitrarily large model uncertainties.

Control Objective. The control objective is specified as a sequence $\mathcal{G} = (\mathcal{G}_0, \mathcal{G}_1, \dots)$ of binary cost functions $\mathcal{G}_t : \mathcal{X} \times \mathcal{U} \mapsto \{0, 1\}$, where each function $\mathcal{G}_t$ encodes a desired condition per time step t: $\mathcal{G}_t(x_t, u_t) = 0$ means that the state x_t and the input u_t meet the requirements at time t. $\mathcal{G}_t(x_t, u_t) = 1$ means that some desired condition is violated at time t and we will say that the system made a *mistake* at t. The performance metric of system trajectories $\boldsymbol{x} := (x_0, x_1, \dots)$ and $\boldsymbol{u} := (u_0, u_1, \dots)$ is the sum of the cost incurred $\mathcal{G}_t(x_t, u_t)$ over the time interval $[0, \infty)$ and we denote this the *total number of mistakes*:

$$\# \text{ mistakes of } \boldsymbol{x}, \boldsymbol{u} = \sum_{t=0}^{\infty} \mathcal{G}_t(x_t, u_t). \tag{6.2}$$

For a state-input trajectory of the system $(\boldsymbol{x}, \boldsymbol{u})$ to achieve an objective $\mathcal{G}$, we want the above quantity to be finite, i.e., eventually the system stops making mistakes and meets the objectives requirements for all time.

One-shot Control Design Goal. The goal is to design an online decision rule $u_t = \mathcal{A}(t, x_t, \dots, x_0)$ such that regardless of the unknown $f^* \in \mathcal{F}$, the online trajectories are guaranteed to be bounded, and to have finite or even explicit upper-bounds on the total number of mistakes (6.2). Thus, we require a strong notion of robustness: $\mathcal{A}$ can control any system (6.1) with the certainty that the objective $\mathcal{G}$ will be achieved after finitely many mistakes. It is suitable to refer to our problem setting as *online control* with *mistake guarantees*.

Motivation and Related Work

The main purpose of this work is to find answers to the following question:

> *How do we learn to control unknown dynamical systems in a systematic and reliable way?*

In one way or another, this is one of the first questions we run into when we approach real-world control problems: Dynamical systems we encounter in the real world are unknown to us; we have ways to find approximate mathematical models (system identification), and use them as a substitute system for control design. That being said, we are always forced to make a leap of faith when we deploy a control system, since we can never be certain how well our system models match up with the dynamics of the real system. Therefore, it is always necessary to collect online data and keep monitoring for potential *inconsistencies* between system behavior and our model/design assumptions, as well as be ready to *adapt* our control algorithms once we *learn* that our models are no longer accurate enough to guarantee performance. Depending on the application, accurate models and/or online adaptation can be of critical importance. This is especially true in safety-critical settings involving physical systems, such as in engineering domains such as aerospace, industrial robotics, automotive, energy plants [123], etc. Throughout the last decades and with the accelerated technological advancement, many important engineering systems have drastically increased in complexity. Finding accurate models of complex systems can quickly become really difficult (or even impossible), which makes the latter problem, of learning to adapt controls from on-line data, a crucial aspect of over control system design. This problem setting is the focus and motivation of our work. The existing literature can be split into two general categories: the more recent literature of what can be described as *System ID, then Robust Control* and the more traditional literature of *Adaptive Control*.

System Identification, then Robust Control

The most common approach in online learning for control literature [42] is to perform system identification [88], then use tools from robust control theory [141]. Robust controller synthesis can provide policies with desired guarantees, so long as one can obtain an approximate model which is "provably close enough" to the real system dynamics. However, estimating a complex system to a desired accuracy level quickly becomes intractable in terms of computational and/or sample complexity. In the adversarial noise setting, system identification of simple linear systems with precision guarantees can be NP-hard [41]. General approaches for nonlinear system identification with precision guarantees are for the most part not available (recently Mania et al. [93] analyzed sample complexity under stochastic noise). Many recent learning approaches for control of dynamical systems have focused on the setting of linear optimal control: One is given a linear system, and the control objective is to minimize a specified cost functional. To relate our problem setting to other approaches in this field, we can view our problem setting as an instance of optimal control where we restrict the cost function to be $\{0, 1\}$ valued. There has been a particular focus on the problem of the Linear Quadratic Regulator (LQR) [1, 37, 42, 43, 51], or linear dynamical system with convex costs [3, 4, 62]. Our work is instead suitable as well for the nonlinear control setting. In addition, even when restricted to the linear system setting, recent line of work on online learning for control differs from our approach in the following aspects:

- Performance Criteria: We focus on bounding the total cost $\sum_{t=0}^{\infty} \mathcal{G}_t(x_t, u_t)$ as defined in Section 6.1 of the main paper. Our notion of control objective is natural to define in control applications, e.g., most popular robotic goals can be formulated as driving the systems towards a desirable set or trajectories. This differs from, but is not incompatible with, the cost-metric formulation that is often seen in optimal control and online learning for control work. Specifically, previous effort on learning LQR has been to improve the regret bound of the learning algorithm [1, 2, 42, 43, 62]. Bounding the regret on the average cost, which is natural for LQR, is not sufficient to guarantee finite mistakes in our problem setting. In Section 6.C, we discuss counterexamples which discuss the relationship between finite mistakes, sublinear regret and asymptotic guarantees. We show that finite mistake guarantees imply sublinear regret, yet sublinear regret does not imply finite mistakes.

- Approach: Our proposed approach does not depend on accurate identification of the online system, which is the focus of several recent works on learning for LQR [37, 42, 51, 62]. As we consider parametric uncertainty, it is plausible to also adopt a system identification approach for the non-linear control settings. However, online system identification with arbitrarily small error is known to be very challenging. As shown by [41], the sample complexity for identifying linear systems under bounded adversarial noises can be exponential in the worst case.

- Assumptions about parameter uncertainty: Some previous work in linear systems [37, 43, 62] assumes knowledge of a stabilizing controller $\pi_{\text{safe}} : \mathcal{X} \mapsto \mathcal{U}$ for the true unknown system parameter θ^*. In our setting, we do not require such an assumption, but merely that for each possible parameter $\theta \in \Omega$ one can find a robust policy $\pi[\theta]$ which stabilizes the small uncertainty model $\mathbb{D}[\theta] \subset \mathcal{F}$.

Robust Adaptive Nonlinear Control

Naturally, our problem setting is of great interest to the adaptive control community, which has had a relatively long history on this topic: [71, 83, 87, 102, 134]. Yet, most of traditional adaptive control approaches can not be applied to the general problem setting we consider without making restrictive assumptions. As an example, in contrast to most adaptive control methods, our framework applies to nonfeedback linearizable nonlinear system (see overview of adaptive control in [65]). Furthermore, many adaptive control techniques can not build on top of methods from other areas of control, like robust control theory, but rather propose separate control algorithms for each problem setting. Additionally, robust stability analysis and thorough empirical validation are largely unavailable for most methods. In fact, most relevant empirical results are only presented for arguably much simpler settings than for the cart-pole swing-up problem, which is considered in this work. In addition, we highlight methodological distinctions.

We provide a modular framework, which allows one to combine robust control tools with online learning algorithms to provide desired guarantees online. We unify the treatment of both uncertain system parameters and unknown disturbance via the construction of confidence sets of candidate systems that are consistent with the historical collected observations. The estimation of such consistent sets is also easily attainable for most robotic systems and allows for non-asymptotic convergence guarantees. Among the relevant adaptive control literature, perhaps the most closely related to ours is Multi-Model Adaptive Control (MMAC) from [9]. The MMAC principle needs to run a high-dimensional Multi-Estimation routine online, which requires the design of nonlinear observers (with the matching and Detectability property - see [63]) for a sufficiently dense covering set of the parameter space. A general construction of such a family is only shown for linear systems (see [63] and references therein), and it is not clear whether designing a tractable Multi-Estimator for the cart-pole system is possible.

6.2   Overview of Approach and Main Results

No Need for SysID and Persistency of Excitation. While accurate models of real systems are hard to obtain, it is often easy to provide more qualitative or rough models of system dynamics *without* performing offline experiments and requiring system identification. Having access to a rough system description, we design a control algorithm in one-shot which can be deployed on the real system with upfront worst-case control-theoretic guarantees on the online performance. Moreover, in contrast to other works, such as [35] for example, we are the first framework that can provide worst-case guarantees *without* requiring the assumption of *persistency of excitation*.

Rough Models as Compactly Parameterisable Uncertainty Sets. In practice, we never have the exact knowledge of f^* in advance. However, for engineering applications involving physical systems, the functional form of f^* can often be derived through first principles and knowledge of the application-specific domain. Conceptually, we can view the unknown parameters of the functional form as conveying both the 'modeled dynamics' and 'unmodeled (adversarial) disturbance' components of the ground truth f^* in the system $x_{t+1} = f^*(t, x_t, u_t)$. It is almost always the case that we can represent the uncertainty in f^* via a collection of parameters in bounded ranges. How we choose to parameterize a given uncertainty set $\mathcal{F}$ is not unique and poses a design choice. We will take this as the starting point for our approach and assume a fixed parameterization of $\mathcal{F}$ in the form of a tuple $(\mathbb{D}, \Omega, d)$, where (Ω, d) is a compact metric space, called *parameter space*, and $\mathbb{D}$ is a map $\Omega \mapsto 2^{\mathcal{F}}$ that defines a collection of models $\{\mathbb{D}[\theta] \mid \theta \in \Omega\}$ which represents a cover of the uncertainty set $\mathcal{F}$. We define this formally as a compact parameterization of $\mathcal{F}$:

Definition 6.1. *A tuple* $(\mathbb{D}, \Omega, d)$, *where* $\mathbb{D} : \Omega \mapsto 2^{\mathcal{F}}$ *is a compact parametrization of* $\mathcal{F}$, *if* (Ω, d) *is a compact metric space and* $\mathcal{F} \subset \bigcup_{\theta \in \Omega} \mathbb{D}[\theta]$.

We will work with candidate parameters $\theta \in \Omega$ of the system and consider a θ^* to be a *true parameter* of f^*, if $f^* \in \mathbb{D}[\theta^*]$. Ideally, each candidate model $\mathbb{D}[\theta]$ has small uncertainty; the precise notion of "small uncertainty" however is problem specific and depends always on the objective. For concreteness, we give several simple examples of common parameter spaces Ω:

1. *Linear time-invariant system*: linear system with matrices A, B perturbed by bounded disturbance sequence $w \in \ell_\infty$, $\|w\|_\infty \leq \eta$:

$$f^*(t, x, u) = Ax + Bu + w_t. \tag{6.3}$$

   The parameter space Ω contains bounded intervals describing the parameters $\theta = (A, B, \eta)$.

2. *Nonlinear system, linear parametrization*: nonlinear system, where dynamics are a weighted sum of nonlinear functions ψ_i perturbed by a bounded disturbance sequence $w \in \ell_\infty$, $\|w\|_\infty \leq \eta$:

$$f^*(t, x, u) = \sum_{i=1}^{M} a_i \psi_i(t, x, u) + w_t. \tag{6.4}$$

   Ω contains bounded intervals that describe $\theta = (\{a_i\}, \eta)$.

3. *Nonlinear system, nonlinear parametrization*: nonlinear system, with function g parameterized by a fixed parameter vector $p \in \mathbb{R}^m$ (e.g., neural networks), perturbed by a bounded disturbance sequence $w \in \ell_\infty$, $\|w\|_\infty \leq \eta$:

$$f^*(t, x, u) = g(t, x, u; p) + w_t. \tag{6.5}$$

   Ω contains bounded intervals that describe $\theta = (p, \eta)$.

Algorithm 2 Meta-Implementation of $\mathcal{A}_\pi(\text{SEL})$

Require: procedures π and SEL
Initialization: $\mathcal{D}_0 \leftarrow \{\}$, x_0 is set to initial condition ξ_0
1: **for** $t = 0, 1, \ldots$ **to** ∞ **do**
2:    $\mathcal{D}_t \leftarrow$ append $(t, x_t, x_{t-1}, u_{t-1})$ to $\mathcal{D}_{t-1}$ (if $t \geq 1$) $\triangleright$ update online history of observations
3:    $\theta_t \leftarrow \text{SEL}[\mathcal{D}_t]$ $\triangleright$ present online data to SEL, get posited parameter θ_t
4:    $u_t \leftarrow \pi[\theta_t](t, x_t)$ $\triangleright$ query π for policy $\pi[\theta_t]$ and evaluate it
5:    $x_{t+1} \leftarrow f^*(t, x_t, u_t)$ $\triangleright$ system transitions with unknown f^* to next state
6: **end for**

In these examples, the uncertainty set $\mathcal{F} \subset \bigcup_{\theta \in \Omega} \mathbb{D}[\theta]$ is covered by models $\mathbb{D}[\theta]$ with smaller uncertainty of the form $\mathbb{D}[\theta] = \{t, x, u \mapsto f_\theta(x, u, w_t) \mid \|w\|_\infty \leq \eta\}$, where f_θ denotes one of the functional forms on the right-hand side of eq. (6.3), (6.4) or (6.5).

Online Robust Control Algorithm. Given a compact parameterization $(\mathbb{D}, \Omega, d)$ for the uncertainty set $\mathcal{F}$, we design a meta-algorithm $\mathcal{A}_\pi(\text{SEL})$ (Algorithm 2) that controls the system (6.1) online by invoking two subroutines π and SEL in each time step.

- *Consistent model chasing.* Procedure SEL receives a finite data set $\mathcal{D}$, which contains state and input observations, and returns a parameter $\theta \in \Omega$.
  Design goal: For each time t, the procedure SEL should select θ_t such that the set of models $\mathbb{D}[\theta_t]$ stays "consistent" with $\mathcal{D}_t$, i.e., candidate models in $\mathbb{D}[\theta_t]$ can *explain* the past data. Moreover posited parameters θ_t should only change when necessary: two posited parameters θ_t and $\theta_{t'}$ should not be very different from each other, if both data sets $\mathcal{D}_t$ and $\mathcal{D}_{t'}$ contain "similar" amount information.

- *Robust oracle.* Procedure π receives a posited system parameter $\theta \in \Omega$ as input and returns a control policy $\pi[\theta] : \mathbb{N} \times \mathcal{X} \mapsto \mathcal{U}$ which can be evaluated at time t to compute a control action $u_t = \pi[\theta](t, x_t)$ based on the current state x_t.
  Design goal: We require that π represents a robust control design subroutine for the collection of models $\mathbb{D}$, in the sense that policy $\pi[\theta]$ could provide mistake guarantees for $\mathcal{G}$ which are robust to bounded noise **if** the uncertainty set $\mathcal{F}$ *were* $\mathbb{D}[\theta]$.

Theoretical Contribution. Our main theoretical results certify safety- and finite mistake guarantees for the online control scheme $\mathcal{A}_\pi(\mathrm{SEL})$ if the sub-routines π and SEL meet the design requirement for "robust oracle" and "consistent model chasing" for a given uncertainty set $\mathcal{F}$ and objective $\mathcal{G}$. We will clarify the consistency and robustness requirements of the sub-routines π and SEL in Section 6.3 and Section 6.5. For now, we present an informal version of the finite mistake guarantees and the worst-case state deviation for the online control scheme $\mathcal{A}_\pi(\mathrm{SEL})$:

Theorem (*Informal*). *For any (adversarial)* $f^* \in \mathcal{F}$, *the online control scheme* $\mathcal{A}_\pi(\mathit{SEL})$ *described in Algorithm 2 guarantees a priori that the trajectories* $\boldsymbol{x}$, $\boldsymbol{u}$ *will achieve the objective* $\mathcal{G}$ *after finitely many mistakes. The total number of mistakes* $\sum_{t=0}^{\infty} \mathcal{G}_t(x_t, u_t)$ *is at most*

$$\text{oracle performance } M_\rho^\pi * \Gamma_1 \left(\frac{\text{size of uncertainty } \mathcal{F}}{\text{efficiency of SEL} * \text{robustness margin } \rho \text{ of } \pi} \right),$$

and the norm of the state $\|x_t\|$ *is at most*

$$\Gamma_2 \left(\frac{\text{size of uncertainty } \mathcal{F}}{\text{efficiency of SEL} * \text{single-step robustness margin of } \pi}, \|x_0\| \right),$$

for some increasing function $\Gamma_1 : \mathbb{R}^+ \mapsto \mathbb{R}^+$ *and some function* $\Gamma_2 : \mathbb{R}^+ \mapsto \mathbb{R}^+$ *which is increasing in the first argument and is linear in the second.*

- *Performance of* π: Assume the worst-possible $f^* \in \mathcal{F}$, but also access to direct online measurements $\theta_t = \theta^* + v_t$ of the a true parameter θ^* with small noise v_t of size ρ; M_ρ^π denotes the worst-case mistakes if we were to apply the almost ideal control law $u_t = \pi[\theta_t](t, x_t)$ in this setting.

- *Efficiency of* SEL: We quantify the efficiency of SEL in the result through competitive analysis of online algorithms. The procedure SEL posits parameters efficiently, if as a function of time, the parameter selection θ_t changes only when necessary; that is, it only changes when new observations are informative and keep a constant value otherwise. We phrase this in terms of a *competitive ratio* γ (with $\gamma \geq 1$) and distinguish here between γ-competitive and (γ, T)-finite-time competitive algorithms. The smaller the constant γ is, the more efficient the algorithm posits parameters. As discussed in Section 6.5, a smaller γ indicates that an online algorithm performs more closely to the ideal optimal algorithm in hindsight.

Remark 31. *If the same procedure* π *serves as a robust oracle for a set of criteria* $\mathcal{G}^{(1)}, \mathcal{G}^{(2)}, \ldots, \mathcal{G}^{(M)}$, *then correspondingly the instantiation* $\mathcal{A}_\pi(\mathit{SEL})$ *provides*

multiple finite mistake guarantees, i.e., one for each corresponding criteria $\mathcal{G}^{(i)}$,
$i = 1, \ldots, M$.

This approach brings several attractive qualities:

- *Generality.* The result applies to a wide range of problem settings. The objective $\mathcal{G}$ and the uncertainty set $\mathcal{F}$ serve as a flexible abstraction to represent a large class of dynamical systems and control-theoretic performance objectives.
- *Robust guarantees in the large uncertainty setting.* Our result applies in settings where only rough models are available. As an example, we can use the result to provide guarantees in control settings with unstable nonlinear systems where stabilizing policies are **not** known a-priori and which are subject to online adversarial disturbances.
- *Decoupling algorithm design for learning and control.* The construction of the "robust oracle" π and the consistent model chasing procedure SEL can be addressed with existing tools from control and learning. More generally, this perspective enables us to decouple learning and control problems in the large uncertainty setting into separate robust control and online learning problems, a novel approach. See discussion in Section 6.3 and Section 6.5.
- *Modular algorithm design for robust learning and control.* The above approach provides a first interface between robust control and online learning, which enables a modular design of learning and control algorithms with versatile worst-case performance guarantees against large model uncertainty.
- *A new tool for performance analysis of learning and control algorithms.* We can view the above theorem also from an analysis point-of view: Many existing certainty-equivalence based learning and control algorithms can be easily represented as an instance of the meta-algorithm $\mathcal{A}_\pi(\text{SEL})$ and thus can be analyzed using the above theorem.

Promising for Design of Efficient Algorithms in pPractice. Besides focusing on providing worst-case guarantees in a general setting, empirical results show that our framework is a promising approach to design efficient algorithms for learning and control in practice. In Section 6.11, we apply our approach to the problem of swinging-up a cartpole with large parametric uncertainty in a realistic and highly challenging setting and show that it achieves consistently (over 900 experiments with different parameter settings) good performance.

As summarized in Algorithm 2, the main ingredients of our approach are a robust control oracle π that returns a robust controller under posited system parameters, and an online algorithm SEL that chases parameter sets that are consistent with the data collected so far. In the following two sections we formulate the formal concept of oracles and model selectors SEL and discuss their respective properties required for the statment of our main results.

6.3 Control Oracle: An Abstraction for Nominal Control Design

It is obvious that our OC-MG problem is only well-posed if the underlying control problem under no uncertainty, i.e., the *nominal control design* problem, is feasible as well. In particular, for any dynamics $f' \in \mathcal{F}$, there has to be a control policy $\kappa' \in \mathcal{K}$ which, in closed-loop, provides worst-case stability and mistake guarantees. To state this formally, let the dynamical model $\mathsf{M}[f, \kappa] \subset (\mathcal{X} \times \mathcal{U})^{\mathbb{N}}$ be the set of all closed-loop trajectories obtained by interconnecting the dynamics f with controller κ, i.e.,

$$\mathsf{M}[f, \kappa] = \left\{ (\boldsymbol{\tau}^{\mathrm{x}}, \boldsymbol{\tau}^{\mathrm{u}})^{\top} \; \middle| \; \forall t \in \mathbb{N} : \begin{array}{ll} \boldsymbol{\tau}^{\mathrm{x}}(t+1) & = f(t, \boldsymbol{\tau}^{\mathrm{x}}(t), \boldsymbol{\tau}^{\mathrm{u}}(t)) \\ \boldsymbol{\tau}^{\mathrm{u}}(t) & = \kappa(t, \boldsymbol{\tau}^{\mathrm{x}}(t)) \end{array} \right\}, \quad (6.6)$$

then our problem OC-MG is only feasible, if for any $f' \in \mathcal{F}$ there exists some $\kappa' \in \mathcal{K}$ such that for all $\boldsymbol{\tau} \in \mathsf{M}[f', \kappa']$ holds $\boldsymbol{\tau} \in \ell_{\infty}^{\mathcal{X} \times \mathcal{U}}$ (bounded closed-loop trajectories) and $\sum_{t=0}^{\infty} \mathcal{G}_t(\boldsymbol{\tau}(t)) < \infty$ (each closed-loop trajectory makes finite mistakes).

The main purpose of the oracle π is to serve as an abstraction for nominal control design which is robust to some small degree of model uncertainty. The procedure π is a map $\Omega \mapsto \mathcal{K}$ from parameter space Ω to the space $\mathcal{K} := \{\kappa : \mathbb{N} \times \mathcal{X} \mapsto \mathcal{U}\}$ of all (non-stationary) control policies of the form $u_t = \kappa(t, x_t)$. The purpose of π is to specify a parametrized collection of "fixed"-model-based controllers suitable for certainty-equivalent control. A desired property of π as an *oracle* is that π returns controllers that satisfy $\mathcal{G}$ if the model uncertainty *were* small. Thus, if we let $\mathsf{CL}_{\pi}[\omega] \subset (\mathcal{X} \times \mathcal{U})^{\mathbb{N}}$ be the dynamic model below,

Definition 6.2. $\mathsf{CL}_{\pi}[\omega] \subset (\mathcal{X} \times \mathcal{U})^{\mathbb{N}}$ *is the dynamic model of the closed-loop interconnection of the dynamics* $\mathbb{D}[\omega]$ *and the oracle policy* $\pi[\omega]$ *and is equivalently written as:*

$$\mathsf{CL}_{\pi}[\omega] = \bigcup_{f \in \mathbb{D}[\omega]} \mathsf{M}[f, \pi[\omega]].$$

We can formulate a necessary requirements for the nominal problem setting to be:

$$\forall \boldsymbol{\tau} \in \bigcup_{\omega \in \Omega} \mathsf{CL}_\pi[\omega] \ : \ \sum_{t=0}^{\infty} \mathcal{G}_t(\boldsymbol{\tau}(t)) < \infty \text{ and } \boldsymbol{\tau} \in \ell_\infty^{\mathcal{X} \times \mathcal{U}}.$$

In other words, if the uncertainty set $\mathcal{F}$ *were* contained in the set $\mathbb{D}[\theta]$, then control policy $\pi[\theta]$ could guarantee to achieve the objective $\mathcal{G}$ with finite mistake guarantees. The above statement is subsumed in the formal oracle requirements, formulated in the next section. Furthermore, in an idealized setting where the true parameter *were* known exactly, the oracle should return a policy such that the system performance is robust to some level of bounded noise. This is a standard notion of *robustness*, which we later define more precisely. Naturally, there exist many control methods in the control literature which are suitable for robust oracle design. Which method to use depends on the control objective $\mathcal{G}$, the specific application, and the system class (linear/nonlinear/hybrid, etc.). For a broad survey, see [118, 119, 142] and references therein. We characterize two general methodologies (which can also be combined):

- *Robust stability analysis focus:* In an initial step, we use analytical design principles from robust nonlinear and linear control design to propose an oracle $\pi[\theta](x)$ in closed-form for all θ and x. In a second step we prove robustness using analysis tools such as for example *Input-to-State Stability* (ISS) stability analysis [75] or robust set invariance methods [106, 107]).
- *Robust control synthesis:* If the problem permits, we can also directly address the control design problem from a computational point of view, by formulating the design problem as an optimization problem and compute for a control law with desired guarantees directly. This can happen partially online, partially offline. Some common nonlinear approaches are robust (tube-based) MPC [30, 96], SOS-methods [100],[21], Hamilton-Jacobi reachability methods [22].

There are different advantages and disadvantages to both approaches, and it is important to point out that robust control problems are not always tractably solvable. See [27, 31] for simple examples of robust control problems which are NP-hard. The computational complexity of robust controller synthesis tends to increase (or even be potentially infeasible) with the complexity of the system of interest; it also further increases as we try optimize for larger robustness margins ρ.

The Dual Purpose of the Oracle. In our framework, access to a robust oracle is a necessary prerequisite to design learning and control agents $\mathcal{A}_\pi(\mathrm{SEL})$ with mistake guarantees. However, this is a mild assumption and is often more enabling than restrictive. First, it represents a natural way to ensure well-posedness of the overall learning and control problem; If robust oracles cannot be found for an objective, then the overall problem is likely intrinsically hard or ill-posed (for example, necessary fundamental system properties like stabilizability/ detectability are not satisfied).

Second, oracle abstraction enables a modular approach to robust learning and control problems and directly leverages existing powerful methods in robust control: Any model-based design procedure π that works well for the small uncertainty setting (i.e., acts as a robust oracle) can be augmented with an online chasing algorithm SEL (with required chasing properties) to provide robust control performance (in the form of mistake guarantees) in the large uncertainty setting via the augmented algorithm $\mathcal{A}_\pi(\text{SEL})$.

Next, we formulate the robustness properties expected from the oracle π.

6.4   Robust Oracle Guarantees

The main requirement on the oracle is that if the model uncertainty were small enough, we could just use the oracle for certainty-equivalent control. We phrase this requirement by describing the performance of the policy in an idealized setting. Let θ^* be a parameter of true dynamics f^*, and assume that online we have access to noisy observations $\boldsymbol{\theta} = (\theta_0, \theta_1, \dots)$, where each measurement θ_t is ρ-close to θ^*, under metric d. The online control algorithm queries π at each time step and applies the corresponding policy $\pi[\theta_t]$. The resulting trajectories obey the equations:

$$x_{t+1} = f^*(t, x_t, u_t), \quad u_t = \pi[\theta_t](t, x_t) \tag{6.7a}$$

$$\theta_t \text{ s.t.: } d(\theta_t, \theta^*) \leqslant \rho, \quad \text{where } f^* \in \mathbb{D}[\theta^*]. \tag{6.7b}$$

To facilitate later discussion, define the set of all feasible trajectories of the dynamic equations (6.7) as the *nominal trajectories* $\mathcal{S}_{\mathcal{I}}[\rho; \theta]$ of the oracle:

Definition 6.3. *For a time-interval* $\mathcal{I} = [t_1, t_2] \subset \mathbb{N}$ *and fixed* $\theta \in \Omega$, *let* $\mathcal{S}_{\mathcal{I}}[\rho; \theta]$ *denote the set of all pairs of finite trajectories* $x_{\mathcal{I}} := (x_{t_1}, \dots, x_{t_2})$, $u_{\mathcal{I}} := (u_{t_1}, \dots, u_{t_2})$ *which for* $\theta^* = \theta$, *satisfy conditions* (6.7) *with some feasible* f^* *and sequence* $(\theta_{t_1}, \dots, \theta_{t_2})$.

Design Specification for Oracles. We will say that π is ρ-robust for some objective $\mathcal{G}$, if all trajectories in $\mathcal{S}_{\mathcal{I}}[\rho; \theta]$ achieve $\mathcal{G}$ after finitely many mistakes. We distinguish between robustness and uniform robustness, which we define precisely below.

ρ-robust	$\forall \theta \in \boldsymbol{\Omega} : \sup_{\gamma \geqslant 0} m_\rho^\pi(\gamma; \theta) < \infty$
uniformly ρ-robust	$M_\rho^\pi := \sup_{\gamma \geqslant 0, \theta \in \Omega} m_\rho^\pi(\gamma; \theta) < \infty$
locally ρ-robust	$\forall \gamma \geqslant 0,\ \theta \in \boldsymbol{\Omega} :\ m_\rho^\pi(\gamma; \theta) < \infty$
locally uniformly ρ-robust	$\forall \gamma \geqslant 0 :\ M_\rho^\pi(\gamma) := \sup_{\theta \in \Omega} m_\rho^\pi(\gamma; \theta) < \infty$

Table 6.1: Notions of oracle-robustness

Definition 6.4 (robust oracle)**.** *Equip $\mathcal{X}$ with some norm $\|\ \|$. For each $\rho, \gamma \geqslant 0$ and $\theta \in \boldsymbol{\Omega}$, define the quantity $m_\rho^\pi(\gamma; \theta)$ as*

$$m_\rho^\pi(\gamma; \theta) := \sup_{\mathcal{I}=[t,t']\ :\ t<t'}\ \sup_{(x_\mathcal{I}, u_\mathcal{I}) \in \mathcal{S}_\mathcal{I}[\rho;\theta], \|x_0\| \leqslant \gamma}\ \sum_{t \in \mathcal{I}} \mathcal{G}_t(x_t, u_t).$$

If $m_0^\pi(\gamma; \theta) < \infty$ for all $\gamma \geqslant 0, \theta \in \boldsymbol{\Omega}$, we call π an oracle for $\mathcal{G}$ w.r.t. parametrization $(\mathbb{D}, \boldsymbol{\Omega}, d)$. In addition, we say that an oracle π is (locally) (uniformly) ρ-robust if the corresponding property shown in Table 6.1 holds. If it exists, M_ρ^π is the mistake constant/function of π.

The constant $\rho > 0$ will be referred to as the robustness margin of π. If we use the above terms without referencing ρ, it should be understood that there exists some $\rho > 0$ for which the corresponding property is feasible. The mistake constant M_ρ^π can be viewed as a robust offline benchmark. It quantifies how many mistakes we would make in the worst case if we could use the oracle π under idealized conditions, that is, described by (6.7).

Invariance Property. On top of ρ-robustness, for some results, we will require the following additional condition from the oracle:

Definition 6.5. *For a fixed objective $\mathcal{G}$, define the set of admissible states at time t as $\mathsf{X}_t = \{x \mid \exists u' :\ \mathcal{G}_t(x, u') = 0\}$, i.e., the set of states for which it is possible to achieve zero cost at the time step t. We call a ρ-robust oracle π cost invariant, if for all $\theta \in \boldsymbol{\Omega}$ and $t \geqslant 0$ the following holds:*

- *For all $x \in \mathsf{X}_t$ holds $\mathcal{G}_t(x, \pi[\theta](t, x)) = 0$.*

- *For all $x \in \mathsf{X}_t$, $f \in \mathbb{D}[\theta]$ and θ' s.t. $d(\theta', \theta) \leqslant \rho$, holds $f(t, x, \pi[\theta'](t, x)) \in \mathsf{X}_{t+1}$.*

Remark 32. *The above condition is related to the well-known notion of positive set/tube invariance in control theory [26]: The above condition requires that the* *oracle policies $\pi[\theta]$ can ensure for their nominal model $\mathbb{D}(\theta)$ the following closed loop condition: $x_t \in X_t \implies x_{t+1} \in X_{t+1}$, $\forall t$.*

Robustness of Single-Step Closed Loop Transitions. To provide worst-case guarantees on the online state trajectory, we need to bound how system uncertainty can affect a single online time-step state transition in the worst case. To this end, consider equipping the state space $\mathcal{X}$ with some norm $\| \cdot \|$ and defining a desired property for the oracle in terms of its performance in the idealized scenario:

Definition 6.6. $\pi : \Omega \mapsto \mathcal{K}$ *is* (α, β)-*single step robust in the space* $(\mathcal{X}, \| \cdot \|)$ *if for any 2-time-steps nominal trajectory* $(x_{t+1}, x_t), (u_{t+1}, u_t) \in \mathcal{S}_{[t,t+1]}[\rho; \theta]$ *holds* $\|x_{t+1}\| \leqslant \alpha\rho\|x_t\| + \beta$.

The above property requires that in the idealized setting (6.7), we can uniformly bound the single-step growth of the state by a scalar linear function in the noise-level ρ and the previous state norm. Equivalently, we can explicitly write out the condition in Def. 6.6 as $\exists \alpha, \beta > 0 : \forall \theta, \theta' \in \Omega, x \in \mathcal{X}, f \in \mathbb{D}[\theta], t \geqslant 0$ s.t.:

$$\|f(t, x, \pi[\theta'](t, x))\| \leqslant \alpha d(\theta, \theta')\|x\| + \beta.$$

We use a simplified problem setting to explain the correspondence between the discussed conditions and standard problems in the control literature.

Mistake Guarantees and Set Convergence Stability

Consider a class of systems of the form $x_{t+1} = g(x_t, u_t; \theta^*) + w_t, \|w_t\| \leqslant 1$, where θ^* is an unknown system parameter that lies in a known compact set $\Omega \subset \mathbb{R}^m$. We represent the uncertainty set as $\mathcal{F} = \cup_{\theta \in \Omega} \mathbb{D}[\theta]$ with $\mathbb{D}[\theta] := \{f^* : t, x, u \mapsto g(x, u; \theta) + w_t \mid \|w\|_\infty \leqslant 1\}$. Let $\pi : \Omega \mapsto \mathcal{K}$ be a procedure which returns state feedback policies $\pi[\theta] : \mathcal{X} \mapsto \mathcal{U}$ for a given $\theta \in \Omega$. Designing an uniformly ρ-robust oracle π can be equivalently viewed as making the closed-loop system (described by (6.7)) of the idealized setting robust to disturbance and noise. For the considered example, the closed loop can be represented by the dynamic model $\mathsf{G}_{\theta*} \in (\mathcal{X} \times \mathcal{U} \times \mathcal{Y} \times \mathcal{V} \times \mathcal{W})^{\mathbb{N}}$, defined as

$$\mathsf{G}_{\theta*} := \{(\boldsymbol{x}, \boldsymbol{u}, \boldsymbol{y}, \boldsymbol{v}, \boldsymbol{w})^\top \text{ such that } (6.8)\}$$

where for all $t \in \mathbb{N}$, $y_t \in \mathcal{Y} = \mathbb{R}$ is an output representing the cost at time t, $v_t \in \mathcal{V} = \mathbb{R}^m$ is an input representing bounded observational noise and $w_t \in \mathcal{W} = \mathcal{X}$

is an input representing a disturbance signal.

$$x_{t+1} = g(x_t, u_t; \theta^*) + w_t \tag{6.8a}$$

$$u_t = \pi[\theta^* + v_t](x_t) \tag{6.8b}$$

$$y_t = \mathcal{G}_t(x_t, u_t) \tag{6.8c}$$

Leaning on our discussion in Chapter 2, we can express the oracle-robustness property in terms of stability of the $\{wv\} \mapsto \{xuy\}$-map, denoted by the operator $\Phi_{\theta*}^{wv \mapsto xuy} \in \mathcal{C}(\ell^{\mathcal{W} \times \mathcal{V}}, \ell^{\mathcal{X} \times \mathcal{U} \times \mathcal{Y}})$ of the dynamic model $G_{\theta*}$. We call π a uniformly ρ-robust oracle if, with respect to the domain $\{(v, w)^\top \mid \|v\|_\infty, \leqslant \rho, \|w\|_\infty \leqslant 1\}$, the family of partial maps $\{\Phi_{\theta*}^{wv \mapsto y}\}_{\theta* \in \Omega}$ is uniformly stable in the $\ell_\infty^{\mathcal{W} \times \mathcal{V}} \mapsto \ell_1$-induced operator norm. The fixed constant M_ρ^π (or function M_ρ^π) which we used in the definitions of oracle robustness quantifies the uniform stability of the family of operators $\{\Phi_{\theta*}^{wv \mapsto y}\}_{\theta* \in \Omega}$ similarly to the concept of gain, which is commonly used in operator theory in the control theory literature [139]. Moreover, if we identify the cost functions $\mathcal{G}_t$ with their level sets $S_t := \{(x, u) \mid \mathcal{G}_t(x, u) = 0\}$, we can also rephrase the former conditions as a form of robust trajectory-tracking problem or a set-point control problem[2] [77]. It is common in control theory to provide guarantees in the form of convergence rates (finite-time or exponential convergence) on the tracking-error; these guarantees can be directly mapped to M_ρ^π and $M_\rho^\pi(\cdot)$, as shown in the next example.

Example 3. *Assume we want to track a desired trajectory x^d within ε precision in some normed state space $(\mathcal{X}, \|\cdot\|)$. We can phrase this as a control objective $\mathcal{G}$ by defining $\mathcal{G}_t(x, u) := 0$, if $\|x_t^d - x\| \leqslant \varepsilon$ and $\mathcal{G}_t(x, u) := 1$, otherwise. Providing an exponential convergence guarantee of type $\|x_t^d - x_t\| \leqslant c\mu^t$ with constants c, $\mu < 1$ is a basic problem studied in control theory. It is easy to see that such a guarantee implies $\sum_{t=0}^\infty \mathcal{G}_t(x_t, u_t) \leqslant \frac{\log(c\|x_0\|) + \log(\varepsilon^{-1})}{\log(\mu^{-1})}$. Hence, if design method π provides a policy $\pi[\theta]$ which guarantees (for any $\|w\|_\infty \leqslant 1$, $\|v\|_\infty \leqslant \rho$) for any trajectory of the closed-loop CL_θ the convergence condition $\forall t : \|x_t^d - x_t\| \leqslant c\mu^t$, then π is a locally ρ-uniformly robust oracle with the mistake function*

$$M_\rho^\pi = \frac{\log(c\|x_0\|) + \log(\varepsilon^{-1})}{\log(\mu^{-1})}.$$

6.5   Consistent Model Chasing

The procedure SEL has the task of efficiently selecting models consistent with the observed online data. In this section, we formulate this as an online learning problem

---

[2]In that case, $\mathcal{G}_t$ would be not time dependent.

of *chasing consistent models* and discuss various design approaches on how to address it. Next, we describe the basic setup of the problem.

Consistent Models and Parameters

Let $\mathbb{O} := \{(d_1, \ldots, d_N) \mid d_i \in \mathbb{N} \times \mathcal{X} \times \mathcal{X} \times \mathcal{U}, N < \infty\}$ be the space of all data sets $\mathcal{D} = (d_1, \ldots, d_N)$ of time-indexed data points $d_i = (t_i, x_i^+, x_i, u_i)$. We will call a data sequence $\mathcal{D} = (\mathcal{D}_1, \mathcal{D}_2, \ldots)$ an online stream if the data sets $\mathcal{D}_t = (d_1, \ldots, d_t)$ are formed from a sequence of observations $(d_1, d_2, \ldots)$. Intuitively, given a data set $\mathcal{D} = (d_1, \ldots, d_N)$ of tuples $d_i = (t_i, x_i^+, x_i, u_i)$, *any* candidate $f \in \mathcal{F}$ which satisfies $x_i^+ = f(t_i, x_i, u_i)$ for all $1 \leqslant i \leqslant N$ is *consistent* with $\mathcal{D}$; Similarly, we will say that f is *consistent* with an online stream $\mathcal{D}$, if at each time-step t, it is consistent with the data set $\mathcal{D}_t$. We will extend this definition to models $\mathbb{D}[\theta]$ and parameters $\theta \in \Omega$: The model $\mathbb{D}[\theta]$ is a consistent model for a data set $\mathcal{D}$ or an online stream $\mathcal{D}$, if it contains at least one function f that is consistent with $\mathcal{D}$ or $\mathcal{D}$, respectively; θ is then called a consistent parameter. Similarly, for some data set $\mathcal{D}$, we define the set of all consistent parameters as $P(\mathcal{D})$:

Definition 6.7 (Consistent Sets). *The map of the consistent set* $P : \mathbb{O} \mapsto 2^{\Omega}$ *returns for each data set* $\mathcal{D} \in \mathbb{O}$ *the corresponding set of consistent parameters* $P(\mathcal{D})$:

$$P(\mathcal{D}) := \mathrm{closure}\Big(\Big\{ \theta \in \Omega \ \Big| \ \exists f \in \mathbb{D}[\theta] : \ \forall (t, x^+, x, u) \in \mathcal{D}, \ x^+ = f(t, x, u) \Big\} \Big).$$

$$(6.9)$$

Some important facts follow from this definition. Since the data $\mathcal{D}_t$ at time t always contain the previous data, it is clear that the constraints defining the consistent set at time t are stricter than those at time $t - 1$. Hence, the set $P(\mathcal{D}_t)$ is contained in the consistent set $P(\mathcal{D}_{t-1})$ from a previous time step. We refer to this as the nestedness property of the sequence of sets $(P(\mathcal{D}_0), P(\mathcal{D}_1), \ldots,)$. Several other important implications are remarked in the corollary below:

Corollary. *Assume* $\mathcal{D}$ *is a data stream with at least one consistent* $f \in \mathcal{F}$. *Then, the following holds for the sequence of consistent sets* $P(\mathcal{D}) = (P(\mathcal{D}_1), P(\mathcal{D}_2), \ldots)$:

  (i) *The sequence of consistent sets is nested in* Ω, *i.e:* $P(\mathcal{D}_t) \supset P(\mathcal{D}_{t+1})$.
  (ii) $P(\mathcal{D}_\infty) := \big(\bigcap_{k=1}^{\infty} P(\mathcal{D}_k) \big) \cap \Omega$ *is non-empty.*
  (iii) *If* $\theta_t \in P(\mathcal{D}_t)$, *and* $\lim_{t \to \infty} \theta_t = \theta_\infty$, *then* $\theta_\infty \in P(\mathcal{D}_\infty)$.

Proof. The first property is clear since $\mathcal{D}_t = \mathcal{D}_{t-1} \cup \{d_t\}$. For the second, notice that $P(\mathcal{D}_t)$ is always a non-empty compact set and recall the basic real analysis fact

[109]: The intersection of a nested sequence of compact sets that are not empty is always non-empty [109]. To verify property (iii), notice that nestedness implies that the subsequence $(\theta_T, \theta_{T+1}, \dots)$ is contained in $P(\mathcal{D}_T)$; since $P(\mathcal{D}_T)$ is closed, we have $\theta_\infty \in P(\mathcal{D}_T)$. We chose T arbitrarily, so we conclude that $\theta_\infty \in P(\mathcal{D}_T)$ for all $T \geqslant 0$, that is: $\theta_\infty \in P(\mathcal{D}_\infty)$. $\qquad\square$

Chasing Consistent Models

Assume a parameterization $(\mathbb{D}, \Omega, d)$ and some fixed data set $\mathcal{D}_T = (d_1, \dots, d_T)$ which is presented to us in an online fashion and which has at least one consistent parameter θ^* (i.e., $P(\mathcal{D}_T)$ is nonempty).

Our goal is to find a consistent parameter $\theta^* \in P(\mathcal{D}_T)$ online, or equivalently, to find the model $\mathbb{D}[\theta^*]$ consistent with all data $\mathcal{D}_T$. However, since we do not have access to all data upfront, the best we can do is posit at each time-step t a parameter $\theta_t \in P(\mathcal{D}_t)$ consistent at least with all so far seen data and make the hypothesis that θ_t is also consistent with $\mathcal{D}_T$ until proven otherwise by new data. Our goal is to posit the parameters θ_t in an efficient way; we would like to change our hypothesis about the consistent parameter θ^* as little as possible, and thus θ_t should change over time as little as possible. This task can be interpreted as a two-player game, where player A (our selection SEL) is trying to chase after player B (the set of consistent models $P(\mathcal{D}_t)$) with the objective of tagging them (selecting a parameter $\theta_t \in P(\mathcal{D}_t)$). It suits to call this problem *consistent models chasing* as our objective is not only selecting from the consistent sets but also accounting for future and potentially adversarial changes in the set $P(\mathcal{D}_t)$. Next, we formulate "chasing" conditions for SEL, defined in Def. 6.9, which can address this problem. The basic requirement for SEL is to output a consistent parameter (if one exists) $\theta = \text{SEL}[\mathcal{D}]$ for a given data set $\mathcal{D}$. Described in the language of set-valued analysis, we require SEL to be a *selection* or *selector* $\mathbb{O} \mapsto \Omega$ of the set-valued map $\mathcal{D} \mapsto P(\mathcal{D})$.

Definition 6.8. *[19]. A function $f : X \mapsto Y$ is a selection/selector of the set-valued map $F : X \mapsto 2^Y$, if $\forall x \in X : f(x) \in F(x)$.*

Since we intend to use SEL in an online manner where we are given a stream of data $\mathcal{D}$, we require that SEL can posit consistent parameters $\theta_t = \text{SEL}[\mathcal{D}_t]$ in an efficient manner online. A fitting notion of "efficiency" can be defined by comparing the variation of the parameter sequence θ and the sequence of consistent sets $P(\mathcal{D}) := (P(\mathcal{D}_1), P(\mathcal{D}_2), \dots)$ over time, where we quantify the latter using the

Hausdorff distance $d_{\mathcal{H}} : 2^{\Omega} \times 2^{\Omega} \mapsto \mathbb{R}^+$ defined as

$$d_{\mathcal{H}}(S, S') = \max \left\{ \max_{x \in S} d(x, S') , \ \max_{y \in S'} d(y, S) \right\} .$$

We phrase this in terms of desired "chasing"-properties (A)-(D) and refer to selectors SEL with such properties as (consistent model) chasers:

Definition 6.9. *Let SEL $: \mathbb{O} \mapsto \Omega$ be a selection of P. Let $\mathcal{D} = (\mathcal{D}_1, \mathcal{D}_2, \dots)$ be an online data stream and let θ be a sequence defined for each time t as $\theta_t = SEL[\mathcal{D}_t]$. Assume that there always exists an $f \in \mathcal{F}$ consistent with $\mathcal{D}$ and consider the following statements:*

(A) asymptotically efficient: $\theta^* = \lim_{t \to \infty} \theta_t$ exists.

(B) asymptotically finite-time (f.t.) efficient: $\lim_{t \to \infty} d(\theta_t, \theta_{t-1}) = 0$.

(C) γ-competitive:

$$t_1 < t_2 \implies \sum_{t=t_1+1}^{t_2} d(\theta_t, \theta_{t-1}) \leqslant \gamma \, d_{\mathcal{H}}(\mathsf{P}(\mathcal{D}_{t_2}), \mathsf{P}(\mathcal{D}_{t_1})).$$

(D) (γ, T)-finite-time (f.t.) competitive:

$$t_2 - t_1 \leqslant T \implies \sum_{t=t_1+1}^{t_2} d(\theta_t, \theta_{t-1}) \leqslant \gamma \, d_{\mathcal{H}}(\mathsf{P}(\mathcal{D}_{t_2}), \mathsf{P}(\mathcal{D}_{t_1})).$$

SEL algorithms that satisfy the above chasing properties as referred to as consistent model chasers.

The desired properties describe a natural notion of efficiency for this problem. The posited consistent parameter θ_t should only change online if new data are also informative. The competitiveness properties (C) and (D) naturally restrict changing θ_t when little new information is available and permit bigger changes in θ_t only when new data is informative. Per time interval $\mathcal{I} = [t_1, t_2]$, the inequalities in (C) and (D) enforce the following: If the consistent sets $\mathsf{P}(\mathcal{D}_{t_2})$ and $\mathsf{P}(\mathcal{D}_{t_1})$ are the same, i.e., $(d_{\mathcal{H}}(\mathsf{P}(\mathcal{D}_{t_2}), \mathsf{P}(\mathcal{D}_{t_1})) = 0)$, the posited parameters $\theta_{t_1}, \dots, \theta_{t_2}$ should all have the same value. On the other hand, if $\mathsf{P}(\mathcal{D}_{t_2})$ is much smaller than $\mathsf{P}(\mathcal{D}_{t_1})$, (i.e. $d_{\mathcal{H}}(\mathsf{P}(\mathcal{D}_{t_2}), \mathsf{P}(\mathcal{D}_{t_1}))$ large), the total variation $\sum_{t=t_1+1}^{t_2} d(\theta_t, \theta_{t-1})$ in the posited parameters is allowed to be at most $\gamma d_{\mathcal{H}}(\mathsf{P}(\mathcal{D}_{t_2}), \mathsf{P}(\mathcal{D}_{t_1}))$.

Properties (C) and (D) are stronger versions of (A) and (B), called *competitiveness/finite-time competitiveness*. The relationship between the chasing properties are summarized in Lemma 51 below. To prove the relation (D) $\implies$ (B) we require the following auxilliary result:

Lemma 50. *Let (Ω, d) be a compact metric space and let $\{S_1, \ldots, S_T\}$, $\forall t : S_{t+1} \subset S_t$, $S_t \subset \Omega$ be a collection of nested subsets in Ω that are ε separated w.r.t. to the Hausdorff metric $d_\mathcal{H}$, that is: $d_\mathcal{H}(S_i, S_j) > \varepsilon, \forall i \neq j$. Then we have $T \leqslant N(\Omega, \varepsilon)$.*

Proof. Assume that $\{S_1, \ldots, S_T\}$ is a ε-separated subset of the metric space $(2^\Omega, d_\mathcal{H})$, where $d_\mathcal{H}$ is the Hausdorff metric. Since for all t we have $d_\mathcal{H}(\Omega_t, \Omega_{t+1}) > \varepsilon$ and $\Omega_{t+1} \subset \Omega_t$, this means that there exists at least one point $p_t \in \Omega_t$ such that $d(p_t, \Omega_{t+1}) > \varepsilon$. Since for all $j > t$ holds $p_j \in \Omega_j \subset \Omega_{t+1}$, we conclude $d(p_t, p_j) \geqslant d(p_t, \Omega_{t+1}) > \varepsilon$ for all $j > t$. This establishes that $\{p_1, \ldots, p_T\}$ is a ε-separated subset of Ω. Therefore, we can bound the size of the set T by the packing number $T \leqslant N(\Omega, \varepsilon)$. $\qquad\square$

Lemma 51. *Then the following implications hold between the properties of Def. 6.9:*

$$
\begin{array}{ccc}
\text{(C)} & \Rightarrow & \text{(A)} \\
\Downarrow & & \Downarrow \\
\text{(D)} & \Rightarrow & \text{(B).}
\end{array}
$$

The reverse (and any other) implications between the properties do not hold in general.

Proof. (A) $\implies$ (B), (C) $\implies$ (D) are obvious. (C) $\implies$ (A) follows by noticing that (C) implies $\sum_{t=1}^\infty d(\theta_t, \theta_{t-1}) \leqslant \gamma \operatorname{diam}(\Omega)$. To prove (D) $\implies$ (B), we use Lem. 50. First notice that (γ, T)-finite time competitiveness (f.t.) implies $(\gamma, 1)$-w.c.. Pick some $\varepsilon > 0$ and let $\mathbb{I} = \{t_1, \ldots, \}$ be all time-steps at which $d(\theta_t, \theta_{t-1}) \geqslant \varepsilon$. Now, for each $t \in \mathcal{I}$ holds $d_\mathcal{H}(\mathsf{P}(\mathcal{D}_t), \mathsf{P}(\mathcal{D}_{t-1})) \geqslant \frac{\varepsilon}{\gamma}$. We can verify that the collection of sets $\{\mathsf{P}(\mathcal{D}_{t-1}) \mid t \in \mathbb{I}\}$ is a $\frac{\varepsilon}{\gamma}$-separated set in the metric space $(2^\Omega, d_\mathcal{H})$; therefore, by Lem. 50 it follows that the index set $\mathbb{I}$ is finite, that is, $|\mathbb{I}| \leqslant N(\Omega, \frac{\varepsilon}{\gamma})$. Since this holds for all $\varepsilon > 0$, we proved $\forall \varepsilon > 0 \exists N$ s.t. $\forall t \geqslant N : d(\theta_t, \theta_{t-1}) < \varepsilon$, i.e., $\lim_{t \to \infty} d(\theta_t, \theta_{t-1}) = 0$. $\qquad\square$

From the above diagram, we can see that γ-competitiveness is the strongest chasing property as it implies all other properties. Furthermore, the condition (γ, T)-finite-time competitiveness strengthens with larger T, as shown below:

Corollary 52. *(γ, T)-finite time competitiveness implies $(k\gamma, kT)$-finite time competitiveness for any $k \in \mathbb{N}$.*

Proof. Assume (γ, T)-f.t.-competitive and pick k consecutive intervals $\mathcal{I}_1, \ldots, \mathcal{I}_k$, $\mathcal{I}_j := [\tau_j, \bar{\tau}_j]$, each of length T. Notice that the nestedness property $\mathsf{P}(\mathcal{D}_{\bar{\tau}_k}) \subset \cdots \subset \mathsf{P}(\mathcal{D}_{\tau_1})$ implies $d_{\mathcal{H}}(\mathsf{P}(\mathcal{D}_{\bar{\tau}_j}), \mathsf{P}(\mathcal{D}_{\tau_j})) \leqslant d_{\mathcal{H}}(\mathsf{P}(\mathcal{D}_{\bar{\tau}_k}), \mathsf{P}(\mathcal{D}_{\tau_1}))$ which leads to:

$$\sum_{j=1}^{k} \sum_{t \in \mathcal{I}_j} d(\theta_{t+1}, \theta_t) \leqslant \gamma \sum_{j=1}^{k} d_{\mathcal{H}}(\mathsf{P}(\mathcal{D}_{\bar{\tau}_j}), \mathsf{P}(\mathcal{D}_{\tau_j})) \leqslant \gamma k d_{\mathcal{H}}(\mathsf{P}(\mathcal{D}_{\bar{\tau}_k}), \mathsf{P}(\mathcal{D}_{\tau_1})).$$

$\square$

In the next section, we discuss that in cases where the consistent set map P returns convex sets, γ-competitive CMC algorithms can be designed via a reduction to the nested convex bodies chasing (NCBC) problem [33]. On the other hand, for $T = 1$ and any $\gamma \geqslant 1$, the weaker chasing property (D) can always be achieved, even if the map P returns arbitrary non-convex sets. We show in Section 6.5 a simple projection-based selection rule, which satisfies $(\gamma, 1)$-finite-time competitiveness.

Competitive Chasing via Competitive Nested Convex Body Chasing

The main difficulty in selecting the parameters θ_t to solve CMC competitively is that, for any time $t < T$, we cannot guarantee the selection of a parameter θ_t that is guaranteed to be in the future consistent set $\mathsf{P}(\mathcal{D}_T)$. The notion of competitiveness is a common performance objective in the design of online learning algorithms [29, 34, 58, 82, 115, 136]. Moreover, it turns out that in the case where a sequence of consistent sets is always convex, we can reduce the CMC problem to a well-known problem of *nested convex body chasing* (NCBC) [33]. This requirement is necessary for the reduction and is stated in Assumption 6.6.

Assumption 6.6. *Given a compact parameterization* $(\mathbb{T}, \Omega, d)$ *of the uncertainty set* $\mathcal{F}$, *the consistent sets* $\mathsf{P}(\mathcal{D})$ *are always convex for any data set* $\mathcal{D} \in \mathbb{O}$.

Constructing consistent sets $\mathsf{P}(\mathcal{D}_t)$ online can be addressed with tools from set-membership identification. For a large collection of linear and nonlinear systems, the sets $\mathsf{P}(\mathcal{D})$ can be constructed efficiently online. Such methods have been developed and studied in the literature of set-membership identification, for a recent survey see [97]. Moreover it is often possible to construct $\mathsf{P}(\mathcal{D})$ as an intersection of finite half-spaces, allowing for tractable representations as LPs. To see a particularly simple example, consider the following nonlinear system with some unknown parameters

$\alpha^* \in \mathbb{R}^M$ and η^*, where w_t is a vector with entries in the interval $[-\eta^*, \eta^*]$:

$$x_{t+1} = \sum_{i=1}^{M} \alpha_i^* \psi_i(x_t, u_t) + w_t, \tag{6.10}$$

where $\psi_i : \mathcal{X} \times \mathcal{U} \mapsto \mathcal{X}$ are M known nonlinear functions. If we represent the above system as an uncertain system $\mathbb{D}$ with parameter $\theta^* = [\alpha^*; \eta^*]$, it is easy to see that the consistent sets $\mathsf{P}(\mathcal{D})$ for some data $\mathcal{D} = \{(x_i^+, x_i, u_i) \mid 1 \leqslant i \leqslant H\}$ takes the form of a polyhedron

$$\mathsf{P}(\mathcal{D}) = \{\theta = [\alpha; \eta] \mid \text{ s.t. } (6.11) \text{ for all } 1 \leqslant i \leqslant H\},$$

defined by the inequalities

$$[\psi_1(x_i, u_i), \dots, \psi_M(x_i, u_i)]\alpha \leqslant x_i^+ + \mathbb{1}\eta, \tag{6.11a}$$

$$[\psi_1(x_i, u_i), \dots, \psi_M(x_i, u_i)]\alpha \geqslant x_i^+ - \mathbb{1}\eta. \tag{6.11b}$$

We can see that any linear discrete-time system can be put into the above form (6.10). Moreover, as shown in Section 6.10 the above representation also applies for a large class of (nonlinear) robotics system.

Nested Convex Body Chasing (NCBC)

In NCBC, we have access to a nested sequence $\mathsf{S}_0, \mathsf{S}_1, \dots, \mathsf{S}_T$ of convex sets online in some metric space $(\mathcal{M}, d)$ (that is: $\mathsf{S}_t \subset \mathsf{S}_{t-1}$). The learner selects at each time t a point p_t from S_t. The goal of competitive NCBC is to produce $p_1, \dots, p_T$ online so that the total cost of moving $\sum_{j=1}^{T} d(p_j, p_{j-1})$ at time T is competitive with the offline optimum, that is, there is some $\gamma > 0$ stain. $\sum_{j=1}^{T} d(p_j, p_{j-1}) \leqslant \gamma \max_{p_0 \in \mathsf{S}_0} \mathrm{OPT}_T(p_0)$, where $\mathrm{OPT}_T(p_0) := \min_{p \in \mathsf{S}_T} d(p, p_0)$.

Remark 33. *NCBC is a special case of the more general convex body chasing (CBC) problem, first introduced by [55], which studied competitive algorithms for metrical goal systems.*

Let the sequence of convex consistent sets $\mathsf{P}(\mathcal{D}_t)$ be the corresponding S_t of the NCBC problem, any γ-competitive agent $\mathcal{A}$ for the NCBC problem can instantiate a γ-competitive selection for competitive model-chasing, as summarized in the following reduction:

Proposition 34. *Consider the setting of Assumption 6.6. Then any γ-competitive algorithm for NCBC in metric space (Ω, d) instantiates via Algorithm 3 a γ-competitive CMC algorithm $\mathsf{SEL}_{\mathrm{NCBC}}$ in the parametrization $(\mathbb{T}, \Omega, d)$.*

Algorithm 3 γ-competitive CMC selection SEL_{NCBC}

Require: γ-competitive NCBC algorithm $\mathcal{A}_{\text{NCBC}}$, consistent set procedure P
 1: **procedure** $\text{SEL}_{\text{NCBC}}(t, x^+, x, u)$
 2:     $\mathcal{D}_t \leftarrow \mathcal{D}_{t-1} \cup (t, x^+, x, u)$
 3:     $S_t \leftarrow P(\mathcal{D}_t)$ $\triangleright$ construct/update new consistent set
 4:     present set S_t to $\mathcal{A}_{\text{NCBC}}$
 5:     $\mathcal{A}_{\text{NCBC}}$ chooses $\theta_t \in S_t$
 6:     **return** θ_t
 7: **end procedure**

Proof. Since we set $S_t = P(\mathcal{D}_t)$, it is clear that $\max_{p_0 \in S_0} \text{OPT}_T(p_0)$ is equal to $d_{\mathcal{H}}(P(\mathcal{D}_T), P(\mathcal{D}_0))$. Therefore, γ-competitive NCBC implies γ-competitive CMC on all time intervals $[t_0, T]$, with $t_0 = 1$. To see that this also holds for any choice of t_0, recall that the NCBC problem requires the competitiveness condition to hold for any sequence of nested convex bodies. Thus, for a fixed sequence S_t the condition must also be satisfied for the shifted sequences $S'_t := S_{t+k}$. $\qquad\square$

Simple Competitive NCBC-Algorithms in Euclidean Space $\mathbb{R}^n$. When (Ω, d) is a compact euclidean finite dimensional space, recent exciting progress on the NCBC problem provides a variety of competitive algorithms [12, 13, 33, 114] that can instantiate competitive selections per Algorithm 3.

We highlight two simple instantiations based on the results in [12] and [33]. Both algorithms can be tractably implemented in the setting of assumption 6.6. The selection criteria for $\text{SEL}_p(\mathcal{D}_t)$ and $\text{SEL}_s(\mathcal{D}_t)$ is defined as:

$$\text{SEL}_p(\mathcal{D}_t) := \underset{\theta \in P(\mathcal{D}_t)}{\arg\min} \|\theta - \text{SEL}_p(\mathcal{D}_{t-1})\|, \tag{6.12a}$$

$$\text{SEL}_s(\mathcal{D}_t) := s(P(\mathcal{D}_t)), \tag{6.12b}$$

where SEL_p defines simply a greedy projection operator and where SEL_s selects according to the *Steiner-Point* $s(P(\mathcal{D}_t))$ of the consistent set $P(\mathcal{D}_t)$ at time t.

Definition 6.10 (Steiner Point). *For a convex body Ω, the Steiner point is defined as the following integral over the $n - 1$ dimensional sphere $\mathbb{S}^{n-1}$:*

$$s(\Omega) = n \int_{v \in \mathbb{S}^{n-1}} \max_{x \in \Omega} \langle v, x \rangle v \, dv. \tag{6.13}$$

Another equivalent definition, which is more useful computationally, is to define the Steiner point as the weighted average of the extreme points of Ω, where the weights are the polar angles of each extreme point [113]. A convenient way to express this in terms of averages of random linear programs: $s(\Omega)$ is the expectation of a random variable $V_t(c)$, where each random variable V_i is the solution to a linear program $\max_{x \in \Omega} c_i^\top x$ with a cost-vector c, whose direction in $\mathbb{R}^n$ is picked uniformly at random. Moreover, in euclidean spaces it suffices to sample c from the n-dimensional standard Gaussian distribution $\mathcal{N}(0, I_n)$:

$$(6.13) \quad \Leftrightarrow \quad s(\Omega) = \mathbb{E}_{c \sim \mathcal{N}(0, I_n)} \arg\max_{v \in \Omega} \langle c, v \rangle.$$

Remark 35. *As shown in [13], the Steiner point can be efficiently approximated by solving randomized linear programs as used in the definition above.*

The competitive analysis presented in [33] applies, and appealing to Proposition 34 we can establish that SEL_p and SEL_s are competitive CMC algorithms:

Corollary 53 (of Theorem 1.3 [12], and Theorem 2.1 [33]). *Assume Ω is a compact convex set in $\mathbb{R}^n$ and $d(x, y) := \|x - y\|_2$. Then, the procedures SEL_p and SEL_s are competitive (CMC)-algorithms with competitive ratio γ_p and γ_s:*

$$\gamma_p = (n - 1)n^{\frac{n+1}{2}}, \qquad\qquad \gamma_s = \frac{n}{2}. \qquad (6.14)$$

The following lemma is needed to translate the results of [12] and Theorem 2.1 [33] into our setting:

Lemma 54 (Steiner Point). *Let $s(\Omega)$ denote the Steiner point of a convex body. The following inequalities hold for a nested sequence of convex bodies $\Omega_0 \supset \Omega_1 \cdots \supset \Omega_T$ and their $d_\mathcal{H}$-pathlength $\Gamma_T = \sum_{t=1}^{T} \|s(\Omega_t) - s(\Omega_{t-1})\|_2$:*

$$\Gamma_T \leqslant \tfrac{n}{2}\text{diam}(\Omega_1), \qquad\qquad \Gamma_T \leqslant nd_\mathcal{H}(\Omega_1, \Omega_T).$$

Proof. Assume $\Omega_1 \supset \Omega_2 \cdots \supset \Omega_T$ is a nested sequence of convex bodies in $\mathbb{R}^n$. Then, as shown in [33] Thm 2.1, we have the following

$$\sum_{t=1}^{T} \|s(\Omega_t) - s(\Omega_{t-1})\|_2 \leqslant \tfrac{n}{2}(w(\Omega_1) - w(\Omega_T)), \qquad (6.15)$$

where $w(\mathsf{S})$ denotes *mean-width* of the set S. $w(\mathsf{S})$ can be written as the average length of a random 1-dimensional projection of S: Let v be uniformly distributed in the sphere $\mathbb{S}^{n-1}$ and let $w(\mathsf{S})$ be defined as the expectation of $l_v(\mathsf{S})$,

$$w(\mathsf{S}) := \mathbb{E}_{v \sim \text{Unif}(\mathbb{S}^{n-1})} l_v(\mathsf{S}), \quad l_v(\mathsf{S}) := \text{diam}(\text{Proj}_v(\mathsf{S})) \qquad (6.16)$$

where $l_v(S)$ is evaluated by first projecting the set S into the subspace spanned by the vector v, denoted $\mathrm{Proj}_v(S)$ and taking its length. By linearity of expectation, we can write

$$\sum_{t=1}^{T} \|s(\mathbf{\Omega}_t) - s(\mathbf{\Omega}_{t-1})\|_2 \leqslant \tfrac{n}{2}\mathbb{E}_{v\sim\mathrm{Unif}(\mathbb{S}^{n-1})}l_v(\mathbf{\Omega}_1) - l_v(\mathbf{\Omega}_T)$$

$$\leqslant \tfrac{n}{2}\mathbb{E}_{v\sim\mathrm{Unif}(\mathbb{S}^{n-1})}2d_{\mathcal{H}}(\mathbf{\Omega}_1, \mathbf{\Omega}_T) = nd_{\mathcal{H}}(\mathbf{\Omega}_1, \mathbf{\Omega}_T) \quad (6.17)$$

where the last inequality comes from noticing that according to the definition of the Hausdorff distance, it holds $\mathbf{\Omega}_1 \subset \mathbf{\Omega}_T \oplus d_{\mathcal{H}}(\mathbf{\Omega}_1, \mathbf{\Omega}_T)\mathsf{B}_{\|\cdot\|_2}$. Similarly, we obtain

$$\sum_{t=1}^{T} \|s(\mathbf{\Omega}_t) - s(\mathbf{\Omega}_{t-1})\|_2 \leqslant \tfrac{n}{2}\mathbb{E}_{v\sim\mathrm{Unif}(\mathbb{S}^{n-1})}l_v(\mathbf{\Omega}_1) \leqslant \tfrac{n}{2}\mathrm{diam}(\mathbf{\Omega}_1).$$

$$\square$$

A General Approach to Finite Time Competitive Consistent Model Chasing

In contrast to the stricter notion of γ-competitiveness, we can give a simple and general selection rule which is always $(\gamma, 1)$-finite time competitive:

Definition 6.11 (Projection-based chasing). *Pick $\theta_t \in \mathsf{P}(\mathcal{D}_t)$ always such that for some fixed $\gamma > 0$, at every time-step t holds $d(\theta_t, \theta_{t-1}) \leqslant \gamma d(\mathsf{P}(\mathcal{D}_t), \theta_{t-1})$.*

This projection-based chasing algorithm might not always be tractable to implement, since it requires solving a potentially non-convex optimization problem with γ relative accuracy. However, it is trivially $(\gamma, 1)$-finite time competitive and describes a simple blueprint to design a general CMC algorithm SEL, that is allowing for potentially *infinite dimensional* metric spaces and *non-convex* consistent sets $\mathsf{P}(\mathcal{D}_t)$. Combined with a suitable oracle π, the resulting online control algorithm $\mathcal{A}_\pi(\mathrm{SEL}_p)$ provides finite mistake guarantees for objectives $\mathcal{G}$ according to Theorem 39.

6.7 Main Results

Assuming that π and SEL meet the required specifications, we can provide the overall guarantees for the algorithm. Let $(\mathbb{D}, \mathbf{\Omega}, d)$ be a compact parametrization of a given uncertainty set $\mathcal{F}$. Let π be robust per Definition 6.3 and SEL return consistent parameters per Definition 6.9. We apply the online control strategy $\mathcal{A}_\pi(\mathrm{SEL})$ described in Algorithm 2 to system $x_{t+1} = f^*(t, x_t, u_t)$ with unknown dynamics $f^* \in \mathcal{F}$ and denote $(\boldsymbol{x}, \boldsymbol{u})$ as the corresponding state and input trajectories.

We organize our results by the strictness of requirements we impose on the model-chaser SEL. The quality of the overall guarantee depends greatly on the type of chasing property that SEL provides. Provided that the oracle guarantees certain robustness requirements, the chasing conditions $\theta \to \theta_\infty$ (A) and $d(\theta_t, \theta_{t-1}) \to 0$ (B) assure that the overall amount of mistakes $\sum_{t=0}^{\infty} \mathcal{G}_t(x_t, u_t)$ is finite. What's more, if SEL provides competitive chasing properties such as γ-competitiveness or (γ, T)-finite-time competitiveness, then we can provide stronger guarantees, which state uniform bounds on the worst-case number of mistakes.

Stability and Boundedness of Closed Loop Trajectories

Regardless of the objective $\mathcal{G}$, we can provide worst-case state norm guarantees for $\mathcal{A}_\pi(\text{SEL})$ in a normed state space $(\mathcal{X}, \| \cdot \|)$, if SEL is a competitive or a finite-time competitive CMC algorithm and π provides sufficient robustness guarantees for a single time-step transition:

Theorem 36. *Assume that* $\pi : \Omega \mapsto \mathcal{K}$ *is* (α, β)-*single step robust in the space* $(\mathcal{X}, \| \cdot \|)$. *Then, the following state bound guarantees hold:*

(i) If SEL is a γ-competitive CMC algorithm, then:

$$\forall t : \quad \|x_t\| \leqslant e^{\alpha\gamma\phi(\Omega)} \left(e^{-t}\|x_0\| + \beta\frac{e}{e-1} \right).$$

(ii) If SEL is a (γ, T)-finite-time competitive CMC algorithm, then:

$$\|x\|_\infty \leqslant \inf_{0<\mu<1} \left(1 + (\alpha\phi(\Omega))^{n^*} \right) \max\{\tfrac{\beta}{1-\mu}, \|x_0\|\} + \beta \sum_{k=0}^{n^*} (\alpha\phi(\Omega))^k$$

where $n^* = N(\Omega, \frac{\mu}{\alpha\gamma})$ *and* $\phi(\Omega)$ *denotes the diameter of* Ω.

Proof. Part 1: In each time-step t, it holds $x_{t+1} = f(t, x_t, \pi[t, \theta_t])$ for some $f \in \mathbb{D}(\theta_{t+1})$. Therefore, the following inequality holds at each time-step:

$$\|x_{t+1}\| = \|f(t, x_t, \pi[\theta_t](t, x_t))\| \leqslant \alpha d(\theta_{t+1}, \theta_t)\|x_t\| + \beta. \tag{6.18}$$

We apply Lem. 55 with the substitution $s_t := \|x_t\|$, $\delta_t := \alpha d(\theta_{t+1}, \theta_t)$ and $c := \beta$, to obtain

$$\|x_t\| \leqslant e^{\alpha L} \left(e^{-t}\|x_0\| + \beta\frac{e}{e-1} \right). \tag{6.19}$$

Part 2: We follow the proof technique used in the main result of [65] to prove boundedness. Given a closed-loop trajectory x, at each time-step t holds $x_{t+1} = f(t, x_t, \pi[\theta_t](t, x_t))$ for some $f \in \mathbb{D}(\theta_{t+1})$. Take an arbitrary time step T. Define $\mathbb{I}$ the set of all indeces $k < T$ for which holds $\|x_{k+1}\| > \mu\|x_k\| + \beta$. Notice that for each $k \notin \mathbb{I}$ holds $\|x_{k+1}\| \leqslant \mu\|x_k\| + \beta$ while for $k \in \mathbb{I}$ we have at least the inequality $\|x_{k+1}\| \leqslant \alpha\mathrm{diam}(\mathbf{\Omega})\|x_k\| + \beta$. Now, for each $k \in \mathbb{I}$ holds

$$\mu\|x_k\| + \beta < \|x_{k+1}\| = \|f(k, x_k, \pi[\theta_k](k, x_k))\| \leqslant \alpha d(\theta_{k+1}, \theta_k)\|x_k\| + \beta,$$

which leads to $d(\theta_{k+1}, \theta_k) > \frac{\mu}{\alpha}$. Now, the (γ, T) finite-time competitive property ensures that $d(\theta_k, \theta_{k+1}) \leqslant \gamma d_{\mathcal{H}}(\mathsf{P}(\mathcal{D}_k), \mathsf{P}(\mathcal{D}_{k+1}))$. We can therefore conclude that:

$$\frac{\mu}{\alpha\gamma} < \frac{1}{\gamma}d(\theta_k, \theta_{k+1}) \leqslant d_{\mathcal{H}}(\mathsf{P}(\mathcal{D}_k), \mathsf{P}(\mathcal{D}_j)), \quad \text{for } j > k. \tag{6.20}$$

Hence, $\{\mathsf{P}(\mathcal{D}_k) \mid k \in \mathbb{I}\}$ is a $\frac{\mu}{\alpha\gamma}$-separated set in $\mathbf{\Omega}$. Therefore, $|\mathbb{I}| \leqslant N(\mathbf{\Omega}, \frac{\mu}{\alpha\gamma})$. Recall again that for each $k \notin \mathbb{I}$ holds $\|x_{k+1}\| \leqslant \mu\|x_k\| + \beta$ while for $k \in \mathbb{I}$ it holds $\|x_{k+1}\| \leqslant \alpha\mathrm{diam}(\mathbf{\Omega})\|x_k\| + \beta$. Following the same arguments as in the Appendix of [65], we obtain the presented $\|\cdot\|_\infty$-bound on x. $\qquad\square$

Lemma 55. *Let $s = (s_0, s_1, \dots)$, $\boldsymbol{\delta} = (\delta_0, \delta_1, \dots)$ be non-negative scalar sequences such that $s_{k+1} \leqslant \delta_k s_k + c$, with $c \geqslant 0$ and $\sum_{t=0}^{\infty} \delta_t \leqslant L$. Then s_t is bounded by:*

$$s_t \leqslant e^L \left(e^{-t}s_0 + c\frac{e}{e-1}\right).$$

Proof. First, we apply the comparison lemma. Therefore, s_k is bounded above by a sequence γ_k with the dynamics $\gamma_{k+1} = \delta_k\gamma_k + c$, where $\gamma_0 = s_0$. Hence, it suffices to bound the sequence γ_k in order to obtain a bound on s_k. Writing out γ_t yields:

$$\gamma_t = \prod_{k=0}^{t-1} \delta_k s_0 + c\left(1 + \sum_{j=1}^{t-1}\prod_{k=j}^{t-1} \delta_k\right).$$

Recall the basic fact $1 + x \leqslant e^x$ and notice that each product $\prod_{k=j}^{t-1} \delta_k$ can be bounded as:

$$\prod_{k=j}^{t-1} \delta_k = \prod_{k=j}^{t-1}(1 + (\delta_k - 1)) \leqslant \prod_{j=0}^{t-1} \exp(\delta_k - 1) = \exp\left(\sum_{j=k}^{t-1}(\delta_k - 1)\right)$$
$$\leqslant \exp(L - t) = e^{-t+k}e^L.$$

Therefore we obtain our desired result by bounding γ_t at each time-step t as:

$$\gamma_t \leqslant \leqslant e^{-t}e^L s_0 + c\left(1 + \sum_{j=1}^{t-1} e^{-j}\right)e^L$$

$$\leqslant e^{-t}e^L s_0 + ce^L \sum_{j=0}^{\infty} e^{-j} \leqslant e^{-t}e^L s_0 + c\frac{e^{L+1}}{e-1}.$$

Notice that the above worst-case guarantee holds for *arbitrarily large* diameter $\phi(\Omega)$ of the parameter space Ω, and applies naturally to scenarios where initially stabilizing control policies do not exist. To this end, it is also important to verify that the required property defined in Def. 6.6 does *not* imply existence of an initially stabilizing control policy.

Finite Mistake Guarantees for Asymptotic CMC

The guarantees stated in Theorem 37 and Theorem 6.7 can be interpreted as global asymptotic stability of the output, since $\mathcal{G}_t(x_t, u_t)$. To provide this guarantee, we only impose weak asymptotic conditions on π and SEL: SEL needs to either satisfy the convergence condition (A) or (B), while π has to be either robust or uniformly robust+cost-invariant for some non-zero margin, respectively.

Theorem 37. *Assume that SEL is a consistent model chaser (CMC) of type (A) and that π is an ρ-robust oracle for an objective $\mathcal{G}$. Then, the amount of mistakes we make online is guaranteed to be finite; i.e., for any closed-loop trajectory $(\boldsymbol{x}, \boldsymbol{u})$ holds $\sum_{t=0}^{\infty} \mathcal{G}_t(x_t, u_t) < \infty$.*

Proof. Denote the online data at time t as the tuple $\mathcal{D}_t := (d_1, \ldots, d_t)$. Per assumption, we know that $\mathsf{P}(\mathcal{D}_\infty)$ is non-empty and that $\lim_{t \to \infty} \theta_t = \theta_\infty \in \mathsf{P}_\infty$. Moreover, there exists some $f' \in \mathbb{D}(\theta_\infty)$ such that the trajectories satisfy for all time $t \geqslant 0$ the dynamics $x_{t+1} = f'(t, x_t, u_t)$. Since $\theta_t \to \theta_\infty$, there exists a time T, such that for all $t \geqslant T$, $d(\theta_t, \theta_\infty) < \rho$, i.e., for $t \geqslant T$, we apply policies $\pi(t; \theta_t)$ with parameters ρ-close to θ_∞. Per definition, this tells us that for the time interval $\mathcal{I} = [T, \infty)$, the tail of the trajectory $x_\mathcal{I}, u_\mathcal{I}$ is contained in $\mathcal{S}_\mathcal{I}[\rho; \theta_\infty]$. Since we assume that π is a ρ-robust oracle for $\mathcal{G}$, (Def. 6.4), we have to conclude that $\sum_{t=T}^{\infty} \mathcal{G}_t(x_t, u_t) \leqslant M$ for some finite number M. This proves the desired claim since $\sum_{t=0}^{\infty} \mathcal{G}_t(x_t, u_t) \leqslant \sum_{t=0}^{T-1} \mathcal{G}_t(x_t, u_t) + M$. $\qquad\square$

The next theorem states that we can guarantee finite mistakes even for the weakest chasing condition (B), provided π satisfies a stronger robustness property.

Theorem. *Assume that SEL is a consistent model chaser (CMC) of type (B) and that π is an uniformly ρ-robust, cost-invariant oracle for an objective $\mathcal{G}$. Then, the amount of mistakes we make online is guaranteed to be finite, i.e., for any closed-loop trajectory $(\boldsymbol{x}, \boldsymbol{u})$ holds $\sum_{t=0}^{\infty} \mathcal{G}_t(x_t, u_t) < \infty$.*

Proof. Pick an arbitrary $0 < \varepsilon < \rho$ and some $T \geqslant M_\rho^\pi$. There exists $N > 0$ such that $\forall t \geqslant N$: $\mathrm{dist}(\mathsf{P}(\mathcal{D}_t), \theta_t) \leqslant \varepsilon/4$ and $d(\theta_t, \theta_{t-1}) \leqslant \varepsilon/(2T)$. Pick an arbitrary time-step $s > N + T$, there exists a $\theta_s' \in \mathsf{P}(\mathcal{D}_s)$ such that $d(\theta_s', \theta_s) \leqslant \varepsilon/4$. Consider now the time steps t in the time window $\mathcal{I}_s = [s - T, s - 1]$. Per assumption and triangle inequality, we have, for all $t \in \mathcal{I}_s$:

$$d(\theta_s', \theta_t) \leqslant d(\theta_s', \theta_s) + \sum_{j=t}^{s-1} d(\theta_{j+1}, \theta_j) \leqslant \frac{\varepsilon}{4} + (s - t)\frac{\varepsilon}{2T} \leqslant \frac{\varepsilon}{4} + T\frac{\varepsilon}{2T} \leqslant \varepsilon < \rho.$$

Since $\theta_s' \in \mathsf{P}(\mathcal{D}_\infty)$, the truncation $(\boldsymbol{x}_{\mathcal{I}_s}, \boldsymbol{u}_{\mathcal{I}_s})$ is contained in the set $\mathcal{S}_{\mathcal{I}_s}[\rho; \theta_s']$. Since we picked T to be larger than the mistake constant M_ρ^π, there has to be at least one time-step $s' \in \mathcal{I}$ prior to s at which $\mathcal{G}_s(x_s, u_s) = 0$; otherwise we would contradict the ρ-robustness property. Now, due to the cost-invariance of π, we have to conclude that for any time-steps $k \in \mathcal{I}_s$, $k \geqslant s'$ after s' holds $\mathcal{G}_k(x_k, u_k) = 0$; this also includes time-step s, hence $\mathcal{G}_s(x_s, u_s) = 0$ is true. Finally, since s was arbitrarily chosen in the interval $[N + T + 1, \infty)$, we know that $\sum_{t=N+T+1}^\infty \mathcal{G}_t(x_t, u_t) = 0$. Therefore, the total cost is $\sum_{t=0}^\infty \mathcal{G}_t(x_t, u_t) = \sum_{t=0}^{N+T} \mathcal{G}_t(x_t, u_t)$ and is finite. $\qquad\square$

It is important to point out that the above asymptotic guarantees do not require ρ to be known, merely that there exist some non-zero ρ for which π is ρ-robust. Next, we discuss that the stronger notion of (finite-time)-competitive chasing allows to bound the total number of mistakes uniformly over all closed-loop trajectories.

Mistake-Bound Guarantees for γ-Competitive CMC

As discussed in Section 6.5, in cases where $\mathsf{P}(\mathcal{D}_t)$ have a convex representation, we can use the Steiner point $s(\mathsf{P}_t)$ or the projection rule $\theta_t = \mathrm{Proj}_{\mathsf{P}_t}(\theta_{t-1})$ to design a γ-competitive CMC-algorithm. As shown in the next theorem, SEL procedures of this type (C) lead to a uniform bound on the total mistakes for the overall adaptive controller $\mathcal{A}_{\pi \times \mathrm{SEL}}$:

Theorem 38. *Assume that SEL is a γ-competitive CMC-algorithm and that π is an uniformly ρ-robust oracle for an objective $\mathcal{G}$. Then, for any trajectory $(\boldsymbol{x}, \boldsymbol{u})$ holds*

$$\sum_{t=0}^\infty \mathcal{G}_t(x_t, u_t) \leqslant M_\rho^\pi \left(\tfrac{2\gamma}{\rho}\mathrm{diam}(\boldsymbol{\Omega}) + 1 \right).$$

Proof. The parameter sequence $\boldsymbol{\theta}$ provided by SEL satisfies $\theta_t \in \mathsf{P}(\mathcal{D}_t)$, $\forall t$ and $\sum_{t=1}^T d(\theta_t, \theta_{t-1}) \leqslant \gamma d_{\mathcal{H}}(\boldsymbol{\Omega}, \mathsf{P}(\mathcal{D}_T))$. Set $t_0 = 0$ and construct the index-sequence

$t_0, t_1, t_2, \ldots, t_N$ as follows:

$$t_k := \begin{cases} \min\left\{t \leqslant T \mid t > t_{k-1} \text{ and } d(\theta_t, \theta_{t_{k-1}}) > \tfrac{1}{2}\rho\right\} & , \text{if } k > 1 \\ 0 & , \text{if } k = 0 \end{cases} \tag{6.21}$$

until for some N, the condition $t \leqslant T, t > t_N, d(\theta_t, \theta_{t_N}) > \tfrac{1}{2}\rho$ becomes infeasible and we terminate the construction. Define the intervals $\mathcal{I}_k := [t_k, \bar{t}_k]$, where $\bar{t}_k := t_{k+1}-1$ for $k < N$ and $\bar{t}_N = T$. The intervals $\mathcal{I}_0, \ldots, \mathcal{I}_N$ are a non-overlapping cover of the time-interval $[0, T]$:

$$\bigcup_{0 \leqslant k \leqslant N} \mathcal{I}_k = [0, T], \quad \mathcal{I}_k \cap \mathcal{I}_{k-1} = \varnothing, \forall k : 1 \leqslant k \leqslant N.$$

Let $(a_0, a_1, \ldots, a_N)$ and $(b_0, b_1, \ldots, b_N)$ be the parameters selected at the start and end of each interval $\mathcal{I}_k$, respectively, $a_k := \theta_{t_k}$ and $b_k := \theta_{\bar{t}_k}$. Per construction, we know that

$$d(a_k, \theta_t) \leqslant \tfrac{1}{2}\rho \text{ for all } t \in \mathcal{I}_k \tag{6.22}$$

$$d(a_k, a_{k-1}) > \tfrac{1}{2}\rho \text{ for all } 1 \leqslant k \leqslant N. \tag{6.23}$$

Inequality (6.22) states that $d(a_k, b_k) \leqslant \tfrac{1}{2}\rho$ and implies via triangle inequality that for all $t \in \mathcal{I}_k$ holds

$$d(\theta_t, b_k) \leqslant d(\theta_t, a_k) + d(a_k, b_k) \leqslant \rho.$$

Since we picked $b_k = \theta_{\bar{t}_k}$ and the procedure SEL assures $\theta_{\bar{t}_k} \in \mathsf{P}(\mathcal{D}_{\bar{t}_k})$, this means that for some $f' \in \mathbb{D}[b_k]$, the partial trajectory $(x_{\mathcal{I}_k}, u_{\mathcal{I}_k})$ satisfies the following equations for the time steps $t \in \mathcal{I}_k$:

$$x_{t+1} = f'(t, x_t, u_t), \quad u_t = \pi[\theta_t](t, x_t). \tag{6.24}$$

We can therefore conclude that $(x_{\mathcal{I}_k}, u_{\mathcal{I}_k}) \in \mathcal{S}_{\mathcal{I}_k}[\rho; b_k]$. Hence, for the time-frame $\mathcal{I}_k$ the trajectory is consistent with the nominal closed-loop w.r.t. to system parameter b_k and therefore the partial trajectory $(x_{\mathcal{I}_k}, u_{\mathcal{I}_k})$ has to obey the conditions implied by ρ-robustness of the oracle, i.e., we have to conclude that $\sum_{t \in \mathcal{I}_k} \mathcal{G}_t(x_t, u_t) \leqslant M_\rho^\pi$. Applying this reasoning to each of the N intervals $\mathcal{I}_k$, we observe that the total mistakes can be at most $M_\rho^\pi(N+1)$:

$$\sum_{t=0}^{T} \mathcal{G}_t(x_t, u_t) = \sum_{k=0}^{N} \sum_{j \in \mathcal{I}_k} \mathcal{G}_j(x_j, u_j) \leqslant M_\rho^\pi(N+1). \tag{6.25}$$

The last step is to bound the number of intervals N, for which we leverage the chasing-property of the selection SEL. We established in (6.23) that over each interval SEL must have changed the consistent parameter by at least $\frac{1}{2}\rho$. From this we notice that N has to scale with the total path-length of our selection and obtain our bound by leveraging the property of γ-competitiveness. The following chain of inequalities

$$\tfrac{1}{2}\rho N \leqslant \sum_{k=1}^{N} d(a_k, a_{k-1}) \leqslant \sum_{k=0}^{N-1} \sum_{t \in \mathcal{I}_k} d(\theta_t, \theta_{t+1}) \leqslant \sum_{t=1}^{T} d(\theta_t, \theta_{t-1}) \leqslant \gamma d_{\mathcal{H}}(\boldsymbol{\Omega}, \mathsf{P}(\mathcal{D}_T))$$

leads to the bound $N \leqslant 2\frac{\gamma}{\rho} d_{\mathcal{H}}(\boldsymbol{\Omega}, \mathsf{P}(\mathcal{D}_T))$. We substitute this into (6.25) to obtain the desired bound on the total number of mistakes:

$$\sum_{t=0}^{T} \mathcal{G}_t(x_t, u_t) \leqslant M_\rho^\pi (2\tfrac{\gamma}{\rho} d_{\mathcal{H}}(\boldsymbol{\Omega}, \mathsf{P}(\mathcal{D}_T)) + 1)$$

$$\leqslant M_\rho^\pi (2\tfrac{\gamma}{\rho} d_{\mathcal{H}}(\boldsymbol{\Omega}, \mathsf{P}(\mathcal{D}_\infty)) + 1) \leqslant M_\rho^\pi (2\tfrac{\gamma}{\rho}\mathrm{diam}(\boldsymbol{\Omega}) + 1). \quad (6.26)$$

We can take the limit $T \to \infty$ and arrive at the desired result. $\qquad\square$

Mistake-Bound Guarantees for Finite-Time Competitive CMC

The strong competitiveness property is not necessary if, instead, stronger conditions on the oracle can be enforced. The next result states that if the oracle π is cost-invariant we can weaken the assumptions on SEL and still provide finite mistake guarantees.

Theorem 39. *Assume that SEL is (γ, T)-finite-time competitive CMC-algorithm (type D) and that π is an uniformly ρ-robust, cost-invariant oracle for an objective $\mathcal{G}$. Then for any closed-loop trajectory holds:*

$$\sum_{t=0}^{\infty} \mathcal{G}_t(x_t, u_t) \leqslant M_\rho^\pi \left(N(\boldsymbol{\Omega}, r^*) + 1 \right), \qquad r^* := \frac{1}{2}\frac{\rho}{\gamma} \frac{T}{M_\rho^\pi + T} \qquad (6.27)$$

where $N(\boldsymbol{\Omega}, r)$ denotes the r packing number of $\boldsymbol{\Omega}$ as defined in (6.12).

Proof. We discuss only some of the main steps and postpone the discussion of the full proof to Section 6.A of the appendix. The first part of the proof follows a construction similar to that in Theorem 38 to arrive at an intermediate bound

$$\sum_{t=0}^{T} \mathcal{G}_t(x_t, u_t) = \sum_{k=0}^{N} \sum_{j \in \mathcal{I}_k} \mathcal{G}_j(x_j, u_j) \leqslant M_\rho^\pi (|\mathbb{S}| + 1),$$

where $\mathbb{S}$ is defined as a subset of intervals defined by the condition:

$$\mathbb{S} := \{\mathcal{I}_k \mid \mathcal{G}_{t_{k+1}}(x_{t_{k+1}}, u_{t_{k+1}}) = 1\}.$$

The second part of the proof is concerned with showing that $\mathbb{S}$ is a finite set and bounding the cardinality $|\mathbb{S}|$. The main argument we use here is similar in spirit to the one used to establish the boundedness of N in the proof of Theorem 38: We show that due to the finite-time competitive chasing property, within each interval $\mathcal{I}_k = [\underline{t}_k, \bar{t}_k]$ our selection θ_t changes at least by a fixed nonzero amount, which leads to a fixed separation of size ε

$$\varepsilon = \frac{1}{2}\frac{\rho}{\gamma}\frac{T}{M_\rho^\pi + T}$$

in Hausdorff metric between consistent sets $\mathsf{P}(\mathcal{D}_{\underline{t}_k})$ at the beginning of the interval $\mathcal{I}_k$ and the consistent set $\mathsf{P}(\mathcal{D}_{\bar{t}_k})$ at the end of the interval $\mathcal{I}_k$. Moreover, it turns out that this observation proves the existence of a ε-separated set of size $|\mathbb{S}|$ that is contained within Ω. Due to the compactness of Ω, this proves that $\mathbb{S}$ is a finite set and that its cardinality is bounded above by the ε packing number of the set $\Omega \backslash \mathsf{P}(\mathcal{D}_\infty) \subset \Omega$. $\qquad\square$

Definition 6.12 ([46]). *Let $(\mathcal{M}, d)$ be a metric space and $\mathsf{S} \subset \mathcal{M}$ a compact set. For $r > 0$, define r-packing number of S, denoted by $N(\mathsf{S}, r)$, as:*

$$N(\mathsf{S}, r) := \max\left\{ n \in \mathbb{N} \,\bigg|\, \exists \theta_1, \ldots, \theta_N \in \mathsf{S} \text{ s.t. } d(\theta_i, \theta_j) > r, \, \forall i \neq j \right\}.$$

The bound of Theorem 39 is much larger than the one of Theorem 38, however it compensates for in generality.

The stronger result Theorem 38 requires SEL to be a γ-competitive CMC-algorithm; however, it is unclear whether γ-competitiveness can be achieved for the case of non-convex consistent sets. On the other hand, the generic projection-based strategy presented in Section 6.5 describes a universal finite-time competitive CMC algorithm SEL_p, regardless of whether the consistents are convex or not. Hence, assuming that we have a suitable oracle π, it suffices to couple the oracle with the simple projection-based CMC SEL_p in order for $\mathcal{A}_{\pi \times \text{SEL}_p}$ to inherit the guarantees of Theorem 39.

Mistake Guarantees with Locally Robust Oracles

The worst-case bound shown in Theorem 36 can be directly used to extend the result of Theorem 38 to problem settings where we only have access to locally robust oracles.

Theorem 40 (Corollary of Thm. 38). *Consider the setting and assumptions of Thm.38 and Thm.39, but relax the oracle robustness requirements to corresponding local versions and enforce the additional oracle assumption stated in Thm.36. Then all guarantees of Thm.38 and Thm.39 still hold, if we replace M_ρ^π in Thm.38 and Thm.39, respectively, by $M_\rho^\pi(R_\infty)$ and $M_\rho^\pi(R_\infty^w)$ and where:*

$$R_\infty = e^{\alpha\gamma\phi(\Omega)}\left(\|x_0\| + \beta\frac{e}{e-1}\right)$$

$$R_\infty^w = \inf_{0<\mu<1}\left(1 + (\alpha\phi(\Omega))^{n^*}\right)\max\{\tfrac{\beta}{1-\mu}, \|x_0\|\} + \beta\sum_{k=0}^{n^*}(\alpha\phi(\Omega))^k$$

and $n^* = N(\Omega, \frac{\mu}{\alpha\gamma})$ and $\phi(\Omega)$ denotes the diameter of Ω.

Writing out the inequalities derived from Theorem 38, i.e., the guarantees for combining γ-competitive model chasers SEL with locally uniformly ρ-robust and (α, β)-single step stable oracles π, we obtain the mistake bound:

$$\sum_{t=0}^{\infty}\mathcal{G}_t(x_t, u_t) \leqslant M_\rho^\pi(R_\infty)\left(\tfrac{2\gamma}{\rho}\mathrm{diam}(\Omega) + 1\right). \tag{6.28}$$

On the other hand, the guarantee of Theorem 39 transforms into the bound:

$$\sum_{t=0}^{\infty}\mathcal{G}_t(x_t, u_t) \leqslant M_\rho^\pi(R_\infty^w)\left(N(\Omega, r^*) + 1\right), \quad r^* := \frac{1}{2}\frac{\rho}{\gamma}\frac{T}{M_\rho^\pi(R_\infty^w) + T} \tag{6.29}$$

6.8 Discussion and Extension of Main Results

Theorem 38 and 39 can be invoked on any learning and control method that instantiates $\mathcal{A}_\pi(\text{SEL})$. It offers a set of sufficient conditions to verify whether a learning agent $\mathcal{A}_\pi(\text{SEL})$ can provide mistake guarantees: We need to show that w.r.t. some compact parametrization $(\mathbb{T}, \Omega, d)$ of the uncertainty set $\mathcal{F}$, π operates as a robust oracle for some objective $\mathcal{G}$, and that SEL satisfies strong enough chasing properties. Theorem 38 also suggests a design philosophy of decoupling the learning and control problem into two separate problems while retaining the appropriate guarantees: (1) design a robust oracle π for a specified control goal $\mathcal{G}$; and (2) design an online selection procedure SEL that satisfies the chasing properties defined in Def. 6.9. Nominal control design methods which have guarantees only in the small uncertainty settings can be naturally extended to the large uncertainty setting. Provided that the design method can be embedded as an oracle sub-routine π_{rc} and that we can find a suitable SEL routine, the meta-algorithm $\mathcal{A}_{\pi_{rc}}(\text{SEL})$ provides a simple extension of the original method to the large uncertainty setting.

ρ-robust	$\forall \theta \in \Omega: \ \sup_{\tau,\gamma \geq 0} m_\rho^\pi(\tau, \gamma; \theta) < \infty$
uniformly ρ-robust	$M_\rho^\pi(\tau) := \sup_{\gamma \geq 0, \theta \in \Omega} m_\rho^\pi(\tau, \gamma; \theta) < \infty$
locally ρ-robust	$\forall \gamma \geq 0, \ \theta \in \Omega: \ \sup_{\tau \geq 0} m_\rho^\pi(\tau, \gamma; \theta) < \infty$
locally uniformly ρ-robust	$\forall \gamma \geq 0: \ M_\rho^\pi(\tau, \gamma) := \sup_{\theta \in \Omega} m_\rho^\pi(\tau, \gamma; \theta) < \infty$

Table 6.2: Notions of oracle-robustness adjusted for finite-time guarantees

Using the theorems of the previous section, any nominal guarantees which can be paraphrased as OC-MG stability and mistake guarantees are carried over by $\mathcal{A}_{\pi_{rc}}(\text{SEL})$ to the large uncertainty setting.

Finite-Time Mistake Guarantees

At the cost of adding notational overhead, we can extend the definitions of the oracle robustness properties, as shown in Def. 6.13 and Table 6.2, to be more granular and allow for finite-time evaluation.

Definition 6.13 (m_ρ^π definition for finite-time guarantees). *Equip $\mathcal{X}$ with some norm* $\|\cdot\|$. *For each* $\rho, \gamma \geq 0$, $\tau \in \mathbb{N}$ *and* $\theta \in \Omega$, *define the quantity* $m_\rho^\pi(\tau, \gamma; \theta)$ *as*

$$m_\rho^\pi(\tau, \gamma; \theta) := \sup_{\mathcal{I}=[t,t+\tau]} \ \sup_{(x_\mathcal{I},u_\mathcal{I}) \in \mathcal{S}_\mathcal{I}[\rho;\theta], \|x_t\| \leq \gamma} \sum_{t \in \mathcal{I}} \mathcal{G}_t(x_t, u_t).$$

With respect to this adjusted set of notations, the proofs of Theorem 38 and Theorem 39 remain, up to some notational substitution, logically the same. Hence, for example, we can state the inequalities (6.28) and (6.29) as the following finite-time guarantees:

$$\sum_{t=0}^{\tau} \mathcal{G}_t(x_t, u_t) \leq M_\rho^\pi(\tau, R_\infty)\left(\tfrac{2\gamma}{\rho}\text{diam}(\Omega) + 1\right). \tag{6.30}$$

$$\sum_{t=0}^{\tau} \mathcal{G}_t(x_t, u_t) \leq M_\rho^\pi(\tau, R_\infty^w)\left(N(\Omega, r^*) + 1\right), \qquad r^* := \frac{1}{2}\frac{\rho}{\gamma}\frac{T}{M_\rho^\pi(R_\infty^w) + T} \tag{6.31}$$

From Mistake Guarantees to Cost Guarantees

The presented mistake guarantees can be used to obtain worst-case guarantees for more general cost-functions. We highlight how to perform this reduction next.

Assume we are given a non-negative and possibly time-dependent cost function $C : \mathbb{N} \times \mathcal{X} \times \mathcal{U} \to \mathbb{R}_0^+$. Then we can define a family $\{\mathcal{G}[\varepsilon]\}_{\varepsilon \in \mathbb{R}^+}$ of objectives $\mathcal{G}[\varepsilon] : \mathbb{N} \times \mathcal{X} \times \mathcal{U} \to \{0, 1\}$ indexed over the open interval $\varepsilon \in (0, \infty)$, such that for a trajectory $(\boldsymbol{x}, \boldsymbol{u})^\top$, the sum $\sum_{t=0}^\infty \mathcal{G}_t[\varepsilon](x_t, u_t)$ represents the total number of time-steps t at which we incurred a cost $C(t, x_t, u_t)$ larger than ε. Suitably, we call this the total number of ε-*mistakes*. We make this definition precise using characteristic functions. Let $\chi_A : \mathbb{R} \mapsto \{0, 1\}$ denote the characteristic function for a subset $A \subset \mathbb{R}$ of the real line and $\overline{\chi}_A := \chi_{A^c}$ be the corresponding complementary function, i.e., $\overline{\chi}_A(s) = 0, \forall s \in A$ and $\overline{\chi}_A(s) = 1, \forall s \notin A$. Then, for each $\varepsilon \in \mathbb{R}, \varepsilon > 0$, the ε-objective $\mathcal{G}[\varepsilon]$ is defined for each $t \in \mathbb{N}$, $x \in \mathcal{X}$, $u \in \mathcal{U}$ as

$$\mathcal{G}[\varepsilon](t, x, u) := \overline{\chi}_{[0,\varepsilon]}(C(t, x, u)).$$

Using the following Lemma, which is proven in the Appendix, we can reformulate the running cost in terms of the mistake bounds.

Lemma 56. *For any scalar sequence $\boldsymbol{a} \in \ell_1$ holds:*

$$\|\boldsymbol{a}\|_1 = \int_0^\infty \sum_{k=0}^\infty \overline{\chi}_{[-s,s]}(a_k)ds = \int_0^{\|\boldsymbol{a}\|_\infty} \sum_{k=0}^\infty \overline{\chi}_{[-s,s]}(a_k)ds \tag{6.32}$$

Proof. See Appendix 6.A $\qquad\qquad\square$

As direct application of the above, we have the following relation between running cost and number of ε-mistakes:

$$\sum_{k=0}^t C(k, x_k, u_k) = \int_0^\infty \sum_{k=0}^t \overline{\chi}_{[0,\varepsilon]}(C(k, x_k, u_k))d\varepsilon = \int_0^\infty \sum_{k=0}^t \mathcal{G}[\varepsilon](k, x_k, u_k)d\varepsilon$$

$$= \int_0^{\max_{k \leqslant t} C(k, x_k, u_k)} \sum_{k=0}^t \mathcal{G}[\varepsilon](k, x_k, u_k)d\varepsilon$$

We can associate the family of objectives $\{\mathcal{G}[\varepsilon]\}_{\varepsilon \in \mathbb{R}^+}$ with a family of corresponding comparison functions $\{m_\rho^\pi[\varepsilon]\}_{\varepsilon \in \mathbb{R}^+}$ and $\{M_\rho^\pi[\varepsilon]\}_{\varepsilon \in \mathbb{R}^+}$ – each $m_\rho^\pi[\varepsilon]$, $M_\rho^\pi[\varepsilon]$ quantifying the nominal mistake bounds of a fixed oracle π. Then, for suitable π and model chasers SEL, we can integrate the finite-time mistake bounds over ε and use the above equivalence, in general, to derive finite-time worst-case cost performance guarantees for $\mathcal{A}_{\pi \times \text{SEL}}$.

As an example, as a corollary of Theorem 38, and its local extension Theorem 40, we obtain the following result:

Theorem 41. *Let $C : \mathbb{N} \times \mathcal{X} \times \mathcal{U} \to \mathbb{R}_0^+$ be a non-negative, possibly time-dependent cost function and let $\{\mathcal{G}[\varepsilon]\}_{\varepsilon \in \mathbb{R}^+}$ be the corresponding family of objectives as defined above. Let $\pi : \Omega \to \mathcal{K}$ be an oracle with the (α, β)-stability property.*

Assume that SEL is a γ-competitive CMC-algorithm and that for each objective $\mathcal{G}[\varepsilon]$, the oracle π is locally uniformly ρ-robust, with corresponding comparison functions $m_\rho^\pi[\varepsilon]$, $M_\rho^\pi[\varepsilon]$. Then, for any closed-loop trajectory $(\boldsymbol{x}, \boldsymbol{u})$ and $\tau \in \mathbb{N}$ holds

$$\sum_{t=0}^{\tau} C(t, x_t, u_t) \leq \left(\tfrac{2\gamma}{\rho}\mathrm{diam}(\boldsymbol{\Omega}) + 1\right) \int_0^\infty M_\rho^\pi[\varepsilon](\tau, R_\infty)d\varepsilon,$$

where $R_\infty := e^{\alpha\gamma\phi(\boldsymbol{\Omega})}(\|x_0\| + \beta\tfrac{e}{e-1})$.

Safety Guarantees Using Families of Lyapunov Functions

Def. 6.6 is closely related to ISS-stability [75] and is favored in our derivations as a simple substitute for generic ℓ_∞-stability conditions. However, Def. 6.6 requires that all nominal closed-loops $\mathsf{CL}_\pi[\omega]$ share $V(x) = \|x\|$ as a common ISS-Lyapunov function, which can seem a rather restrictive condition. Nevertheless, we can replace Def. 6.6 with a more flexible definition without significant repercussions on the theoretical results.

Assume that for each $\omega \in \boldsymbol{\Omega}$, there exists a family $\{V(\,\cdot\,|\omega)\}_{\omega\in\Omega}$ of non-negative functions $V(\,\cdot\,;\omega) : \mathbb{N} \times \mathcal{X} \to \mathbb{R}^+$ and a scalar positive-definite, increasing bijective function $h : \mathbb{R}_0^+ \to \mathbb{R}_0^+$ such that the following conditions are met:

1. For all $\omega \in \boldsymbol{\Omega}, t, x \in \mathcal{X}$:

$$V(t, x \mid \omega) \leq h(\|x\|)$$

2. For some $\mu > 0, \alpha > 0$ and all $\omega_1, \omega_2 \in \boldsymbol{\Omega}, f \in \mathbb{D}[\omega_2], t \in \mathbb{N}, x \in \mathcal{X}$:

$$V_{t+1}(f(t, x_t, \pi[\omega_1]) \mid \omega_2) \leq (e^{-\mu} + \alpha d(\omega_1, \omega_2))V_t(x_t \mid \omega_1) + 1$$

If we assume an oracle π to satisfy the above set of conditions for some μ, α, $\{V(a;\omega) \mid \omega \in \boldsymbol{\Omega}\}$, h, and couple it with a model chaser SEL, we can obtain theoretical safety guarantees analogous to Theorem 36 and finite mistake guarantees analogous to Theorem 40. A thorough discussion of these extensions is being prepared for publication.

Extension to Unbounded Uncertainties

Surprisingly, knowledge of a compact parametrization of $\mathcal{F}$ is not a fundamental necessity for the PixSel framework. In fact, we can extend the approach to a large class of problem settings with unbounded uncertainty sets, and still retain the innate ability to provide worst-case guarantees.

Algorithm 4 Meta-Model Chaser SEL_∞

Require: Family of consistent model chasers $\{\text{SEL}_i\}_{i \in \mathbb{N}}$ corresponding to a filtration $\{\Omega_i\}$
Initialization: $\mathcal{D}_0 \leftarrow \{\}$, $i = 0$, x_0 is set to initial condition ξ_0

```
1: for t = 0, 1, … to ∞ do
2:     𝒟ₜ ← append (t, xₜ, xₜ₋₁, uₜ₋₁) to 𝒟ₜ₋₁ (if t ≥ 1)          ▷ update history
3:     θₜ ← SELᵢ[𝒟ₜ]
4:     while θₜ = ∅ do
5:         i ← i + 1                                              ▷ switch to next CMC
6:         θₜ ← SELᵢ[𝒟ₜ]                                          ▷ reattempt selection
7:     end while
8:     Output consistent parameter θₜ
9: end for
```

Consider the more general setup, where we have a (non-compact) metric space (Ω_∞, d) representing the space of parameters, and a set-valued map $\mathbb{D} : \Omega_\infty \to 2^{\mathcal{F}}$ representing the set of dynamics $\mathbb{D}[\omega] \subset \mathcal{F}$ associated with each parameter $\omega \in \Omega$, such that $\{\mathbb{D}[\omega]\}_{\omega \in \Omega_\infty}$ covers the uncertainty $\mathcal{F}$.

Now, suppose we can construct a filtration $\{\Omega_i\}_{i \in \mathbb{N}}$ of compact subsets $\Omega_i \subset \Omega_\infty$ which covers Ω_∞, i.e., $i < j \implies \Omega_i \subset \Omega_j$ and $\bigcup_{i \in \mathbb{N}} \Omega_i = \Omega_\infty$, and assume that for each Ω_i we have a consistent model chaser SEL_i, which has the ability to notify us, for example by returning the empty set $\emptyset$, the moment the set of consistent parameters $\mathsf{P}_i(\mathcal{D}_t) \subset \Omega_i$ becomes empty at time t.

Remark. *As a simple example, consider $\mathbb{P} = \mathbb{R}$, $d(x, y) = |x - y|$ and define Ω_i as the interval $[-i, i] \subset \mathbb{R}$. Then $\{\Omega_i\}_{i \in \mathbb{N}}$ is a filtration of compact subsets which covers $\mathbb{R}$.*

Given such a family of model chasers $\{\text{SEL}_i\}$, we can construct a Meta-chasing-algorithm SEL_∞, which initializes as SEL_0, and thereafter switches to the implementation of the next model chaser SEL_{i+1}, once the consistent set of the previous model chaser $\mathsf{P}_i(\mathcal{D}_t)$ becomes empty. This process is diagrammed in Algorithm 4.

Despite its simplicity, the Meta-chasing algorithm SEL_∞ inherits any chasing-properties which hold uniformly over the family of model chasers $\{\text{SEL}_i\}_{i \in \mathbb{N}}$. This inheritance relationship rests on a simple observation. If f^* is contained in $\mathcal{F}$, then there has to exist some $i^* \in \mathbb{N}$ for which $f^* \in \mathbb{D}[\theta^*]$ for some $\theta^* \in \Omega_{i^*}$; it is immediately clear that asymptotic chasing properties, i.e., statements (A) and (B) of Def. 6.9, hold true for the model-chaser SEL_∞ if the same properties are true for the entire CMC-family $\{\text{SEL}_i\}_{i \in \mathbb{N}}$. In regards to the competitive chasing properties (C) and (D), we can easily see that if each model chaser in the family $\{\text{SEL}_i\}_{i \in \mathbb{N}}$ is

γ-competitive / (γ, T)-f.t.-competitive, then SEL_∞ is at the very least $i^*\gamma$-competitive / $(i^*\gamma, T)$-f.t.-competitive. Unfortunately, since i^* is not known in general, we can not expect to have knowledge of the former constants in advance. Nevertheless in the case of convex consistent sets, there are cases in which the competitive ratio is preserved. A thorough theoretical discussion of this case is being prepared for future publication.

Oracle Policies with Memory and System Level Controllers

The previous results assume that π returns static policies of the type $(t, x) \mapsto u$. However, this assumption is only made for ease of exposition. All previous results also hold in the case where π returns policies that have an internal state, as long as we can define the internal state to be shared among all oracle policies; namely, as part of the oracle implementation online, we update the state z_t at each step t according to some fixed update rule h

$$z_t = h(t, z_{t-1}, x_t, u_t, \ldots, x_0, u_0),$$

and control policies $\pi[\theta]$, $\theta \in \Omega$ are maps $(t, x, z) \mapsto u$ which we evaluate at time t as $u_t = \pi[\theta](t, x_t, z_t)$.

The results presented in Chapter 4 provide a natural and seamless way to encapsulate families of system level controllers as oracles and instantiate corresponding PixSel Algorithms with worst-case guarantees. A powerful application, which fuses the results of Chapter 4 and 6, is presented in our recent work [9]. In said publication, we provide the first fully distributed and scalable control algorithm capable of learning to stabilize unknown large-scale linear systems in the adversarial setting.

To summarize our findings and recap some of the main results, we review our design framework for a simple, yet non-trivial example.

6.9 Design Example: Control of Uncertain Scalar Linear System

Let us consider a very basic problem setting, wherein we are given an unknown scalar linear system

$$x_{k+1} = \alpha^* x_k + \beta^* u_k + w_k =: f^*(k, x_k, u_k),$$

s.t. $|w_k| \leqslant \gamma^* \leqslant \eta < 1$ and $\alpha^* \in [-a, a]$, $\beta^* \in [1, 1 + 2b_\Delta]$, and our goal is to reach the target interval $\mathcal{X}_T = [-1, 1]$ and remain there. We can equivalently phrase this as to achieve the objective $\mathcal{G} = (\mathcal{G}_0, \mathcal{G}_1, \dots)$ with cost functions

$$\mathcal{G}_t(x, u) := \begin{cases} 0, & \text{if } |x| \leqslant 1 \\ 1, & \text{else} \end{cases}, \quad \forall t \geqslant 0$$

after finitely many mistakes.

Compact Parametrization of Uncertainty Set. We define our parameter space as $\Omega = [-a, a] \times [1, 1+2b_\Delta]$, and define the true parameter as $\theta^* = (\theta_x^*, \theta_u^*) = (\alpha^*, \beta^*)$ and parametrize the uncertainty set as $\mathcal{F} = \cup_{\theta \in \Omega} \mathbb{D}[\theta]$ with

$$\mathbb{D}[\theta] := \{t, x, u \mapsto \theta_x x + \theta_u u + w_t \mid \|w\|_\infty \leqslant \eta\}.$$

We choose the metric as $d(\theta, \theta') := |\theta_x - \theta_x'| + a|\theta_u - \theta_u'|$. The diameter of the metric space (Ω, d) is $\phi(\Omega) = d((-a, 1), (a, 1 + 2b_\Delta)) = 2(a + b_\Delta)$.

A Locally Uniformly Robust Oracle. As an oracle, we take the simple deadbeat controller: $\pi[\theta](t, x) := -(\theta_x/\theta_u)x$. It can be easily shown that π is a locally ρ-uniformly robust oracle for $\mathcal{G}$ for any margin in the interval $(0, \bar{\rho})$, $\bar{\rho} := 1 - \eta$, by noticing the inequality:

$$|x_{t+1}| \leqslant |\theta_x^* x_t + \theta_u^* \pi[\theta_t](t, x_t)| + \eta = |((\theta_x^* - \theta_{x,t}) - (\theta_u^* - \theta_{u,t})\tfrac{\theta_{x,t}}{\theta_{u,t}})x_t| + \eta \tag{6.33}$$

$$\leqslant (|\theta_x^* - \theta_{x,t}| + |\theta_u^* - \theta_{u,t}||\tfrac{\theta_{x,t}}{\theta_{u,t}}|)|x_t| + \eta \leqslant d(\theta^*, \theta_t)|x_t| + \eta. \tag{6.34}$$

To obtain the mistake function M_ρ^π for a fixed $\rho \in (0, 1-\eta)$, notice that if $d(\theta^*, \theta_t) \leqslant \rho$, then

$$|x_{t+1}| \leqslant \rho|x_t| + \eta = \rho|x_t| + (1 - \rho)\tfrac{1}{1-\rho}\eta \quad \Leftrightarrow \quad |x_{t+1}| - \tfrac{\eta}{1-\rho} \leqslant \rho(|x_t| - \tfrac{\eta}{1-\rho})$$

$$\Rightarrow \quad |x_t| \leqslant \tfrac{\eta}{1-\rho} + \rho^t(|x_0| - \tfrac{\eta}{1-\rho}).$$

Notice that

$$\tfrac{\eta}{1-\rho} + \rho^t|x_0| < 1 \Leftrightarrow t > \tfrac{\log(|x_0|)}{\log(\rho^{-1})} + \underbrace{\tfrac{\log(1-\rho)-\log(1-\rho-\eta)}{\log(\rho^{-1})}}_{c(\rho)}$$

which implies the mistake function

$$M_\rho^\pi(\gamma) \leqslant \tfrac{\log(\gamma)}{\log(\rho^{-1})} + c(\rho). \tag{6.35}$$

Construction of Consistent Sets via LPs. The consistent set for the data set $\mathcal{D}_N$ of N observed system transitions (x_i^+, x_i, u_i) can be written as an intersection of Ω with $2N$ halfspaces:

$$P(\mathcal{D}_N) = \left\{\theta \in \Omega \mid \text{s.t.: } \forall 1 \leqslant i \leqslant N : x_i^+ - \eta \leqslant \theta_x x_i + \theta_u u_i \leqslant x_i^+ + \eta\right\}.$$

It can be constructed online and is convex.

Competitive Consistent Model Chasing via Steiner Point. We can construct a competitive CMC-algorithm by using algorithms for competitive NCBC. Assume we use the Steiner point and denote the selection procedure SEL_s as in (6.12). SEL_s is a $\frac{n}{2} = 1$-competitive CMC algorithm in euclidean space, and since the euclidean norm is bounded above by the 1-norm, SEL_s is also 1-competitive w.r.t. the metric space (Ω, d).

Mistake Guarantee for $\mathcal{A}_\pi(\text{SEL}_s)$. We apply the extension of the results in Theorem 40. It is easy to see that our π satisfies the extra condition with $\alpha = 1$, $\beta = \eta$. Assuming $|x_0| = 0$, the constant γ_∞ takes the value

$$\gamma_\infty = e^{\alpha\phi(\Omega)}(\|x_0\| + \beta\tfrac{e}{e-1}) = \tfrac{\eta e}{e-1}e^{\phi(\Omega)}. \tag{6.36}$$

For ease of exposition, assume that $\eta = e^{-1}$ and that we picked $\rho = e^{-1}$. This gives us $M_\rho^\pi(\gamma_\infty) = \phi(\Omega) - \log(e - 2)$ and substituting all constants gives us a finite mistake guarantee for the objective $\mathcal{G}$:

$$\sum_{t=0}^{\infty} \mathcal{G}_t(x_t, u_t) \leqslant M_\rho^\pi(\gamma_\infty)\left(\tfrac{2L}{\rho} + 1\right) \approx \phi(\Omega)(1 + 2e\phi(\Omega))$$

$$= 8e(a + b_\Delta)^2 + 2(a + b_\Delta).$$

The above inequality shows that the worst-case total number of mistakes grows quadratically with the size of the initial uncertainty in the system parameters θ_x and θ_u. Notice, however, that the above inquality holds for arbitrary large choices of a and b_Δ. Thus, $\mathcal{A}_\pi(\text{SEL}_s)$ gives finite mistake guarantees for this problem setting for arbitrarily large system parameter uncertainties.

This small-scale example serves as a warm-up for the the next section where we discuss applications for learning and control of uncertain robotic systems.

6.10 Application to Uncertain Robotic Systems

We walk through an example of how to design adaptive controllers $\mathcal{A}_\pi(\text{SEL})$ for a class of robotic systems with the goal of learning to follow a trajectory. We discuss how to embed well-known control methods in robotics such as robust oracles π and couple it with SEL selections based on competitive algorithms (NCBC). Consider a general case of online control of uncertain fully-actuated robotic systems. Most robotic systems can be modeled via the robotic equation of motion [98]:

$$\mathbf{M}_\eta(q)\ddot{q} + \mathbf{C}_\eta(q,\dot{q})\dot{q} + \mathbf{N}_\eta(q,\dot{q}) = \tau + \tau_d \tag{6.37}$$

where $q \in \mathbb{R}^n$ is the multi-dimensional generalized coordinates of the system, $\dot{q}$ and $\ddot{q}$ are its first and second (continuous) time derivatives, $\mathbf{M}_\eta(q), \mathbf{C}_\eta(q,\dot{q}), \mathbf{N}_\eta(q,\dot{q})$ are matrix and vector-value functions that depend on the parameters $\eta \in \mathbb{R}^m$ of the robotic system, i.e., η comes from a parametric physical model. Often, τ is the control action (e.g., torques and forces of actuators), which acts as input of the system. Disturbances and other uncertainties present in the system can be modeled as additional torques $\tau_d \in \mathbb{R}^n$ that perturb the equations. Moreover, one can derive from first principles [98], that for many robotic systems (for example robot manipulators) the following two properties hold:

$$\dot{\mathbf{M}}_\eta(q) - 2\mathbf{C}_\eta(q,\dot{q}) \text{ is skew-symmetric} \tag{6.38a}$$

$$\mathbf{M}_\eta(q)\ddot{q} + \mathbf{C}_\eta(q,\dot{q})\dot{q} + \mathbf{N}_\eta(q,\dot{q}) = \mathbf{Y}(q,\dot{q},\ddot{q})\eta = \tau + \tau_d. \tag{6.38b}$$

The second equation says that the left-hand-side of equation (6.37) can always be factored into a $n \times m$ matrix of known functions $\mathbf{Y}(q,\dot{q},\ddot{q})$ and a constant vector $\eta \in \mathbb{R}^m$. Assume that the disturbances are bounded at each time t, as $|\tau_d(t)| \leqslant \omega$, $\omega \in \mathbb{R}^n$ and where the inequality should be read entry-wise. Consider that we are given a system with unknown η^*, ω^*, where the parameter $\theta^* = [\eta^*; \omega^*]$ is known to be contained in a bounded set Ω. Assume that our goal is to follow a desired trajectory q_d, which is given as a function of time $q_d : \mathbb{R} \mapsto \mathbb{R}^n$, within the precision ϵ. Denoting $x = [q^\top, \dot{q}^\top]^\top$ as the state vector and $x_d = [q_d^\top, \dot{q}_d^\top]^\top$ as the desired state, we want the state trajectory of the system $x(t)$ to satisfy:

$$\limsup_{t \to \infty} \|x(t) - x_d(t)\| \leqslant \epsilon. \tag{6.39}$$

As is common in practice, we assume we can observe the sampled measurements $x_k := x(t_k)$, $x_k^d := x^d(t_k)$ and apply a constant control action (zero-order-hold actuation) $\tau_k := \tau(t_k)$ at the discrete time-steps $t_k = kT_s$ with small enough sampling-time T_s to allow for continuous-time control design and analysis.

Control Objective $\mathcal{G}^\epsilon$. We phrase trajectory tracking as a control objective $\mathcal{G}^\varepsilon$ with the cost functions

$$\mathcal{G}_k^\epsilon(x, u) := \begin{cases} 0, & \text{if } \|x - x_k^d\| \leq \epsilon \\ 1, & \text{else} \end{cases}, \quad \forall k \geq 0,$$

which we wish to achieve online with finite mistake guarantees against the uncertainty set $\mathcal{F} = \bigcup_{\theta \in \Omega} \mathbb{D}[\theta]$, where:

$$\mathbb{D}[\theta] := \{k, x_k, \tau_k \mapsto f^*(x_k, \tau_k, \tau_d(\,\cdot\,); \theta) \mid \tau_d : [0, T_s] \mapsto \mathbb{R}^n, \|\tau_d\|_\infty \leq \omega\}.$$

The function f^* denotes the discretized dynamics of (6.37) w.r.t. the sampling time T_s.

Robust Oracle Design. We outline how to design a robust oracle based on a well-established robust control method for robotic manipulators proposed in [118]. Define v, a and r as the quantities

$$v = \dot{q}_d - \Lambda\tilde{q}, \quad a = \dot{v}, \quad r = \dot{\tilde{q}} + \Lambda\tilde{q}, \quad \tilde{q} = q - q_d \tag{6.40}$$

and denote $\mathbf{Y}'(q, \dot{q}, v, a)$ as the corresponding $n \times m$ matrix which allows the factorization:

$$\mathbf{M}_\eta(q)a + \mathbf{C}_\eta(q, \dot{q})v + \mathbf{N}_\eta(q, \dot{q}) = \mathbf{Y}'(q, \dot{q}, v, a)\eta. \tag{6.41}$$

Based on the control law presented in [118], we define the oracle $\pi[\theta](k, x_k)$ for $x = [q; \dot{q}]$ and $\theta = [\eta; \omega]$ through the equations:

$$\pi[\theta](k, x_k) = \mathbf{Y}'(q_k, \dot{q}_k, v_k, a_k)(\eta + u_k) - K_\omega r_k, \tag{6.42}$$

$$u = \begin{cases} -\rho\dfrac{\mathbf{Y}'^\mathsf{T} r_k}{\|\mathbf{Y}'^\mathsf{T} r_k\|_2} & \text{if } \|\mathbf{Y}'^\mathsf{T} r_k\|_2 > \varepsilon \\ -\dfrac{\rho}{\varepsilon}\mathbf{Y}'^\mathsf{T} r_k & \text{if } \|\mathbf{Y}'^\mathsf{T} r_k\|_2 \leq \varepsilon \end{cases} \tag{6.43}$$

where $\Lambda, K_\omega > 0$ are diagonal positive definite design and where ρ, ε are design variables. Following the analysis in [118] and [40] one can design a suitable gain K_ω in terms of ω, such that π is a uniformly ρ robust oracle for $\mathcal{G}^\epsilon$ in the compact parametrization $(\mathbb{D}, \Omega, d)$.

Remark 42. *The analysis in [118] shows that uniform ultimate boundedness properties of the tracking error $\tilde{x} = [q - q_d; \dot{q} - \dot{q}_d]$ are preserved, if we replace η in equation (6.42) with some perturbation $\eta + \delta(t)$, $\|\delta(t)\|_2 \leq \rho$ for all t. In [118], the disturbance τ_d is assumed to be zero, i.e., the $\omega = 0$ case, and the gain K_0 is left* *as a tuning variable. However, with standard Lyapunov arguments, the analysis of [118] can be extended to consider the nonzero disturbance case and specify gains K_ω for each ω such that the above oracle π becomes a uniformly ρ robust oracle for the above objective $\mathcal{G}^\epsilon$: For each ω, increase the gain K_ω until the uniform ultimate boundedness guarantee implies the desired ϵ-tracking behavior described by $\mathcal{G}^\epsilon$.*

Constructing Consistent Sets. The linear factorization property (6.38b) can be exploited to construct convex consistent sets. Denote $\mathbb{D}$ as an uncertain robotic system (6.38) with some convex compact uncertainty Ω in euclidean space $(\mathbb{R}^{m+n}, \| \cdot \|_2)$. Recall that we parameterize the bound on the disturbance by $\omega \in \mathbb{R}^n$, i.e., $|\tau_d| \leqslant \omega$ holds entry-wise and that our system parameter is represented by $\theta = [\eta^\top, w^\top]^\top \in \Omega$. At the sampled time-steps t_k, equations (6.38b) say that measurements $q_k, \dot{q}_k, \ddot{q}_k, \tau_k$ enforce the following entry-wise condition on consistent parameters η and ω:

$$\tau_k - \omega \leqslant \mathbf{Y}(q_k, \dot{q}_k, \ddot{q}_k)\eta \leqslant \tau_k + \omega. \tag{6.44}$$

In matrix form, the consistent set is captured via the following relationship:

$$\underbrace{\begin{bmatrix} \mathbf{Y}(q_k, \dot{q}_k, \ddot{q}_k) & -\mathbf{I}_n \\ -\mathbf{Y}(q_k, \dot{q}_k, \ddot{q}_k) & -\mathbf{I}_n \end{bmatrix}}_{\mathbf{A}_k} \begin{bmatrix} \eta \\ \omega \end{bmatrix} \leqslant \underbrace{\begin{bmatrix} \tau_k \\ -\tau_k \end{bmatrix}}_{\mathbf{b}_k}. \tag{6.45}$$

Consequently, we have a concrete construction of consistent set at each time t:

$$\mathsf{P}(\mathcal{D}_t) = \left\{ \theta = \begin{bmatrix} \eta \\ w \end{bmatrix} \in \mathbb{R}^{m+n} \middle| \mathbf{A}_t \begin{bmatrix} \eta \\ w \end{bmatrix} \leqslant \mathbf{b}_t \right\} \cap \mathsf{P}(\mathcal{D}_{t-1}), \quad \mathsf{P}(\mathcal{D}_0) = \Omega \tag{6.46}$$

where $\mathbf{A}_k = \mathbf{A}_k(x_k, u_k)$ and $\mathbf{b}_k = \mathbf{b}_k(x_k, u_k)$ are matrix and vector of "features" constructed from current control policy and state at time t via the known functional form of $\mathbf{Y}$. Data sets $\mathcal{D}_k$ are tuples of the form $\mathcal{D}_k = (d_1, \ldots, d_k)$, $d_k = (q_k, \dot{q}_k, \ddot{q}_k, \tau_k)$.

Designing a Competitive Chasing Selection. The above consistent sets are simply an intersection of halfspaces, hence we are in the setting of Assumption 6.6 and we can instantiate competitive selections from the (NCBC) competitive greedy and Steiner point algorithm algorithms:

- **Greedy Projection.** SEL_p selects $\theta_t = \mathrm{SEL}_p(\mathcal{D}_t)$ as the solution to the following convex optimization problem, which can be solved efficiently:

$$\theta_t = \underset{\theta \in \mathbb{R}^p \cap \Omega}{\arg\min} \quad \frac{1}{2} \|\theta - \theta_{t-1}\|^2,$$

$$\text{s.t:} \quad \mathbf{A}_i \theta \leqslant \mathbf{b}_i, \forall i = 1, \ldots, t.$$

Algorithm 5 design of $\mathcal{A}_\pi(\mathrm{SEL}_{p/s})$ for ϵ-trajectory tracking for fully actuated robots

1: **for** $t = 0, T_s, \ldots, kT_s$ **to** ∞ **do**
2:       measure $q_k, \dot{q}_k, \ddot{q}_k$
3:       update polyhedron $\mathsf{P}(\mathcal{D}_k)$ as in (6.46)
4:       select according to (6.12a) or (6.12b) $\triangleright$ selection SEL_p or SEL_s
5:       choose $\tau_k = \mathbf{Y}'(q_k, \dot{q}_k, v_k, a_k)(\eta_k + u_k) - K_{\omega_k} r_k$ using (6.40), (6.42) $\triangleright$ use oracle $\pi(x_k; \theta_k)$
6: **end for**

- **Steiner-Point.** Alternatively, SEL_s outputs the Steiner point of the polyhedron $\mathsf{P}_\mathbb{D}(\mathcal{D}_t)$, which in principle requires calculating an integral over multidimensional sphere. Fortunately, as shown in [13], the Steiner point can be efficiently approximated by solving randomized linear programs; an approach we take in our empirical validation.)

Mistake Guarantee for $\mathcal{A}_\pi(\mathbf{SEL}_{p/s})$ Since π is a robust oracle for $\mathcal{G}$ and both SEL_p and SEL_s are γ-competitive CMC algorithms in $(\mathbb{D}, \mathbf{\Omega}, \|\cdot\|)$ for some $\gamma > 0$, our result Theorem 38 tells us that $\mathcal{A}_\pi(\mathrm{SEL}_p)$ and $\mathcal{A}_\pi(\mathrm{SEL}_s)$ guarantees upfront finiteness of the total number of mistakes $\sum_{k=0}^\infty \mathcal{G}_k^\epsilon(x_k, \tau_k)$, which implies the desired tracking behavior guarantee $\limsup_{k\to\infty} \|x - x^d\| \leq \epsilon$. Moreover, if we can provide a bound M on the mistake constant $M_\rho^\pi < M$, we obtain from Theorem 38 an explicit performance bound for the tracking performance in the form of the mistake guarantee

$$\sum_{k=0}^\infty \mathcal{G}_k^\epsilon(x_k, \tau_k) \leq M\big(\tfrac{2\gamma}{\rho}\mathrm{diam}(\mathbf{\Omega}) + 1\big).$$

6.11   Empirical Validation: Cart-Pole Swing-Up on a Constrained Track

We illustrate the practical potential for of our approach on a challenging cart-pole swing-up goal from limited amount of interaction. Compared to the standard cart-pole domain commonly used in RL [32], we introduce modifications motivated by real-world concerns in several important ways:

1. *Goal specification*: the goal is to swing up and balance the cart-pole from a down position, which is significantly harder than balancing from the up-right position (the standard RL benchmark).
2. *Realistic dynamics*: we use a high-fidelity continuous-time nonlinear model, with noisy measurements of discrete-time state observations.
3. *Safety*: cart position has to be kept in a bounded interval for all time. Furthermore, the acceleration should not exceed a specified maximum limit.

$\pi[\theta^*]$	0	0.4	0.99	1	1
$\mathcal{A}_\pi(\text{SEL})$	0	0.2	0.8	0.95	1
T	$3\,s$	$6\,s$	$12\,s$	$30\,s$	$50\,s$

Table 6.3: Fraction of experiments completing the swing up before time T: ideal policy $\pi[\theta^*]$ vs. $\mathcal{A}_\pi(\text{SEL})$

4. *Robustness to adversarially chosen system parameters*: We evaluate 900 uncertainty settings, each with a different θ^* reflecting mass, length, and friction. The tuning parameter remains the same for all experiments. This robustness requirement amounts to a generalization goal in contemporary RL.

5. *Other constraints*: no system reset is allowed during learning (i.e., a truly continuous goal).

Our introduced modification makes this goal significantly more challenging from both on-line learning and adaptive control perspective. Table 6.3 summarizes the results for 900 different parameter conditions (corresponding to 900 adversarial settings). It compares the online algorithm to the corresponding ideal oracle policy $\pi[\theta^*]$ shows that the online controller is only marginally slower. See Appendix 6.11 and [5] for detailed description of our setup and results.

We employ well-established techniques to synthesize model-based oracles. Expert controllers are a hybrid combination of a linear state-feedback LQR around the upright position, a so-called energy-based swing-up controller (see [17]) and a control barrier function to respect the safety constraints [7]. As also described in [48], adding constraints on state and acceleration makes learning the swing-up of the cart-pole a significantly harder goal for state-of-the-art learning and control algorithms.

Table 6.3 compares the on-line algorithm with the corresponding ideal oracle policy $\pi[\theta^*]$ showing that the on-line controller is only marginally slower.

6.A Proofs

Theorem. *Assume procedure SEL is (γ, T)-finite-time competitive for some $\gamma > 0$, $T \geq 1$ and that procedure π is a uniform ρ-robust, cost-invariant oracle for $\mathcal{G}$. Then, the total number of mistakes is guaranteed to be bounded above by:*

$$\sum_{t=0}^{\infty} \mathcal{G}_t(x_t, u_t) \leq \left(N(\mathbf{\Omega}^\circ, r^*) + 1\right) M_\rho^\pi \leq \left(N(\mathbf{\Omega}, r^*) + 1\right) M_\rho^\pi,$$

where $\Omega^\circ := \Omega\backslash\mathrm{int}(\mathrm{P}(\mathcal{D}_\infty))$ and $r^* := \frac{1}{2}\frac{\rho}{\gamma}\frac{T}{M_\rho^\pi + T}$.

Proof. Denote x, u to be some fixed online trajectories and denote θ as the corresponding parameter sequence selected by procedure SEL. The sequence θ satisfies $\theta_t \in \mathrm{P}(\mathcal{D}_t)$ and $\sum_{t=t_1+1}^{t_2} d(\theta_t, \theta_{t-1}) \leqslant \gamma d(\mathrm{P}(\mathcal{D}_{t_2}), \mathrm{P}(\mathcal{D}_{t_1}))$ for all $t_2 - t_1 \leqslant T$. For some time-step $\tau > 0$, we derive bounds on the mistakes $\sum_{t=0}^{\tau} \mathcal{G}_t(x_t, u_t)$. Set $t_0 = 0$ and construct the index-sequence $t_0, t_1, t_2, \ldots, t_N$ as follows:

$$t_k := \begin{cases} \min\left\{t \leqslant \tau \mid t > t_{k-1} \text{ and } d(\theta_t, \theta_{t_{k-1}}) > \frac{1}{2}\rho\right\} & , \text{if } k > 1 \\ 0 & , \text{if } k = 0 \end{cases} \tag{6.47}$$

until for some N, the condition $t \leqslant \tau, t > t_N, d(\theta_t, \theta_{t_N}) > \frac{1}{2}\rho$ becomes infeasible and we terminate the construction. Define the intervals $\mathcal{I}_k := [t_k, \bar{t}_k]$, where $\bar{t}_k := t_{k+1} - 1$ for $k < N$ and $\bar{t}_N = \tau$. The intervals $\mathcal{I}_0, \ldots, \mathcal{I}_N$ are a non-overlapping cover of the time-interval $[0, \tau]$:

$$\bigcup_{0 \leqslant k \leqslant N} \mathcal{I}_k = [0, \tau], \quad \mathcal{I}_k \cap \mathcal{I}_{k-1} = \varnothing, \forall k : 1 \leqslant k \leqslant N.$$

Let $(a_0, \ldots, a_N)$ and $(b_0, \ldots, b_N)$ be the parameters selected at the start and end of each interval $\mathcal{I}_k$, respectively: $a_k := \theta_{t_k}$ and $b_k := \theta_{\bar{t}_k}$. Per construction, we know that

$$d(a_k, \theta_t) \leqslant \tfrac{1}{2}\rho \text{ for all } t \in \mathcal{I}_k \tag{6.48}$$

$$d(a_k, a_{k-1}) > \tfrac{1}{2}\rho \text{ for all } 1 \leqslant k \leqslant N. \tag{6.49}$$

Inequality (6.22) states that $d(a_k, b_k) \leqslant \frac{1}{2}\rho$ and implies via triangle inequality that for all $t \in \mathcal{I}_k$ holds

$$d(\theta_t, b_k) \leqslant d(\theta_t, a_k) + d(a_k, b_k) \leqslant \rho.$$

Since we picked $b_k = \theta_{\bar{t}_k}$ and the procedure SEL assures $\theta_{\bar{t}_k} \in \mathrm{P}(\mathcal{D}_{\bar{t}_k})$, it means that for some $f_k \in \mathbb{D}[b_k]$, the partial trajectory $(x_{\mathcal{I}_k}, u_{\mathcal{I}_k})$ satisfies the following equations for the time-steps $t \in \mathcal{I}_k$:

$$x_{t+1} = f'(t, x_t, u_t), \quad u_t = \pi[\theta_t](t, x_t). \tag{6.50}$$

We can therefore conclude that $(x_{\mathcal{I}_k}, u_{\mathcal{I}_k}) \in \mathcal{S}_{\mathcal{I}_k}[\rho; b_k]$. We apply to conclude that

$$\sum_{t \in \mathcal{I}_k} \mathcal{G}_t(x_t, u_t) \leqslant M_\rho^\pi \tag{6.51}$$

for each $k \in \{0, \dots, N\}$. Now, define $\mathbb{S}$ as the following collection of intervals

$$\mathbb{S} := \{\mathcal{I}_k \mid \mathcal{G}_{t_{k+1}}(x_{t_{k+1}}, u_{t_{k+1}}) = 1\}, \tag{6.52}$$

i.e., all intervals $\mathcal{I}_k$ where at the start of the *next* interval $\mathcal{I}_{k+1}$ the cost is 1. Combining this with the former bound (6.51), we can decompose the total mistake sum as

$$\sum_{t=0}^{T} \mathcal{G}_t(x_t, u_t) = \sum_{t \in \mathcal{I}_0} \mathcal{G}_t(x_t, u_t) + \sum_{\mathcal{I}_j \in \mathbb{S}} \sum_{t \in \mathcal{I}_{j+1}} \mathcal{G}_t(x_t, u_t) + \underbrace{\sum_{\mathcal{I}_j \notin \mathbb{S}} \sum_{t \in \mathcal{I}_{j+1}} \mathcal{G}_t(x_t, u_t)}_{0}$$

$$= \sum_{t \in \mathcal{I}_0} \mathcal{G}_t(x_t, u_t) + \sum_{\mathcal{I}_j \in \mathbb{S}} \sum_{t \in \mathcal{I}_{j+1}} \mathcal{G}_t(x_t, u_t) \leq M_\rho^\pi \left(|\mathbb{S}| + 1\right). \tag{6.53}$$

Notice that the last term $\sum_{\mathcal{I}_j \notin \mathbb{S}} \sum_{t \in \mathcal{I}_{j+1}} \mathcal{G}_t(x_t, u_t)$ in the first equation is zero because $\mathcal{I}_j \notin \mathbb{S}$ implies that the next interval $\mathcal{I}_{j+1}$ start with zero cost; due to the cost-invariance property it follows that $\sum_{t \in \mathcal{I}_{j+1}} \mathcal{G}_t(x_t, u_t) = 0$. The remainder of the proof is concerned with bounding the cardinality of the collection $\mathbb{S}$.

Bounding $|\mathbb{S}|$: We know that for each l in the range $1 \leq l \leq |\mathcal{I}_k|$, there exists at least one sub-interval $\mathcal{I}'_l \subset \mathcal{I}_k$, $|\mathcal{I}'_l| = l$ of length l, such that

$$\sum_{t \in \mathcal{I}'} d(\theta_t, \theta_{t+1}) > \tfrac{1}{2} \rho \tfrac{l}{(|\mathcal{I}_k|+l)}. \tag{6.54}$$

The above has to be true, since otherwise we would contradict (6.49):

- Let $\mathcal{I}'_1, \dots, \mathcal{I}'_m$, $m = \lceil |\mathcal{I}_k|/l \rceil$, $|\mathcal{I}'_i| = l$, $\mathcal{I}'_i \subset \mathcal{I}_k$ be an overlapping cover of $\mathcal{I}_k$, then

$$d(a_k, a_{k+1}) \leq \sum_{t \in \mathcal{I}_k} d(\theta_{t+1}, \theta_t) \leq \sum_{j=1}^{m} \sum_{t \in \mathcal{I}'_j} d(\theta_{t+1}, \theta_t) \leq \tfrac{1}{2} \rho \left\lceil \tfrac{|\mathcal{I}_k|}{l} \right\rceil \tfrac{l}{|\mathcal{I}_k|+l} \tag{6.55}$$

$$\leq \tfrac{1}{2} \rho \left(\tfrac{|\mathcal{I}_k|}{l} + 1 \right) \tfrac{l}{|\mathcal{I}_k|+l} = \tfrac{1}{2} \rho, \qquad \text{(recall that } a_{k+1} = \theta_{\bar{t}_k+1}) \tag{6.56}$$

which is a contradiction to (6.49)

Hence, we can always pick a sequence of sub-intervals $[t_k^l, \bar{t}_k^l] = \mathcal{I}_k^{(l)} \subset \mathcal{I}_k$ (either of length l or identical to $\mathcal{I}_k$ if $|\mathcal{I}_k| \leq l$) such that

$$\sum_{t \in \mathcal{I}_k^{(l)}} d(\theta_t, \theta_{t+1}) > \tfrac{1}{2} \rho \tfrac{l}{|\mathcal{I}_k|+l}. \tag{6.57}$$

Notice that if $|\mathcal{I}_k| \leq l$, we pick $\mathcal{I}_k^{(l)} = \mathcal{I}_k$, and therefore the above inequality is vacuously true since, $\sum_{t \in \mathcal{I}_k} d(\theta_t, \theta_{t+1}) \geq d(a_k, a_{k+1}) > \frac{1}{2}\rho \geq \frac{1}{2}\rho\frac{l}{|\mathcal{I}_k|+l}$. Now, the (γ, T)-finite-time competitiveness property ensures that for all k and all $t \geq \bar{t}_k + 1$ holds:

$$\frac{1}{2}\rho\frac{T}{|\mathcal{I}_k|+T} < \sum_{h \in \mathcal{I}_k^{(T)}} d(\theta_h, \theta_{h+1}) \leq \gamma d_{\mathcal{H}}(\mathsf{P}(\mathcal{D}_{t_k}), \mathsf{P}(\mathcal{D}_{\bar{t}_k+1})) \leq \gamma d_{\mathcal{H}}(\mathsf{P}(\mathcal{D}_{t_k}), \mathsf{P}(\mathcal{D}_t))$$

$$\implies d_{\mathcal{H}}(\mathsf{P}(\mathcal{D}_{t_k}), \mathsf{P}(\mathcal{D}_t)) > \frac{1}{2}\frac{\rho}{\gamma}\frac{T}{|\mathcal{I}_k|+T}$$

where $d_{\mathcal{H}}(\mathsf{P}(\mathcal{D}_{\tau_k}), \mathsf{P}(\mathcal{D}_{\bar{\tau}_k+1})) \leq d_{\mathcal{H}}(\mathsf{P}(\mathcal{D}_{\tau_k}), \mathsf{P}(\mathcal{D}_t))$ follows from nestedness. From now on, we use the abbreviation P_t to refer to the sets $\mathsf{P}(\mathcal{D}_t)$.

Recall the definition $\mathbb{S} := \{\mathcal{I}_k \mid \mathcal{G}_{t_{k+1}}(x_{t_{k+1}}, u_{t_{k+1}}) = 1\}$ and let k_j denote the j-th interval that belongs to $\mathbb{S}$, i.e., $\mathcal{I}_{k_j} \subset \mathbb{S}$. Now set $l = T$ and define S_j as a subsequence of $\mathsf{P}_1, \mathsf{P}_2, \dots$ as follows:

$$\mathsf{S}_j := \begin{cases} \mathsf{P}_{\bar{t}_{k_j}} & \text{if } x_{\bar{t}_{k_j}} \in \mathsf{X}_{\bar{t}_{k_j}} \\ \mathsf{P}_{t_{k_j}^T} & \text{if } x_{\bar{t}_{k_j}} \notin \mathsf{X}_{\bar{t}_{k_j}}. \end{cases} \tag{6.58}$$

We will show that this collection $P = \{\mathsf{S}_1, \mathsf{S}_2, \dots\}$ of sets S_j is a $\frac{1}{2}\frac{\rho}{\gamma}\frac{T}{M_\rho^\pi+T}$ - separated set in the metric space $(2^\Omega, d_{\mathcal{H}})$ via the following inequality:

$$\forall j < i : d_{\mathcal{H}}(\mathsf{S}_j, \mathsf{S}_i) > \frac{1}{2}\frac{\rho}{\gamma}\frac{T}{M_\rho^\pi+T}.$$

This is proven below:

- Recall SEL is defined to always pick $\theta_t \in \mathsf{P}(\mathcal{D}_t)$ and π is ρ-uniformly robust and cost-invariant. Due to the SEL property, there always exists a function $f' \in \mathbb{D}[\theta_{t_{k+1}}]$ such that $x_{t_{k+1}} = f'(\bar{t}_k, x_{\bar{t}_k}, \pi[\theta_{\bar{t}_k}](\bar{t}_k, x_{\bar{t}_k}))$. On the other hand, because of the π property, the statement $x_{t_{k+1}} \notin \mathsf{X}_{t_{k+1}}$ implies that one of the following two has to hold at time $\bar{t}_k$:

  1. Assume $x_{\bar{t}_k} \in \mathsf{X}_{\bar{t}_k}$, then it has to hold that $d(\theta_{\bar{t}_k}, \theta_{t_{k+1}}) > \rho$. Notice due to (γ, H)-w.c. property, that $d(\theta_{\bar{t}_k}, \theta_{t_{k+1}}) \leq \gamma d_{\mathcal{H}}(\mathsf{P}_{\bar{t}_k}, \mathsf{P}_{t_{k+1}})$ which gives us

  $$d_{\mathcal{H}}(\mathsf{P}_{\bar{t}_k}, \mathsf{P}_{t_{k+1}}) > \frac{1}{2}\frac{\rho}{\gamma}.$$

  2. Assume $x_{\bar{t}_k} \notin \mathsf{X}_{\bar{t}_k}$, then by Def. 6.5, it follows that $|\mathcal{I}_k| \leq M_\rho^\pi$, which then implies that $d_{\mathcal{H}}(\mathsf{P}_{t_k^T}, \mathsf{P}_{t_{k+1}}) > \frac{1}{2}\frac{\rho}{\gamma}\frac{T}{|\mathcal{I}_k|+T} \geq \frac{1}{2}\frac{\rho}{\gamma}\frac{T}{M_\rho^\pi+T}$.

- Taking the minimum of both cases we can see that $d_{\mathcal{H}}(\mathsf{S}_j, \mathsf{P}_{t_{k_j}+1}) > \frac{1}{2}\frac{\rho}{\gamma}\frac{T}{M_\rho^\pi+T}$. Due to nestedness, it holds for $i > j$ that $\mathsf{S}_i \subset \mathsf{P}_{t_{k_j}+1} \subset \mathsf{S}_j$. Thus, it holds $d_{\mathcal{H}}(\mathsf{S}_j, \mathsf{S}_i) \geq d_{\mathcal{H}}(\mathsf{S}_j, \mathsf{P}_{t_{k_j}+1})$ and we arrive at the separation condition:

  $$\forall j < i : d_{\mathcal{H}}(\mathsf{S}_j, \mathsf{S}_i) > \frac{1}{2}\frac{\rho}{\gamma}\frac{T}{M_\rho^\pi+T}.$$

We conclude from Lem. 50, that $|\mathbb{S}| = |P|$ is bounded by the packing number $N(\Omega, \frac{1}{2}\frac{\rho}{\gamma}\frac{T}{M_\rho^\pi+T})$. Substituting into the bound (6.53) and taking the limit $\tau \to \infty$, we get the total number of mistakes as:

$$\sum_{t=0}^{\infty} \mathcal{G}_t(x_t, u_t) \leqslant M_\rho^\pi \left(N(\Omega, \tfrac{1}{2}\tfrac{\rho}{\gamma}\tfrac{T}{M_\rho^\pi+T}) + 1 \right).$$

A Tighter Bound. We can define $\Omega^\circ = \Omega\backslash\text{int}(\mathsf{P}(\mathcal{D}_\infty))$ and $\mathsf{S}_j^\circ = \mathsf{S}_j\backslash\text{int}(\mathsf{P}(\mathcal{D}_\infty))$ and notice that S_j° is non-empty for all j: S_j° and $\mathsf{P}(\mathcal{D}_\infty)$ are closed, so $\mathsf{S}_j \subset \mathsf{P}(\mathcal{D}_\infty)$ implies that S_j° contains at least the boundary of $\mathsf{P}(\mathcal{D}_\infty)$. Moreover, we can verify that the corresponding collection $P^\circ = \{\mathsf{S}_1^\circ, \mathsf{S}_2^\circ, \dots\}$ of sets S_j° is still a $\frac{1}{2}\frac{\rho}{\gamma}\frac{H}{M_\rho^\pi+H}$-separated set in the compact metric space $(2^{\Omega^\circ}, d_\mathcal{H})$. Therefore we can improve the previous mistake guarantee and state the tighter inequality:

$$\sum_{t=0}^{\infty} \mathcal{G}_t(x_t, u_t) \leqslant \left(N(\Omega\backslash\text{int}(\mathsf{P}(\mathcal{D}_\infty)), \tfrac{1}{2}\tfrac{\rho}{\gamma}\tfrac{H}{M_\rho^\pi+H}) + 1 \right) M_\rho^\pi.$$

$\square$

Let $\chi_A : \mathbb{R} \mapsto \{0, 1\}$ denote the characteristic function for a subset $A \subset \mathbb{R}$ of the real line and $\overline{\chi}_A := \chi_{A^c}$ be the corresponding complementary function, i.e., $\overline{\chi}_A(s) = 0, \forall s \in A$ and $\overline{\chi}_A(s) = 1, \forall s \notin A$.

Lemma 57. *For any scalar sequence $a \in \ell_1$ holds:*

$$\|a\|_1 = \int_0^\infty \sum_{k=0}^\infty \overline{\chi}_{[-s,s]}(a_k)ds = \int_0^{\|a\|_\infty} \sum_{k=0}^\infty \overline{\chi}_{[-s,s]}(a_k)ds \qquad (6.59)$$

Proof. We first prove the result for the finite sequence case and for non-negative sequences, which is stated in equation (6.61); the general result then follows as a corollary. Let $a \in \mathbb{R}^\mathbb{N}$ be a non-negative scalar sequence and define $Q_a(t, r)$, for $r \geqslant 0$ as the number of indeces $i \leqslant t$ for which the sequence is larger than r, i.e.,

$$Q_a(t, r) := |\{i \mid i \leqslant t \text{ and } a_i > r\}|, \qquad (6.60)$$

then our goal is to show that

$$\sum_{k=0}^{t} a_k = \int_0^\infty Q_a(t, r)dr = \int_0^{\max_{k\leqslant t} a_k} Q_a(t, r)dr. \qquad (6.61)$$

Let $a_0^\uparrow, \dots, a_t^\uparrow$ denote the sequence we obtain from rearranging the sequence $a_{[0,t]}$ in an increasing order. Thus, $a_0^\uparrow \leqslant a_1^\uparrow \leqslant \cdots \leqslant a_t^\uparrow$. It is clear that rearranging does not change the sum, hence it holds: $\sum_{i=0}^{t} a_i = \sum_{i=0}^{t} a_i^{\uparrow}$. Now notice that we can rewrite the sum in terms of the increments $a_i^{\uparrow} - a_{i-1}^{\uparrow}$ as follows:

$$\sum_{k=0}^{t} a_i^{\uparrow} = a_0^{\uparrow} + \sum_{k=1}^{t} \left(a_0^{\uparrow} + \sum_{j=1}^{k} (a_j^{\uparrow} - a_{j-1}^{\uparrow}) \right)$$

$$= a_0^{\uparrow}(t+1) + \sum_{j=1}^{t} \sum_{k=j}^{t} (a_j^{\uparrow} - a_{j-1}^{\uparrow})$$

$$= a_0^{\uparrow}(t+1) + \sum_{j=1}^{t} \left(a_j^{\uparrow} - a_{j-1}^{\uparrow} \right)(t+1-j) \tag{6.62}$$

where we obtain the last equality through changing the order of summation. Observe that the function $Q_a(t, \cdot) : r \mapsto Q_a(t, r)$ is the following piece-wise constant non-increasing function:

$$Q_a(t, r) = \begin{cases} t+1 & \text{if } r \in [0, a_0^{\uparrow}) \\ t+1-j & \text{if } r \in [a_j^{\uparrow}, a_{j-1}^{\uparrow}) \text{ for } j \geq 1 \\ 0 & \text{if } r \in [a_t^{\uparrow}, \infty) \end{cases}$$

Integrating over the domain $[0, \infty)$ shows that $\int_0^{\infty} Q_a(t, r)dr$ is equal to the right-hand side of (6.62) and concludes the proof of the partial result (6.61).

For the final step, notice that for $Q_a(t, r) + Q_{-a}(t, r) = \sum_{k=0}^{t} \overline{X}_{[-r,r]}(a_k)$ and therefore by (6.61), we have

$$\sum_{k=0}^{t} |a_k| = \int_0^{\max_{k \leq t} |a_k|} \sum_{k=0}^{t} \overline{X}_{[-s,s]}(a_k)ds \tag{6.63}$$

Furthermore, notice that $\sup_t \sum_{k=0}^{t} \overline{X}_{[-s,s]}(a_k)$ is finite for all $s > 0$, because otherwise it would contradict $a \in \ell_1$:

- Assume there exists some $s' > 0$ for which $\sum_{k=0}^{t} \overline{X}_{[-s',s']}(a_k)$ grows unbounded in t, then it means that there are infinitely many elements $\{a_j\}$ which are bounded below as $a_j > s'$; this contradicts $\sum_{k=0}^{\infty} |a_k| < \infty$.

It is also clear that $\sum_{k=0}^{\infty} \overline{X}_{[-s,s]}(a_k) = 0$ for any $s \geq \|a\|_{\infty}$ and that the right-hand side of (6.63) is always bounded above $\|a\|_1$. Hence, by taking the limit $t \to \infty$ in (6.63), we obtain our final result (6.59). $\qquad\square$

6.B Oracle for Cartpole-Swing up with Constraints

Next, we describe the oracle we used to instantiate our approach $\mathcal{A}_\pi(\text{SEL})$ for the test of swinging-up the cart pole on a constrained track. The overall control strategy is a hybrid combination of a linear state-feedback LQR around the upright position, a so-called energy-based swing-up controller (See [17]) and a control barrier-function to respect the safety constraints [7].

For convenience of the reader we recall the nonlinear dynamics of the cart pole system:

$$(M + m)\ddot{x} - ml\ddot{\phi}\cos(\phi) + ml\dot{\phi}^2\sin(\phi) - b_x\dot{x} = F \tag{6.64}$$
$$l\ddot{\phi} - g\sin(\phi) - b_\phi\dot{\phi} = \ddot{x}\cos(\phi).$$

Let x and $\dot{x}$ be the position and velocity of the cart and ϕ, $\dot{\phi}$ the angle and angular velocity of the pole. F is the force onto the cart pole and serves as our control input to the system. Throughout the discussion $\bar{a}$ and $\bar{d}$ are design parameters, where $\bar{a}$ denotes the maximal cart acceleration allowed, and $\bar{d}$ be the maximum distance the cart is allowed to move from the center.

Model-Based Oracle for Cart-Pole Swing Up

The outermost layer of the control strategy is partial feedback linearization. Let $F_d(\ddot{x}, \dot{x}, \phi, \dot{\phi}, \dot{x})$ be the force F we need to apply at time t in order to achieve a desired cart acceleration of $\ddot{x}_d$. Multiply the second equation of (6.64) by $ml\cos(\phi)$ and add it to the first, to see that F_d has to be chosen as:

$$F_d(\ddot{x}_d, \dot{x}, \phi, \dot{\phi}, \dot{x}) = (M + m\sin(\phi)^2)\ddot{x}_d - mg\cos(\phi)\sin(\phi) + ml\dot{\phi}^2\sin(\phi) - b_x\dot{x}.$$
$$\tag{6.65}$$

By choosing $F = F_d(\ddot{x}_d, \dot{x}, \phi, \dot{\phi}, \dot{x})$, we can now treat the desired acceleration $\ddot{x}_d$ as our new control input. With respect to our new input $\ddot{x}_d$, we can simplify the original equations (6.64) to

$$\ddot{x} = \ddot{x}_d \tag{6.66}$$
$$l\ddot{\phi} - g\sin(\phi) - b_\phi\dot{\phi} = \ddot{x}_d\cos(\phi). \tag{6.67}$$

The swingup controller consists now of three separate control laws that are later combined.

- *Around up-right position: static linear LQR controller* If the pole has small enough kinetic energy and is close to the upright position, we simply choose $\ddot{x}_d$ to be LQR-state feedback controller based on the system (6.66) linearized around the equilibrium position $x = 0, \dot{x} = 0, \phi = 0, \dot{\phi} = 0$. The policy then takes the form

$$\ddot{x}_{d,LQR} = -K_{LQR}(\theta)z.$$

- *Swing up-controller: energy-based controller.* The Swing-Up controller is based on an elegant energy-based approach by [17], in which we simply choose $\ddot{x}_d$ to control the total normalized energy

$$E(\phi, \dot{\phi}) = \frac{l}{2g}\dot{\phi}^2 + cos(\phi) \qquad (6.68)$$

of the pole. In-depth derivation can be found in [17] and $\ddot{x}_d$ takes the form:

$$\ddot{x}_{d,swing} = -\mathbf{Sat}_{\bar{a}}\left(\frac{1}{2}\gamma|\cos(\phi)|(E(\phi, \dot{\phi}) - 1)\mathrm{sign}(\dot{\phi}\cos(\phi))\right) \qquad (6.69)$$

where $\mathbf{Sat}_{\bar{a}}$ is the saturation function which saturates at the max specified acceleration $\bar{a}$.

- *Wrapping a safety controller.* As part of our oracle policy, we also use a control barrier function controller that prevents us from triggering the safety policy. We do this simply by internally overriding our swing-up $\ddot{x}_{d,swing}$ or balancing $\ddot{x}_{d,LQR}$ terms, if we get too close to the boundary of $[-x_{max}, x_{max}]$. To this end, define $B(x, \dot{x})$ as the barrier function

$$B(x, \dot{x}) = \frac{1}{2\bar{a}}\dot{x}|\dot{x}| + x$$

and define $\ddot{\phi}_{max} := \bar{a}g/l * \sin(30°)$.

The full description of the oracle policy is mapped out in Algorithm (6).

The controller switches to an LQR if the system is close to the upright position, and otherwise defaults to the swing up controller that brings the pendulum to the right energy level. A correction is performed to the previous control action depending on the barrier-function value $|B(x, \dot{x})|$. As $|B(x, \dot{x})|$ gets closer to the boundary $\bar{d} - \epsilon_{safe}$, the controller prioritizes safety and overwrites the previous planned control action. If x exceeds the buffer $\bar{d} - \epsilon_{safe}$, then a safe policy is called, which brings the cart position back to the region $[-\bar{d} + \epsilon_{safe}, \bar{d} - \epsilon_{safe}]$.

Selection Process SEL

We apply the approach presented in the main paper Section 6.10 to obtain polytopes of consistent parameters of $\mathbf{P}_t$ for the lumped parameters $p = [m_c +$

Algorithm 6 oracle policy $\pi[\theta]$ under potential safety policy π_{safe} override

---

Input: $z = [x, \phi, \dot{x}, \dot{\phi}]$, parameters $\theta := [M, m, l, b_x, b_\theta]$, ϵ_{safe}
Output: F
if $|x| < \bar{d} - \epsilon_{safe}$ **then**
    if $|\dot{\phi}^2/\ddot{\phi}_{max}| < 60°$ **and** $\cos(30°/\ddot{\phi}_{max} + \phi\,\text{sign}(\dot{\phi})) > \cos(30°)$ **and** $|-$
$K_{LQR}(\theta)z| \leq \bar{a}$ **then**
        $\ddot{x}_d = -K_{LQR}(\theta)z$
    else
        $\ddot{x}_d = -\text{Sat}_{\bar{a}}[\frac{1}{2}\gamma|\cos(\phi)|(E(\phi, \dot{\phi}) - 1)\text{sign}(\dot{\phi}\cos(\phi))]$
    end if
    $\ddot{x}_{d,back} = -\bar{a}\,\text{sign}(\dot{x})$
    $\lambda = \frac{|B(x,\dot{x})|}{\bar{d}-\epsilon_{safe}}$
    if $B(x, \dot{x}) \geq 0$ **then**
        $\ddot{x}_d \leftarrow (1 - \lambda^2)\ddot{x}_d + \lambda^2 \min\{\ddot{x}_d, \ddot{x}_{d,back}\}$
    else
        $\ddot{x}_d \leftarrow (1 - \lambda^2)\ddot{x}_d + \lambda^2 \max\{\ddot{x}_d, \ddot{x}_{d,back}\}$
    end if
    $F = F_d(\ddot{x}_d, z)$
else
    $F = \pi_{\text{safety}}(z)$
end if

$m_p, m_p l, b_x, l, b_\theta, \tau_{d,x}, \tau_{d,\theta}]$. We use randomized LPs [13], to approximate the Steiner point of the polytope P_t and select the corresponding oracle policy $\pi[\theta_t]$ as described in the meta-algorithm 2.

6.C Mistake Guarantees vs Sublinear Regret

A common performance metric in online learning for control is phrased in terms of the regret $R(T)$. For our general problem setting, we show that sublinear regret does not imply finite mistake guarantees, however finite mistake guarantees do imply sublinear regret.

Regret Definition

Assume we are given some cost function $C : \mathcal{X} \times \mathcal{U} \mapsto \mathbb{R}^+$ and the system $x_{t+1} = f^*(t, x_t, u_t)$, $x_0 = \xi_0$. Assume that some "ideal" policy π^* would generate the trajectory x_t^*, u_t^*, while the online algorithm $\mathcal{A}$, characterized by the sequence of policies, produces x_t, u_t. The optimal total cost of $J^*(T)$ at time T is defined as $J^*(T) := \sum_{k=0}^{T} C(x_k^*, u_k^*)$. The regret $R(T)$ of $\mathcal{A}$ usually refers to the sum of costs of the online algorithm up to time T minus the sum of costs that the optimal policy

π^* would achieve [3]:

$$R(T) := \sum_{k=0}^{T} C(x_k, u_k) - J^*(T). \tag{6.70}$$

Sublinear regret is defined as follows.

Definition 6.14. *The regret $R(T)$ is called sublinear if $R(T) = o(T)$, or equivalently:*

$$\lim_{T\to\infty} \frac{1}{T} R(T) = \lim_{T\to\infty} \frac{1}{T} \sum_{k=0}^{T} (C(x_k, u_k) - C(x_k^*, u_k^*)) = 0. \tag{6.71}$$

The slower $R(T)$ grows with T, (for example $O(\log(T))$), the faster convergence we can guarantee to the above limit.

Sublinear Regret Does Not Imply Bounded Cost

Sublinear regret is a common way to measure the performance of online learning and control algorithms. Ideally, we would expect the sublinear regret to subsume some more basic performance criteria such as the boundedness of the online cost, that is, $\sup_k |C(x_k, u_k) - C(x_k^*, u_k^*)| < \infty$. However, simple derivations show that without additional assumptions, sublinear regret growth is not sufficient to show cost boundedness. The reason for that is intrinsic to the very definition of regret, and simple real-analysis arguments will suffice to demonstrate that.

Abbreviate $c_k := C(x_k, u_k)$, $c_k^* := C(x_k^*, u_k^*)$ and define the sequences

$$s_k := c_k - c_k^* \tag{6.72}$$

$$m_k := \frac{1}{k} \sum_{j=0}^{k} (c_k - c_k^*) = \frac{1}{k} \sum_{j=1}^{k} s_k. \tag{6.73}$$

Now, sublinear regret is defined as the condition $\lim_{k\to\infty} m_k = 0$, while bounded cost considers the statement $\sup_k |s_k| < \infty$.

The next counter examples show that these statements are not related; The first example shows that sublinear regret does *not* imply finiteness of the sequence $|s_k|$; the second shows that the boundedness of $|s_k|$ does *not* imply sublinear regret.

1. $\lim_{k\to\infty} m_k = 0 \;\;\not\Longrightarrow\;\; \sup_k |s_k| < \infty.$

---

[3]Not the most widespread definition, but the most suited for adaptive control setting. See, for example, [143].

Proof. Consider s_k, where $s_{e^n} = n$ and otherwise 0. Define $\bar{n}(k) := \lfloor \log(k) \rfloor$, then

$$m_k \leqslant m_{\bar{n}(k)} = \frac{1}{e^{\bar{n}(k)}} \sum_{i=1}^{\bar{n}(k)} i = \frac{\bar{n}(k)(\bar{n}(k)+1)}{2e^{\bar{n}(k)}}.$$

This shows $\lim_{k\to\infty} m_{\bar{n}(k)} = 0$, but s_k is unbounded. $\square$

2. For all $\varepsilon > 0$: $\sup_k s_k - \inf_k s_k < \varepsilon \implies m_k$ converges.

Proof. Define s_k as the sequence

$$(1, \delta, 1, 1, \delta, \delta, \underbrace{1, \dots, 1, \delta, \dots, \delta}_{}, \underbrace{1, \dots, 1, \delta, \dots, \delta}_{}, \dots, \underbrace{1, \dots, 1, \delta, \dots, \delta}_{}, \dots$$

with $\delta = 1 - \varepsilon$. From the above pattern, it becomes apparent that the corresponding m_k is satisfied for all n:

$$m_{2\times 3^n} = 1 - \varepsilon/2 \qquad\qquad m_{4\times 3^n} = 1 - \varepsilon/4, \qquad (6.74)$$

hence m_k does not converge, yet $\sup_k s_k - \inf_k s_k < \varepsilon$. $\square$

Remark 43. *The above arguments still hold if we change s_k and m_k to the definitions*

$$s'_k := |C(x_k, u_k) - C(x_k^*, u_k^*)| \quad m'_k := \frac{1}{k} \sum_{j=0}^{k} |C(x_k, u_k) - C(x_k^*, u_k^*)| = \frac{1}{k} \sum_{j=1}^{k} s'_k.$$

Sublinear Regret Does Not Imply Finite Mistakes, Finite Mistakes Imply Sublinear Regret

In our problem setting, the costs $C(x, u)$ are represented by $\mathcal{G}(x, u)$, which are $\{0, 1\}$-valued cost functions. Moreover, we compare to an oracle-policy π^* which guarantees at most M_ρ^π mistakes i.e., $\sum_{k=0}^{\infty} c_k^* \leqslant M_\rho^\pi$. Trivially, finite mistakes implies sublinear regret: If it holds that $\sum_{k=0}^{\infty} c_k < M$, then $\frac{1}{T} \sum_{k=0}^{T} (c_k - c_k^*) \leqslant \frac{M}{T} \to 0$. However, the opposite is not true. Consider as an example the following sequence for c_k:

$$c = \underbrace{0, 1}_{2}, \underbrace{0, 0, 0, 1}_{4}, \underbrace{0, 0, 0, 0, 0, 0, 0, 1}_{8}, \dots, \underbrace{0, \dots, 1}_{2^k}, \dots$$

The total mistakes $\sum_{t=1}^{T} c_k = \mathcal{O}(\log(T))$ grow unbounded, but we still have $\lim_{t\to\infty} m_k = 0$.

9 783384 257444